Springer Monographs in Mathematics

For further volumes:
http://www.springer.com/series/3733

Johannes Ueberberg

Foundations of Incidence Geometry

Projective and Polar Spaces

 Springer

Johannes Ueberberg
SRC Security Research & Consulting GmbH
Graurheindorfer Strasse 149A
53117 Bonn
Germany
johannes.ueberberg@gmx.de

ISSN 1439-7382
ISBN 978-3-642-26960-8 ISBN 978-3-642-20972-7 (eBook)
DOI 10.1007/978-3-642-20972-7
Springer Heidelberg Dordrecht London New York

Mathematics Subject Classification (2010): 51E24, 51A50, 51A05

Cover design: deblik, Berlin

Printed on acid-free paper

Springer is part of Springer Science+Business Media (www.springer.com)

To Monika, Vera and Philipp

Preface

Incidence geometry has a long tradition starting with Euclid [25] and his predecessors and is at the same time a central part of modern mathematics. The main progress in incidence geometry until about 1900 are the axioms of affine and projective geometries and the two fundamental theorems of affine and projective geometries as described in David Hilbert's "Grundlagen der Geometrie" [27]. The specific value of the fundamental theorems is due to the fact that a few geometric axioms give rise to algebraic structures as vector spaces and groups.

Starting from these results, modern incidence geometry has been developed under the strong influence of Jacques Tits and Francis Buekenhout. Jacques Tits developed the fascinating theory of buildings and classified the buildings of spherical types. The main representatives of buildings of spherical type are projective spaces and polar spaces.

Starting from the theory of buildings, Francis Buekenhout laid the foundations for modern incidence geometry, also called diagram geometry or Buekenhout–Tits-geometry, and contributed a number of outstanding results such as the classification of quadratic sets or (together with Ernest Shult) the Theorem of Buekenhout–Shult about polar spaces.

The objective of this book is to give a comprehensible insight into this fascinating and complex matter. The book provides an introduction into affine and projective geometry, into polar spaces and into quadratic sets and quadrics.

The book is mainly directed to students who are interested in a clear and comprehensible introduction into modern geometry and to lecturers who offer courses and seminars about incidence geometry. However, the book may also be of interest for researchers since there are many results in this book which were up to now only available in the original research literature.

The book is organized as follows:

Chapter I introduces the terminology of geometries and diagrams and deals with projective and affine spaces. The main results of Chap. I are:

- The assignment of subspaces and dimensions to projective and affine spaces (see Sects. 4 and 5).

- The relation between affine and projective spaces (see Sect. 5).
- The characterization of affine spaces by Buekenhout (see Sect. 6).
- The introduction of diagrams and the Theorem of Tits about the existence of specific paths in residually connected geometries (see Sect. 7).
- The classification of projective and affine geometries by their diagrams (see Sect. 7).
- The introduction of finite geometries (see Sect. 8).

Chapter 2 is devoted to the study of isomorphisms and collineations. Among the collineations, the central collineations are of particular importance. They are investigated in detail. The main results of Chap. II are:

- The fact that (parallelism-preserving) collineations of projective and affine spaces are isomorphisms (see Sect. 4).
- The fact that collineations admitting a centre or an axis are already central collineations (see Sect. 5).
- The fact that every projective space of dimension at least 3 satisfies the Theorem of Desargues (see Sect. 6).
- The fact that every Desarguesian projective space admits as many central collineations as possible (see Sect. 6).

Chapter III provides the introduction of coordinates for affine and projective spaces and the classification of all Desarguesian affine and projective spaces. The main results of Chap. III are:

- The introduction of projective and affine spaces defined by means of vector spaces (see Sects. 2, 3 and 5).
- The introduction of collineations of affine and projective spaces defined by linear and semilinear transformations of the underlying vector space (see Sects. 4 and 6).
- The first fundamental theorem of affine and projective spaces classifying all Desarguesian projective and affine spaces (see Sect. 7).
- The second fundamental theorem of affine and projective spaces determining the automorphism group of Desarguesian projective and affine spaces (see Sect. 8).

Chapter IV deals with polar spaces and polarities. The classification of buildings of spherical type by Tits [50] and the work of Veldkamp [56] (for Char $K \neq 2$) also provides a classification of all polar spaces of rank at least 3 such that all planes are Desarguesian. In Chap. IV these polar spaces are introduced. Unfortunately, an inclusion of the proof of the classification theorem would go beyond the scope of this book. The main results of Chap. IV are:

- The Theorem of Buekenhout–Shult providing the current definition of polar spaces (see Sect. 2).
- The classification of polar spaces by means of their diagrams (see Sect. 3).
- The discussion of the Neumaier geometry (see Sect. 3).
- The introduction of polar spaces defined by a polarity (see Sect. 4).

- The Theorem of Birkhoff and von Neumann classifying the polarities by means of sesquilinear forms (see Sect. 5).
- The introduction of polar spaces defined by a pseudo-quadratic form (see Sect. 6).
- The introduction of the Kleinian polar space defined by the lines of a 3-dimensional projective space (see Sect. 7).
- The Theorem of Buekenhout and Parmentier classifying projective spaces as linear spaces with polarities (see Sect. 8).

Chapter V is devoted to the study of quadrics and quadratic sets. Quadrics are special cases of pseudo-quadrics (polar spaces defined by a pseudo-quadratic form). They have been studied long before polar spaces appeared. The main subject of Chap. V is the Theorem of Buekenhout stating that every quadratic set of a d-dimensional projective space is either an ovoid or a quadric. In fact, this theorem is a special case of the classification of polar spaces mentioned above. The main results of Chap. V are:

- The fact that every quadratic set defines a polar space (see Sect. 2).
- The fact that every quadric defines a quadratic set (see Sect. 3).
- The classification of quadratic sets in a 3-dimensional projective space (see Sect. 4).
- The fact that every quadratic set is perspective (see Sect. 5).
- The Theorem of Buekenhout stating that every quadratic set of a d-dimensional projective space is either an ovoid or a quadric (see Sect. 6).
- The relation between the Kleinian quadric and the Kleinian polar space (see Sect. 7).
- The Theorem of Segre stating that every oval of a *finite* Desarguesian projective plane of odd order is a conic (see Sect. 8).

Acknowledgements

I want to thank Albrecht Beutelspacher who introduced me into projective and finite geometries and to Francis Buekenhout who introduced me into polar spaces and buildings.

Many thanks to one of the referees who read the manuscript carefully and improved the text considerably.

I also want to thank Springer Verlag, in particular Catriona Byrne for her encouragement to limit the material of the book, Marin Reizakis and Federica Corradi Dell'Acqua for their continuous support and Ms Priyadharshini and her team for transferring a Word manuscript into a Tex book.

Finally, I want to thank Monika for a lot of things.

Contents

1 Projective and Affine Geometries 1
  1 Introduction.. 1
  2 Geometries and Pregeometries 1
  3 Projective and Affine Planes.. 4
  4 Projective Spaces.. 10
  5 Affine Spaces.. 23
  6 A Characterization of Affine Spaces 31
  7 Residues and Diagrams 39
  8 Finite Geometries 52

2 Isomorphisms and Collineations 57
  1 Introduction.. 57
  2 Morphisms.. 57
  3 Projections.. 59
  4 Collineations of Projective and Affine Spaces 61
  5 Central Collineations... 67
  6 The Theorem of Desargues 73

3 Projective Geometry Over a Vector Space 83
  1 Introduction.. 83
  2 The Projective Space $P(V)$... 83
  3 Homogeneous Coordinates of Projective Spaces 88
  4 Automorphisms of $P(V)$ 90
  5 The Affine Space $AG(W)$ 97
  6 Automorphisms of $A(W)$.. 104
  7 The First Fundamental Theorem 107
  8 The Second Fundamental Theorem 119

4 Polar Spaces and Polarities 123
  1 Introduction.. 123
  2 The Theorem of Buekenhout–Shult..................................... 124
  3 The Diagram of a Polar Space.. 140

4   Polarities 153
5   Sesquilinear Forms 157
6   Pseudo-Quadrics 166
7   The Kleinian Polar Space 173
8   The Theorem of Buekenhout and Parmentier 179

5   Quadrics and Quadratic Sets 185
1   Introduction 185
2   Quadratic Sets 186
3   Quadrics 191
4   Quadratic Sets in $PG(3, K)$ 197
5   Perspective Quadratic Sets 210
6   Classification of the Quadratic Sets 219
7   The Kleinian Quadric 229
8   The Theorem of Segre 234
9   Further Reading 243

References 245

Index 247

Chapter 1
Projective and Affine Geometries

1 Introduction

The present chapter deals mainly with projective and affine geometries. In Sect. 2, the notions pregeometry and geometry are introduced. In Sects. 3–6, affine and projective spaces are discussed in detail. In particular, the affine and projective spaces are endowed with a structure of subspaces, and the relation between affine and projective spaces is explained.

One can assign a little pictogram to a geometry often indicating important information about this geometry. These pictograms are called diagrams. They are introduced in Sect. 7. Finally, in Sect. 8, we will turn to finite geometries, that is, geometries consisting of finitely many elements.

2 Geometries and Pregeometries

The notions geometry and pregeometry are fundamental in the theory of modern incidence geometry. In the present section, we shall define these two notions, and we shall illustrate them by considering simple geometries like n-gons or the geometry of a cube. The whole strength of the concept of geometries will become clearer in the forthcoming sections and chapters.

Definition. Let I be a non-empty set whose elements are called **types**. A **pregeometry** over I is a triple $\Gamma = (X, *, \text{type})$ fulfilling the following conditions:

(i) X is a non-empty set whose elements are called the **elements** of the pregeometry Γ.

(ii) type is a surjective function from X onto I. It is called the **type function** of Γ.

(iii) $*$ is a reflexive and symmetric relation on X, the so-called **incidence relation**. It fulfils the condition: If x and y are incident elements of the same type, that is, $x * y$ and $\text{type}(x) = \text{type}(y)$, we have $x = y$.

J. Ueberberg, *Foundations of Incidence Geometry*, Springer Monographs in Mathematics, DOI 10.1007/978-3-642-20972-7_1, © Springer-Verlag Berlin Heidelberg 2011

The **rank** of Γ is the cardinality of I. If $|I| = n$, Γ is called a **pregeometry of rank n**.

Definition. (a) Let Δ_n be an n-gon with the set V of vertices and the set E of edges. The geometry $\Gamma_n = (X_n, *, \text{type})$ over $I = \{0, 1\}$ or $I = \{\text{vertex, edge}\}$ of an n-gon is defined as follows: The set X_n consists of the vertices and the edges of Δ_n. A vertex x and an edge k are called incident if the vertex x lies in Δ_n on the edge k.

   The pregeometry Γ_n is called an **n-gon**.

(b) Let Δ be a polyhedron with the set V of vertices, the set E of edges and the set F of faces. The geometry $\Gamma = (X, *, \text{type})$ over $I = \{0, 1, 2\}$ or $I = \{\text{vertex, edge, face}\}$ is defined as follows: The set X consists of the vertices, the edges and the faces of Δ. Two elements of X are called incident if, in Δ, one of the two elements is contained in the other one.

The geometry of the cube with notations as in the figure has the following vertices, edges and faces:

   Vertices: $\{1, 2, 3, 4, 5, 6, 7, 8\}$
   Edges: $\{12, 23, 34, 14, 15, 26, 37, 48, 56, 67, 78, 58\}$
   Faces: $\{1234, 1256, 2367, 3478, 1458, 5678\}$

Definition. Let $\Gamma = (X, *, \text{type})$ be a pregeometry over a type set I.

(a) A **flag** of Γ is a set of pairwise incident elements of Γ. The empty set is a flag of Γ as well.

(b) Two flags F and G are called **incident** if every element of F is incident with every element of G.

(c) Let F be a flag of Γ. The set $\{\text{type}(x) \mid x \in F\}$ is called the **type** of F and is denoted by $\text{type}(F)$. The **rank** of F is the cardinality of the set $\text{type}(F)$.

(d) The **corank** of F is the cardinality of the set $I \setminus \text{type}(F)$.

(e) A **chamber** is a flag of type I.

The geometry of the cube with notations as in the figure has the following flags through the vertex 1:

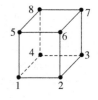

   $\{1\}, \{1, 12\}, \{1, 14\}, \{1, 15\},$
   $\{1, 1234\}, \{1, 1256\}, \{1, 1458\},$
   $\{1, 12, 1234\}, \{1, 12, 1256\}, \{1, 14, 1234\},$
   $\{1, 14, 1458\}, \{1, 15, 1256\}, \{1, 15, 1458\}.$

An incident vertex-face-pair is an example of a flag of the pregeometry defined by the cube. The chambers of the cube are the sets of pairwise incident elements consisting of exactly one vertex, one edge and one face.

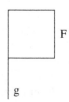

If a line g and a face F of a flag $\{g, F\}$ are drawn as in the above figure, the origin of the name **flag** becomes clear.

2.1 Theorem. *Let $\Gamma = (X, *, \text{type})$ be a pregeometry, and let F be a flag of Γ.*

(a) *The sets F and $\text{type}(F)$ are of the same cardinality, that is, $|F| = |type(F)|$. In particular, the cardinality of F equals the rank of F.*

(b) *If C is a chamber of Γ, C and I are of the same cardinality, that is, $|C| = |I|$. If Γ is of finite rank n, we have $|C| = n$.*

Proof. The proof is obvious.   □

Definition. A pregeometry Γ over a type set I is called a **geometry** if every flag of Γ is contained in a chamber of Γ.

The n-gons and the polyhedra are examples of geometries. Often, a geometry can be constructed as a set of points and a set of subspaces (lines, planes, etc.) which are subsets of the point set. This approach is reflected by the definition of a **set geometry**.

Definition. Let $\Gamma = (X, *, \text{type})$ be a pregeometry over the type set $I = \{0, 1, \ldots, d-1\}$ where the elements of type 0 are called points and all elements are called subspaces. For a subspace U of Γ, denote by U_0 the set of points incident with U.

(a) Γ is called a **set pregeometry** if any two subspaces U and W of Γ with $\text{type}(U) < \text{type}(W)$ are incident if and only if U_0 is contained in W_0 and if $U_0 \neq W_0$ whenever $U \neq W$.

(b) Γ is called a **set geometry** if Γ is a set pregeometry and a geometry.

Remark. Let Γ be a set pregeometry over the type set $I = \{0, 1, \ldots, d-1\}$. The subspaces of Γ can be seen as subsets of the point set of Γ. The incidence relation is defined by set-theoretical inclusion. For set pregeometries, we often use the terminology "U is contained in W" or "W is contained in U" instead of "U and W are incident".

2.2 Examples. (a) The geometries of a polyhedron or an n-gon are examples of set geometries.

(b) Consider the tiling of the Euclidean plane with black and white squares such that two squares with a common edge have different colours. We shall define a geometry Γ of rank 3 over the type set {point, white, black} as follows: The point set of Γ consists of the vertices of the squares. The white elements are the white squares, the black elements are the black squares.

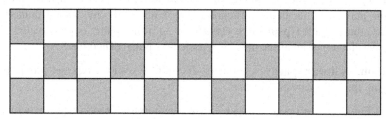

A point and a square are incident if the point is a corner of the square. A black and a white element are incident if they share an edge. The so-defined geometry is a geometry, but not a set geometry.

In this book, we shall mostly deal with set geometries.

Definition. Let $\Gamma = (X, *, \text{type})$ be a geometry over a type set I.

(a) Let F be a flag of Γ. F is called **co-maximal** if there exists an element x of $X \setminus F$ such that $F \cup \{x\}$ is a chamber.
(b) The geometry Γ is called **thin** (respectively **thick**, respectively **firm**) if every co-maximal flag of Γ is contained in exactly two (respectively in at least three, respectively in at least two) chambers.

2.3 Theorem. *Let Γ be a geometry of finite rank n over a type set I. A flag F of Γ is co-maximal if and only if F is of rank $n - 1$.*

Proof. The proof is obvious. $\square$

The n-gons and the geometries of the polyhedra are thin geometries. We shall illustrate this fact by the geometry of the cube:

The co-maximal flags of Γ consist of a vertex and an edge, a vertex and a face or an edge and a face. Considering as an example the co-maximal flag $\{1, 12\}$, we see that there are the two chambers $\{1, 12, 1234\}$ and $\{1, 12, 1256\}$ through this flag.

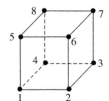

Definition. Let $\Gamma = (X, *, \text{type})$ and $\Gamma' = (X', *', \text{type}')$ be two geometries over the type sets I and I', respectively. Γ' is called a **subgeometry** of Γ if the following two conditions are fulfilled:

(i) We have $I' \subseteq I$, $X' \subseteq X$ and $\text{type}' = \text{type}|_{X'}$.
(ii) Two elements x' and y' of Γ' are incident in Γ' if and only if they are incident in Γ.

3   Projective and Affine Planes

The main result of the present section is the following relation between projective and affine planes:

Starting from a projective plane, an affine plane can be constructed by deleting one line and all of its points (Theorem 3.7). Conversely, by adding points and one line, the so-called **projective closure** of an affine plane is constructed. The projective closure of an affine plane is a projective plane (Theorem 3.8).

Definition. A **linear space** is a geometry L of rank 2 over the type set $\{\text{point}, \text{line}\}$ satisfying the following conditions:

(L_1) For every two points x and y of L, there exists exactly one line which is incident with x and y.
(L_2) Every line is incident with at least two points; there are at least two (distinct) lines.

3.1 Theorem. *Let L be a linear space. For every two lines g and h, there is at most one point of L which is incident with g and with h.*

Proof. The proof is obvious. □

For linear spaces, we use the following terminology: If a point x is incident with a line g, we say that x **is on** g or that g **goes through** x.

A line, being incident with two points x and y, is called the line **through** x **and** y. It is denoted by xy.

Conversely, a point, which is incident with two lines g and h, is called the **intersection point** of g and h. It is denoted by $g \cap h$.

Definition. Let L be a linear space, and let U be a set of points of L. U is called a **subspace** of L if for any two points x and y of U, all points of the line xy are contained in U.

Examples of subspaces of a linear space L are the empty set, a single point, a line or the whole point set of L. In linear spaces, we identify the point set of a subspace with the subspace itself. If all points of a line g are contained in a point set M, we say that the line g itself is contained in M.

3.2 Theorem. *Let L be a linear space. The intersection of an arbitrary family of subspaces of L is a subspace of L.*

Proof. Let $(U_i)_{i \in I}$ be a family of subspaces of L, and let x and y be two points contained in U_i for all i of I. Since U_i is a subspace for all i of I, the line xy is contained in U_i for all i of I. □

Definition. Let L be a linear space.

(a) A set M of points of L is called **collinear** if all points of M lie on a common line of L.
(b) Let M be a set of points of L, and let

$$\langle M \rangle := \cap \, U \,|\, U \text{ is a subspace of } L \text{ containing } M.$$

$\langle M \rangle$ is the smallest subspace of L containing M. It is called **the subspace generated by** M.
(c) Let x, y and z be three non-collinear points of L. The subspace $\langle x, y, z \rangle$ generated by x, y and z is called a **plane** of L.

3.3 Lemma. *Let L be a linear space, and let U be a subspace of L. Furthermore, let M be a set of points of U. Then, $\langle M \rangle$ is contained in U.*

Proof. The proof is obvious. □

Definition. Let L be a linear space. A maximal proper subspace of L is called a **hyperplane** of L.

3.4 Theorem. *Let L be a linear space, and let H be a subspace of L with the property that every line of L has at least one point in common with H. Then, H is a hyperplane of L.*

Proof. Assume that there exists a proper subspace U of L such that H is a proper subspace of U. Let x be a point of U outside of H, and let y be a point of L outside of U. Finally, let g be the line joining x and y. By assumption, the line g and H meet in a point $z \neq x$. It follows that the line $g = xz$ is contained in U, in contradiction to the fact that the point y is on g but not contained in U. $\qquad\square$

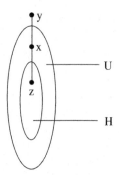

Definition. (a) A **projective plane** is a geometry P of rank 2 over the type set {point, line} satisfying the following conditions:

(PP_1) Every two points of P are incident with exactly one line.
(PP_2) Every two lines of P meet in exactly one point.
(PP_3) Every line is incident with at least three points. There are at least two (distinct) lines.

(b) If the geometry P satisfies the conditions (PP_1), (PP_2) and the weaker condition

(PP_3') Every line is incident with at least two points, and there are at least two lines,

the geometry P is called a **generalized projective plane**.

Remark. Obviously, a projective plane is a linear space. In order to verify Axiom (PP_2) of a projective plane, it suffices to show that any two lines meet in a point. The uniqueness of the intersection point then follows from Theorem 3.1.

3.5 Theorem. *Let P be a generalized projective plane. Then, one of the following cases occurs:*

(i) *P consists of three points and three lines. Every line has exactly two points.*

(ii) *All lines except one line have exactly two points The point set of P consist of the points on the "long" line and one further point.*[1]

(iii) *P is a projective plane.*

Proof. Suppose that P has at least two lines g and h with at least three points. We shall show that every line of P is incident with at least three points:

---

[1]This structure is sometimes called a **near-pencil**.

Let z be the intersection point of g and h, and let x_1 and x_2 be two further points on g and y_1 and y_2 be two further points on h. The lines x_1y_1 and x_2y_2 meet in a point p which is neither incident with g nor with h.

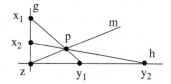

Let m be the line through z and p. If l is a line of P not incident with z, the line l intersects the lines g, h, and m in three different points. If l is a line of P through z different from g, the line l is incident with the three different points z, $l \cap x_1y_1$ and $l \cap x_1y_2$.   □

Definition. (a) An **affine plane** is a geometry A of rank 2 over the type set {point, line} satisfying the following conditions:

(AP_1) Every two points of A are incident with exactly one line.

(AP_2) **Parallel axiom.** Let g be a line and let x be a point not on g. Then, there exists exactly one line h through x which has no point in common with g.

(AP_3) Every line is incident with at least two points. There are at least two (distinct) lines.

(b) Two lines g and h of an affine plane are called **parallel** if $g = h$ or if g and h have no point in common. The parallelism of two lines g and h is denoted by $g \parallel h$.

(c) Let g be a line of an affine plane, and let $\pi(g)$ be the set of lines parallel to g. Then, every set π of the form $\pi = \pi(g)$ is called a **parallel class** of A.

Obviously, an affine plane is a linear space.

3.6 Theorem. *Let A be an affine plane. The relation $\parallel$ is an equivalence relation.*

Proof. Since every line is parallel to itself, the relation $\parallel$ is reflexive. Obviously, the relation $\parallel$ is symmetric. In order to show the transitivity, we consider three lines g, h and l such that g and h are parallel and such that h and l are parallel.

Assume that g and l are not parallel. Then, g and l meet in a point x. Therefore, x is incident with two lines parallel to h, namely g and l, in contradiction to the parallel axiom.   □

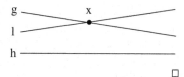

Remark. It follows from Theorem 3.6 that a parallel class $\pi(g)$ is independent of the choice of the line g, that is, $\pi(g) = \pi(h)$ for all h of $\pi(g)$.

The following two theorems show that projective and affine planes are strongly related.

3.7 Theorem. *Let P be a projective plane, and let l be a line of P. Let A be the geometry of rank 2 over the type set {point, line} defined as follows:*

The points of A are the points of P not on l.
The lines of A are the lines of P different from l.
The incidence and the type function of A are induced by the incidence and the type function of P.

The geometry A is an affine plane.

Proof. Axiom (AP_1) follows from Axiom (PP_1).

For the verification of the parallel axiom (AP_2), let g be a line and let x be a point of A not on g. Let $y := g \cap l$ be the intersection point of g and l in P, and let $h := xy$ be the line joining x and y.

(i) Existence: By construction, the point x lies on the line h, and g and h do not have a point of A in common.

(ii) Uniqueness: Let m be a line through x which has (in A) no point in common with g. Since m and g intersect in P, the intersection point lies on l. It follows that $m \cap g = g \cap l = y$. Hence, $m = xy = h$.

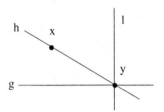

Axiom (AP_3) finally follows from Axiom (PP_3). Note that Axiom (PP_3) provides the existence of a third line of P.   □

Definition. Let P be a projective plane, and let A be the affine plane obtained from P by deleting a line l and the points on l as described in Theorem 3.7. The affine plane A is denoted by $P \setminus l$.

In general, the affine plane $P \setminus l$ depends on the choice of the line l. It can occur that for two distinct lines l and m of a projective plane P, the affine planes $P \setminus l$ and $P \setminus m$ are not isomorphic.[2]

Every parallel class of $P \setminus l$ corresponds to the set of lines of P through a point of l. This fact motivates the following definition and the subsequent theorem.

Definition. Let A be an affine plane, and let $P(A)$ be the geometry of rank 2 over the type set {point, line} defined as follows:

The points of $P(A)$ are the points of A and the parallel classes of A.

The lines of $P(A)$ are the lines of A and an additional line l. This line is called **the line at infinity**.

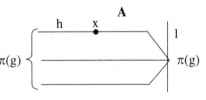

The incidence between points and lines is defined as follows:

_____

	Line h of A	Line l at infinity
Point x of A	Incidence as in A	x and l are not incident
Parallel class $\pi(g)$	$\pi(g)$ and h are incident if and only if h is contained in $\pi(g)$	$\pi(g)$ and l are for every parallel class $\pi(g)$ incident (The points of l are exactly the parallel classes of A)

$P(A)$ is called the **projective closure** of A.

3.8 Theorem. *Let A be an affine plane. The projective closure $P(A)$ of A is a projective plane.*

Proof. Verification of (PP_1): Let x and y be two points of $P(A)$.

If x and y both are contained in A, the line joining x and y in A is the unique line joining x and y in $P(A)$.

If x is a point of A and if $y = \pi(g)$ is a parallel class of A, by the parallel axiom, there exists exactly one line h of $\pi(g)$ through x. This line h is the unique line joining x and y in $P(A)$.

If $x = \pi(g)$ and $y = \pi(h)$ are two parallel classes of A, the line l is the unique line joining x and y in $P(A)$.

Verification of (PP_2): Let g and h be two lines of $P(A)$.

If g and h both are lines of A, one has to distinguish whether g and h are parallel or not. If they are not parallel, there exists an intersection point in A and therefore also in $P(A)$. If g and h are parallel, g and h both are contained in $\pi(g) = \pi(h)$, therefore, g and h intersect in $P(A)$ in the point $\pi(g)$.

If g is a line of A and if $h = l$ is the line at infinity of $P(A)$, g and h meet in the point $\pi(g)$ of $P(A)$.

Verification of (PP_3): Since there are at least two lines in A, there are at least two lines in $P(A)$ as well.

If g is a line of A, $\pi(g)$ is an additional point on g in $P(A)$, therefore, g is incident with at least three points.

It remains to show that the line l has at least three points. For, let g and h be two lines of A intersecting in a point x. Let y and z be two further points (different from x) on g and h, respectively, and let m be the line joining y and z. Then, the lines g, h and m mutually intersect, hence, the parallel classes $\pi(g)$, $\pi(h)$ and $\pi(m)$ are pairwise distinct.

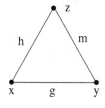

Consequently, the line l has at least three points. □

4   Projective Spaces

The present section is devoted to the discussion of projective spaces. Besides the
definition of projective spaces, the notions subspace, basis and dimension will be
introduced.

The main results of this section are the description of a subspace generated
by a subspace and a point (Theorem 4.2), the existence of bases (Theorem 4.7),
the Exchange Lemma (Theorem 4.9), the Exchange Theorem of Steinitz (Theorem 4.10) and the dimension formula (Theorem 4.15).

Finally, it is shown that projective spaces are thick geometries (Theorem 4.20).

Definition. (a) A **projective space** is a geometry P of rank 2 over the type set
{point, line} satisfying the following conditions:

(PS_1) Any two points are incident with exactly one line.

(PS_2) **Axiom of Veblen–Young.** If p, x, y, a
and b are five points of P such that the lines
xy and ab meet in the point p, the lines xa and
yb also intersect in a point.

(PS_3) Every line is incident with at least three
points. There are at least two (distinct) lines.

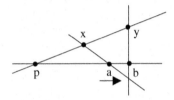

(b) If the geometry P satisfies the conditions (PS_1), (PS_2) and the weaker
condition

(PS_3') Every line is incident with at least two points, and there are at least
two lines,

the geometry P is called a **generalized projective space**.

Sometimes, the empty set, a point, or a line together with the points on that line are
also called projective spaces.

4.1 Theorem. *Let P be a projective space, and let U be a subspace of P such that
there are at least two lines in U. Then, U is a projective space.*

Proof. Verification of (PS_1): If x and y are two points of U, there exists the line
xy in P. By definition of a subspace, this line is contained in U.

Verification of (PS_2): Let p, x, y, a and b be five points such that the lines
xy and ab meet in the point p. Since P is a projective space, the lines xa and yb
intersect in a point s. Since x and a are points of U, it follows that the point s on xa
is contained in U.

Verification of (PS_3): Since every line in P is incident with at least three points,
every line of U is incident with at least three points. By assumption, U contains at
least two lines.   □

The following theorem describes the subspace generated by a subspace and a
point. It is an essential step on the way to introduce the notions dimension and basis.

4.2 Theorem. *Let P be a projective space, let U be a subspace of P, and let p be a
point of P outside of U. We have $\langle U, p \rangle = \cup\{pu \mid u \in U\}$.*

In other words, the subspace $\langle U, p \rangle$ consists of the point p and all points z such that the line pz meets the subspace U in a point.

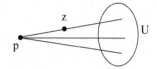

Proof. Let $M := \cup\{pu \mid u \in U\}$.

(i) M is a subspace of P: Let x and y be two points of M, and let z be an arbitrary point on the line xy. We need to show that the point z is contained in M.

1st case. Suppose that the points x and y are contained in U. Then, the line xy is contained in U. Since U is contained in M, it follows that $z \in xy$ is contained in M.

2nd case. Suppose that the point x is contained in U and that the point y is not contained in U. If y is incident with the line px, z is incident with px and therefore contained in M.

Let $y \notin px$. Since y is a point of M, the line py intersects the subspace U in a point b. Since the lines pb and xz meet in the point y, it follows from the axiom of Veblen–Young that the lines pz and xb intersect in a point. Since xb is contained in U, the point z is a point of M.

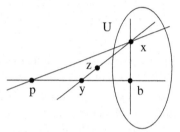

3rd case. Suppose that neither the point x nor the point y is contained in U. If the points p, x and y are collinear,[3] the line $pz = py$ meets the subspace U since y is contained in M. It follows that z is contained in M.

Let p, x and y be non-collinear. Since x and y are contained in M, the lines px and py meet the subspace U in two points a and b. The line $g := ab$ is contained in U. We shall apply the axiom of Veblen–Young twice.

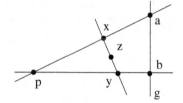

First, we shall construct an auxiliary point s: Since the lines pb and xz meet in the point y, it follows from the axiom of Veblen–Young that the lines px and bz intersect in a point s.

In particular, the lines pa and bz intersect in s, thus, it follows again from the axiom of Veblen–Young that the lines pz and $ab = g$ intersect. Since g is contained in U, it follows that z is a point of M, therefore, M is a subspace.

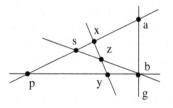

---

[3]This includes the case $x = p$ or $y = p$. W.l.o.g., let $y \neq p$.

(ii) We have $M = \langle U, p \rangle$: Since M is a subspace of P containing p and U, it
follows from Theorem 3.3 that $\langle U, p \rangle$ is contained in M.

In order to show that M is contained in $\langle U, p \rangle$, we consider a point z of M. If
$z = p$, z is contained in $\langle U, p \rangle$. If $z \neq p$, the line pz intersects the subspace U
in a point c. Since z is incident with pc, it follows that z is contained in $\langle U, p \rangle$.
$\square$

4.3 Theorem. *Let P be a projective space, and let $W = \langle U, p \rangle$ be a subspace of
P which is generated by a subspace U of P and a point p of P outside of U.*

Let q be a further point of W outside of U. Then, we have

$$W = \langle U, p \rangle = \langle U, q \rangle.$$

Proof. (i) Since U and q are contained in W, it follows that $\langle U, q \rangle$ is contained
in W.

(ii) Since q is a point of W, it follows from Theorem 4.2 that the line pq intersects
the subspace U. By Theorem 4.2, it follows that p is contained in the subspace
$\langle U, q \rangle$. Thus, $W = \langle U, p \rangle$ is contained in $\langle U, q \rangle$.
$\square$

Definition. Let P be a projective space, and let U be a subspace of P. Furthermore,
let B be a set of points of U.

(a) B is called a **point set generating** U if $\langle B \rangle = U$.
(b) B is called **independent** if for any point x of B, we have: $\langle B \setminus \{x\} \rangle \neq \langle B \rangle$.
(c) B is called a **basis** of U if B is independent and if B is generating U.

4.4 Theorem. *Let P be a projective space, and let U be a subspace of P. Furthermore, let B be a set of points of U. The following conditions are equivalent:*

  (i) The set B is a basis of U.
 (ii) The set B is a minimal point set generating U.
(iii) The set B is a maximal independent subset of U.

Proof. (i) $\Rightarrow$ (ii): If B is a basis of U, B generates U. Assuming that B is not a
minimal point set generating U, there would exist a proper subset B' of B with
$\langle B' \rangle = U$. Let x be a point of $B \setminus B'$. It follows that $\langle B \setminus \{x\} \rangle$ contains $\langle B' \rangle = U$,
in contradiction to the independency of B.

(ii) $\Rightarrow$ (iii): Let B be a minimal point set generating U. Since B is minimal, it
follows that B is independent. The assumption that B is not maximal implies the
existence of an independent subset B' of U such that $\langle B \rangle$ is strictly contained in
$\langle B' \rangle$, contradicting the fact that $\langle B \rangle = U$.

(iii) $\Rightarrow$ (i): Since B is maximal, it follows that B is generating U. Since B is
independent, it follows that B is a basis of U.
$\square$

4.5 Theorem. *Let P be a projective space, and let U be a subspace of P with a
basis B.*

(a) *Let B' be a subset of B, and let $W := \langle B' \rangle$.*

  (i) B' is a basis of W.
 (ii) The point x is not contained in W for all x of $B \setminus B'$.

(b) *If p is a point of P outside of U, $B \cup \{p\}$ is a basis of $\langle U, p \rangle$.*

Proof. (a) (i) By definition, the subspace W is generated by B'. Assume that B' is dependent. Then, there exists a point z of B' with $\langle B' \setminus \{z\}\rangle = W$. It follows that

$$\langle B \setminus \{z\}\rangle = \langle (B' \setminus \{z\}) \cup B \setminus B'\rangle = \langle W, \; B \setminus B'\rangle \supseteq \langle B', \; B \setminus B'\rangle = \langle B\rangle = U,$$

in contradiction to the independency of B.

(ii) Assume that there is an element x of $B \setminus B'$ contained in W. Then, it follows that x is contained in $\langle B'\rangle$. Since $\langle B'\rangle$ is a subset of $\langle B \setminus \{x\}\rangle$, it follows that $\langle B \setminus \{x\}\rangle = \langle B \setminus \{x\}, \; x\rangle = \langle B\rangle = U$, in contradiction to the independency of B.

(b) Let $X := \langle U, \; p\rangle$, and let $C := B \cup \{p\}$. We need to show that C is a basis of X. Obviously, we have $\langle C\rangle = \langle B \cup \{p\}\rangle = \langle U, \; p\rangle = X$.
In order to show that C is independent, let x be a point of C.
If $x = p$, it follows that $\langle C \setminus \{x\}\rangle = \langle B\rangle = U \neq X = \langle C\rangle$.

Let $x \neq p$. It follows from the independency of B that $W := \langle B \setminus \{x\}\rangle \neq \langle B\rangle = U$. Let z be a point of U outside of W. We will show that z is not contained in $\langle C \setminus \{x\}\rangle$. Then, it will follow that $\langle C \setminus \{x\}\rangle \neq X = \langle C\rangle$, that is, the independency of C.

The line pz meets the subspace U in the point z. Since z is not contained in W, the line pz and the subspace W do not have a point in common.
    Because $C = B \cup \{p\}$ and $W = \langle B \setminus \{x\}\rangle$, it follows that $\langle C \setminus \{x\}\rangle = \langle W, \; p\rangle$.
By Theorem 4.2, it follows that z is not contained in $\langle W, \; p\rangle = \langle C \setminus \{x\}\rangle$. □

In order to prove the existence of a basis in an arbitrary projective space, we need the Lemma of Zorn. For details about the Lemma of Zorn, see for example Lang [34].

Lemma of Zorn 4.6. *Let $(M, \leq)$ be a non-empty partially ordered set such that every chain, that is, every totally ordered subset of M, admits an upper bound in M. Then, M has a maximal element.*

4.7 Theorem. *Let P be a projective space, and let U be a subspace of P. Let X be a set of points of U generating U. Then, there exists a basis of U which is contained in X. In particular, there exists a basis of U.*

Proof. We shall apply the Lemma of Zorn. Let M be the set of independent subsets of X. Since M contains at least the empty set, M is not empty. Obviously, M is partially ordered by inclusion.
    Let $(C_i)_{i \in I}$ be a chain of independent subsets of X, and let $A := \cup \{C_i \mid i \in I\}$. We claim that A is independent: Assume that A is dependent. There exists an element x of A such that $\langle A \setminus \{x\}\rangle = A$.

Step 1. The point x is not contained in $\langle C_j \setminus \{x\}\rangle$ for all j of I.

1st case. Suppose that x is contained in C_j for some j of I. Since C_j is independent, it follows that $\langle C_j \setminus \{x\}\rangle \neq C_j$. In particular, x is not contained in $\langle C_j \setminus \{x\}\rangle$.

2nd case. Suppose that x is contained in $\langle C_j \setminus \{x\}\rangle$, but not contained in C_j for some j of I. Since x is contained in $A = \cup\{C_i \mid i \in I\}$, there exists an index k of I such that x is contained in C_k. Since $(C_i)_{i \in I}$ is a chain, it follows that C_j is contained in C_k. It follows that

$$\langle C_j \rangle = \langle C_j \setminus \{x\}\rangle \subseteq \langle C_k \setminus \{x\}\rangle.$$

In view of Case 1, the point x is not contained in $\langle C_k \setminus \{x\}\rangle$, hence x is not contained in $\langle C_j \setminus \{x\}\rangle$.

Step 2. Let $(U_i)_{i \in I}$ be a chain of subsets of $\mathbf{P}$. Then, $U := \cup\{U_i \mid i \in I\}$ is also a subspace of $\mathbf{P}$:

For, let x and y be two elements of U. There exist two indices i and j of I such that x is contained in U_i and y is contained in U_j. Since $(U_i)_{i \in I}$ is a chain, we may assume w.l.o.g. that the subspace U_i is contained in U_j. Hence the line xy is contained in U_j and therefore in U.

Step 3. The set A is independent.

We have

$$A = \langle A \setminus \{x\}\rangle$$

$$= \langle (\bigcup_{i \in I} C_i) \setminus \{x\}\rangle \qquad \text{(Definition of } A\text{)}$$

$$= \langle \bigcup_{i \in I}(C_i \setminus \{x\})\rangle$$

$$= \langle \bigcup_{i \in I}\langle C_i \setminus \{x\}\rangle\rangle \qquad \text{(By Step 1, the point } x \text{ is not contained}$$
$$\qquad\qquad\qquad\qquad\qquad \text{in}\langle C_j \setminus \{x\}\rangle\text{for all } j \text{ of } I.)$$

$$= \bigcup_{i \in I}\langle C_i \setminus \{x\}\rangle \qquad \text{(By Step 2, the union of a chain of sub-}$$
$$\qquad\qquad\qquad\qquad\qquad \text{spaces is a subspace.)}$$

Since x is contained in A, but not in $\bigcup_{i \in I}\langle C_i \setminus \{x\}\rangle$, this equation yields a contradiction. Hence, A is independent.

Step 4. There exists a basis of U which is contained in X:

In view of Step 3, we can apply the Lemma of Zorn. Hence, there exists a maximal independent set B in X. Since B is a maximal independent subset of X, it follows that X is contained in $\langle B \rangle$. Hence, $U = \langle X \rangle$ is contained in $\langle B \rangle$, that is, B generates U. Hence, B is a basis of U. $\square$

In the following, we shall show that all bases of a subspace U of a projective space either have infinitely many elements or that all bases of U have finitely many elements. In the finite case, we shall show that any two bases are of the same cardinality. In the infinite case, this is also true, but we do not give a formal proof. This number of elements of a base will be called the **dimension** of U.

Major steps to prove this assertion are the Exchange Lemma (Theorem 4.9) and the Exchange Theorem of Steinitz (Theorem 4.10).

4.8 Theorem. *Let P be a projective space, and let U be a subspace of P with basis B. For every point x of U, there exists a finite subset C of B such that x is contained in $\langle C \rangle$.*

Proof. W.l.o.g., we may assume that B is infinite. Let

$$A := \cup \left(\langle C \rangle \mid C \subseteq B, \ C \text{ is finite} \right)$$

and assume that $A \neq \langle B \rangle \ (= U)$.

A is a subspace of U: For, let x and y be two points of A. There exist two finite subsets C and D of B such that x is contained in $\langle C \rangle$ and y is contained in $\langle D \rangle$. Hence, x and y are contained in $\langle C \cup D \rangle \subseteq A$. Hence, the line xy is contained in A.

There exists a point x of B not contained in A: Otherwise, $\langle B \rangle$ would be contained in A, a contradiction.

Finally, the subspace A contains the subspace $\langle x \rangle$, a contradiction. $\square$

4.9 Theorem (Exchange Lemma). *Let P be a projective space, and let U be a subspace of P admitting a basis B. For every point p of U, there exists an element x of B such that $B \setminus \{x\} \cup \{p\}$ is a basis of U.*

"In the basis B, the point x is exchanged by the point p."

Proof. Let p be a point of U, and let r be the minimal number of points $x_1, \ldots, x_r$ of B such that p is contained in $W := \langle x_1, \ldots, x_r \rangle$. The number r exists in view of Theorem 4.8.

By definition of r, the point p is not contained in $W' := \langle x_1, \ldots, x_{r-1} \rangle$.

Since p is not contained in W', it follows from Theorem 4.3 that $W = \langle p, W' \rangle$. In particular, the point x_r is contained in $\langle x_1, \ldots, x_{r-1}, p \rangle$. Let $X := \langle B \setminus \{x_r\} \rangle$. Note that the point p is not contained in X: Otherwise, the point x_r would be contained in X, contradicting the fact that $\langle B \setminus \{x_r\} \rangle \neq \langle B \rangle$.

It follows from Theorem 4.5 that $(B \setminus \{x_r\}) \cup \{p\}$ is a basis of $\langle X, p \rangle = U$. $\square$

4.10 Exchange Theorem of Steinitz. *Let P be a projective space, and let U be a subspace with a finite basis B. Furthermore, let C be an independent set of points of U. If $r := |B|$ and $s := |C|$, it holds:*

(a) C is finite, and we have $s \leq r$.

(b) There is a subset B' of B with $|B'| = r - s$ such that $B' \cup C$ is a basis of U.

Proof. Let C be a **finite** set. We shall prove the assertions (a) and (b) by induction on s.

(i) If $s = 1$, U contains at least one point. It follows that $|B| \geq 1 = s$. This shows Part (a). Part (b) is a consequence of the Exchange Lemma.

(ii) $s - 1 \rightarrow s$: Let p be a point of C, and let $C' := C \setminus \{p\}$. By induction (applied on the set C'), we have $s - 1 \leq r$, and there is a subset B'' of B with $|B''| = r - (s - 1)$ such that $B'' \cup C'$ is a basis of U.

(a) We have $s \leq r$: Since, by induction, we have $s - 1 \leq r$, we have to show that $s - 1 \neq r$. Assuming $s - 1 = r$, it follows that $|B''| = r - (s - 1) = 0$, that is, $B'' = \varnothing$. Hence, $C' = B'' \cup C'$ is a basis of U, and it follows that

$$p \in \langle C' \rangle = \langle C \setminus \{p\} \rangle,$$

in contradiction to the independency of C.

(b) By induction, $B'' \cup C'$ is a basis of U where $|B''| = r - (s - 1)$ and $C' = C \setminus \{p\}$. We need to show that there exists a point x of B'' such that $(B'' \setminus \{x\}) \cup C$ is a basis of U.

For, let $t := |B''| = r - (s - 1)$, and let $B'' = \{x_1, \ldots, x_t\}$. For $i = 1, \ldots, t$, let

$$W_i := \langle C', x_1, \ldots, x_i \rangle \text{ and } W_0 := \langle C' \rangle.$$

Then, $\langle C' \rangle = W_0 \subseteq W_1 \subseteq \ldots \subseteq W_t = U$ is a sequence of subspaces. Each subspace W_i has the basis $C' \cup \{x_1, \ldots, x_i\}$, and we have $W_{i+1} = \langle W_i, x_{i+1} \rangle$. Since p is not contained in $\langle C' \rangle$ and since p is contained in U, there exists an index j of $\{0, \ldots, t-1\}$ such that p is not contained in W_j, but p is contained in W_{j+1}.

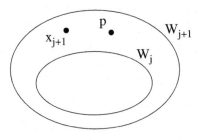

Since $W_{j+1} = \langle W_j, x_{j+1} \rangle$, it follows from Theorem 4.3 that $W_{j+1} = \langle W_j, p \rangle$. It follows from Theorem 4.5 that $C' \cup \{x_1, \ldots, x_j\} \cup \{p\}$ is a basis of W_{j+1}. Repeated application of Theorem 4.5 yields that

$$C' \cup \{x_1, \ldots, x_j, x_{j+2}, \ldots, x_t\} \cup \{p\} = C \cup (B'' \setminus \{x_{j+1}\})$$

is a basis of $W_t = U$.

It remains to show that the set C is finite. Otherwise, C would contain an independent subset C' with $r + 1$ elements. It follows from Part (a) that $r + 1 = |C'| \leq r$, a contradiction.   □

Note that the Exchange Theorem of Steinitz is also valid for a subspace U with an infinite basis, but we do not include a formal proof.

4.11 Theorem. *Let P be a projective space, and let U be a subspace of P. One of the following two cases occurs:*

(i) *All bases of U have infinitely many elements.*

(ii) *All bases of U have finitely many elements. Every two bases of U have the same cardinality.*

Proof. Suppose that U has a basis B with finitely many elements. Let B' be a second basis of U. Since B' is an independent set in U, it follows from the Exchange Theorem of Steinitz (Theorem 4.10) that B' is finite and that the relation $|B'| \leq |B|$ holds. Since B' is a finite basis, the same argument shows that $|B| \leq |B'|$. □

As mentioned earlier, note that any two bases of U are of the same cardinality, not only in the finite case.

Definition. Let P be a projective space, and let U be a subspace of P.

(a) Let B be a basis of U. If B has $d+1$ elements, d is called the **dimension of** U and is denoted by dim U. If B has infinitely many elements, the subspace U is called **infinite-dimensional**.
(b) A d-**dimensional projective space** is a projective space of the **finite** dimension d.

For the empty set $\emptyset$, a point x, a line g and a plane E, we have dim $\emptyset = -1$, dim $x = 0$, dim $g = 1$, and dim $E = 2$.

From now on, we will mainly consider projective spaces of **finite** dimension d.

4.12 Theorem. *Let P be a projective space, and let U be a finite-dimensional subspace of P. If W is a subspace of U with a basis C, the set C can be extended to a basis of U.*

Proof. Let B be a basis of U. Since C is a basis of $W \subseteq U$, the set C is an independent set in U. It follows from the Exchange Theorem of Steinitz (Theorem 4.10) that there is a subset B' of B such that $(B \setminus B') \cup C$ is a basis of U. □

Note that the above theorem is also valid for an infinite-dimensional subspace U of P.

4.13 Theorem. *Let P be a projective space.*

(a) If U and W are two finite-dimensional subspaces of P with $U \subseteq W$, we have dim $U \leq$ dim W with equality if and only if $U = W$.
(b) Let U be a subspace of P, and let x be a point outside of U. Then, dim $\langle U, x \rangle =$ dim $U + 1$.

Proof. (a) The assertion follows from the fact that every basis of U can be extended to a basis of W (Theorem 4.12).
(b) Let B be a basis of U. Then, by Theorem 4.5, $B \cup \{x\}$ is a basis of $\langle U, x \rangle$. It follows that dim $\langle U, x \rangle =$ dim $U + 1$. □

Note that Theorem 4.13 is not valid in the infinite case: If W is an infinite-dimensional subspace with basis $\{x_i \mid i \in N\}$, the subspace $U := \langle x_{2i} \mid i \in N \rangle$ is strictly contained in W, however, the dimension of U equals the dimension of W.

4.14 Theorem. *Let P be a projective space, and let U and W be two non-empty subspaces of P. If $U = \{x\}$ and $W = \{y\}$ are two points, suppose that $x \neq y$. Then,*

$$\langle U, W \rangle = \cup \{uw \mid u \in U, w \in W, u \neq w\}.$$

Proof. Let $M := \bigcup \{uw \mid u \in U, w \in W, u \neq w\}$. Obviously, M is contained in $\langle U, W \rangle$.

It remains to show that $\langle U, W \rangle$ is contained in M: For, let B be a basis of W. For every integer $s \geq -1$, let A_s be the set of points x of $\langle U, W \rangle$ such that x is contained in $\langle U, w_1, \ldots, w_s \rangle$ for some points $w_1, \ldots, w_s$ of B.

Step 1. We have $\langle U, W \rangle = \bigcup_{s \in N} A_s$: Let x be a point of $\langle U, W \rangle$, and let C be a basis of U.

Since $\langle C, B \rangle = \langle U, W \rangle$, by Theorem 4.7, there exist a subset C' of C and a subset B' of B such that $C' \cup B'$ is a basis of $\langle U, W \rangle$. By Theorem 4.8, there exist a finite subset $\{u_1, \ldots, u_r\}$ of C' and a finite subset $\{w_1, \ldots, w_s\}$ of B' such that x is contained in $\langle u_1, \ldots, u_r, w_1, \ldots, w_s \rangle$. In particular, x is contained in $\langle U, w_1, \ldots, w_s \rangle$.

Step 2. The subspace $\langle U, W \rangle$ is contained in M: In view of Step 1, we shall show that the set A_s is contained in M for every integer $s \geq -1$.

For $s = -1$, we have $A_{-1} = U$, and the assertion is obvious.

For $s = 0$, the assertion follows from Theorem 4.2.

$s-1 \rightarrow s$: Let x be a point of A_s not contained in A_{s-1}. By definition of A_s, there exist some elements $w_1, \ldots, w_s$ of B such that x is contained in $\langle U, w_1, \ldots, w_s \rangle$. Let $X := \langle w_1, \ldots, w_s \rangle$ and $X' := \langle w_1, \ldots, w_{s-1} \rangle$. Since x is not contained in A_{s-1}, x is not contained in $\langle U, X' \rangle$.

Since $\langle U, X \rangle = \langle \langle U, X' \rangle, w_s \rangle$, it follows from Theorem 4.2 that there is a point y of $\langle U, X' \rangle$ such that x lies on the line yw_s.

If y is contained in X', it follows that x is contained in $\langle X', w_s \rangle = X$ and hence in M.

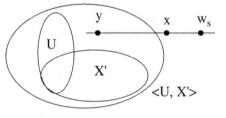

If y is contained in U, it also follows that x is contained in M.

If y is neither contained in U nor in X', there exist by induction two points u of U and x' of X such that y is incident with ux'. The lines ux' and xw_s intersect in the point y.

By the axiom of Veblen–Young, the lines xu and $x'w_s$ meet in a point z.

Since the point z lies on the line $x'w_s$, it follows that z is contained in X. Since x lies on the line zu, it follows that z is contained in M. $\square$

4.15 Theorem (Dimension Formula). *Let P be a projective space, and let U and W be two finite-dimensional subspaces of P. Then,*

$$\dim \langle U, W \rangle = \dim U + \dim W - \dim (U \cap W).$$

Proof. Let $B := \{x_1, \ldots, x_r\}$ be a basis of $U \cap W$. By Theorem 4.12, the base B can be extended to two bases $C := \{x_1, \ldots, x_r, y_1, \ldots, y_a\}$ of U and $D := \{x_1, \ldots, x_r, z_1, \ldots, z_b\}$ of W.

(i) We shall prove by induction on a that $\{x_1, \ldots, x_r, y_1, \ldots, y_a, z_1, \ldots, z_b\}$ is a basis of $\langle U, W \rangle$.

If $a = 0$, $U = U \cap W$, that is, $U \subseteq W$ and $\langle U, W \rangle = W$. It follows that $\langle U, W \rangle = W$ has the basis $\{x_1, \ldots, x_r, z_1, \ldots, z_b\}$.

Let $a > 0$, and let $U' := \langle \{x_1, \ldots, x_r, y_1, \ldots, y_{a-1}\} \rangle$. By induction, the subspace $\langle U', W \rangle$ has the basis $\{x_1, \ldots, x_r, y_1, \ldots, y_{a-1}, z_1, \ldots, z_b\}$.

The point y_a is not contained in $\langle U', W \rangle$: Otherwise, by Theorem 4.14, there is a point u of U' and a point w of W such that $u \neq w$ and such that y_a lies on the line uw. This point w is not contained in U since, otherwise, it would follow that

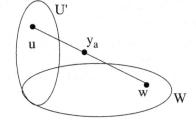

$$w \in U \cap W = U' \cap W \subseteq U',$$

implying that the point $y_a \in uw$ is contained U'.

On the other hand, the point w lies on the line $y_a u$ contained in U.

It follows from this contradiction that y_a is not contained in $\langle U', W \rangle$.

By Theorem 4.5, the set $\{x_1, \ldots, x_r, y_1, \ldots, y_a, z_1, \ldots, z_b\}$ is a basis of $\langle U', W, y_a \rangle = \langle U, W \rangle$.

(ii) We have $\dim \langle U, W \rangle = \dim U + \dim W - \dim (U \cap W)$: For the different subspaces, we have:

Subspace	Basis	Dimension
$U \cap W$	$\{x_1, \ldots, x_r\}$	$r - 1$
U	$\{x_1, \ldots, x_r, y_1, \ldots, y_a\}$	$r + a - 1$
W	$\{x_1, \ldots, x_r, z_1, \ldots, z_b\}$	$r + b - 1$
$\langle U, W \rangle$	$\{x_1, \ldots, x_r, y_1, \ldots, y_a, z_1, \ldots, z_b\}$	$r + a + b - 1$

It follows that

$$\dim\langle U, W \rangle = r + a + b - 1$$
$$= (r + a - 1) + (r + b - 1) - (r - 1)$$
$$= \dim U + \dim W - \dim (U \cap W). \qquad \square$$

4.16 Theorem. *Let P be a projective space. Then, every plane of P is a projective plane.*

Proof. Let E be a plane of P. Axiom (PP_1) follows from Axiom (PS_1).

In order to verify Axiom (PP_2), let g and h be two lines of E. Since $\langle g,\ h \rangle = E$, it follows from the dimension formula (Theorem 4.15):

$$\dim(g \cap h) = \dim g + \dim h - \dim \langle g,\ h \rangle$$
$$= 1 + 1 - 2 = 0.$$

Thus, $g \cap h$ is a point.

Axiom (PP_3) follows from the fact that a plane of P contains at least two lines and that already in P, every line is incident with at least three points.   □

4.17 Theorem. *Let P be a linear space. P is a projective space if and only if every plane of P is a projective plane.*

Proof. (i) Let P be a projective space. In view of Theorem 4.16, every plane of P is a projective plane.

(ii) Let P be a linear space such that every plane of P is a projective plane. We just have to verify the axiom of Veblen–Young.

For, let p, x, y, a and b be five points of P such that the lines xy and ab meet in the point p.

The lines xa and yb are contained in the plane $E := \langle p,\ x,\ y \rangle$. Since E is a projective plane, the lines xa and yb meet in a point.   □

Hyperplanes play a crucial role among the subspaces of a projective space. They are characterized by the fact that every line of the projective space is either contained in the hyperplane or meets the hyperplane in a point (Theorem 4.18).

Definition. Let P be a projective space, and let U and W be two subspaces of P.

(a) U and W are called **complementary** if $U \cap W = \emptyset$ and $\langle U,\ W \rangle = P$.

(b) Suppose that U and W are complementary and that W is of finite dimension r. Then, U is called a subspace of **codimension** $r + 1$.

4.18 Theorem. *Let P be a projective space, and let W be a proper subspace of P.*

(a) The subspace W is a hyperplane if and only if every line of P contains at least one point of W.

(b) The subspace W is a hyperplane of P if and only if W is of codimension 1. If P is of finite dimension d, then every hyperplane of P is of dimension $d - 1$.

(c) Let H be a hyperplane of P. If U is a t-dimensional subspace of P, which is not contained in H, then $\dim (U \cap H) = t - 1$.

Proof. (a) Step 1. Let W be a hyperplane of P. Every line of P contains at least one point of W: Assume that there exists a line g disjoint to W. Let x and y be two arbitrary points on g.

The point y is not contained in the subspace $\langle W,\ x \rangle$: Otherwise, by Theorem 4.2, the line $g = xy$ would intersect W in a point, contradicting the fact that W and g are disjoint.

The fact that $\langle W, x \rangle$ is a proper subspace of P contradicts the maximality of W.

Step 2. Suppose that every line of P contains at least one point of W. By Theorem 3.4, W is a hyperplane of P.

(b) Step 1. If W is maximal, we have $\langle W, x \rangle = P$ for all points x outside of W, hence W is of codimension 1.

Step 2. If W is a subspace of codimension 1, W is obviously maximal.

Step 3. If dim $P = d$ and if H is a hyperplane of P, we have dim $H = d - 1$: Consider a basis B of H and a point x outside of H. It follows from Theorem 4.5 that $B \cup \{x\}$ is a basis of P.

(c) Let g be a line of U. By (a), either the line g is contained in $U \cap H$, or the line g meets $U \cap H$ in a point. It follows from Part (a) that $U \cap H$ is a maximal subspace of U. Hence, we can apply Part (b).   □

4.19 Theorem. *Let P be a d-dimensional projective space, and let U and W be two subspaces of P with $U \cap W = \varnothing$. Then, there exists a subspace W' through W such that U and W' are complementary.*

Proof. Let $\{b_1, \ldots, b_n\}$ and $\{b_{n+1}, \ldots, b_{n+m}\}$ be two bases of U and W, respectively. It follows from the dimension formula (Theorem 4.15) that

$$\dim \langle U, W \rangle = \dim U + \dim W - \dim(U \cap W) = n - 1 + m - 1 - (-1) = n + m - 1,$$

hence, $\{b_1, \ldots, b_{n+m}\}$ is a basis of $\langle U, W \rangle$. By Theorem 4.12, this basis can be extended to a basis $\{b_1, \ldots, b_{d+1}\}$ of P.

Let $W' := \langle b_{n+1}, \ldots, b_{d+1} \rangle$. Then, W' contains the subspace W, and we have

$$\dim (U \cap W') = \dim U + \dim W' - \dim \langle U, W' \rangle = n - 1 + (d + 1 - n - 1) - d = -1,$$

that is, $U \cap W' = \varnothing$.   □

We will conclude this section with Theorem 4.20 stating that the subspaces of a d-dimensional projective space define a thick geometry of rank d over the type set $\{0, 1, \ldots, d - 1\}$.

Definition. Let P be a d-dimensional projective space. Let X be the set of the subspaces of P with $X \neq \varnothing$ and $X \neq P$. Furthermore, let $I := \{0, 1, \ldots, d - 1\}$. The type function type: $X \to I$ is defined by type$(U) := \dim U$. Two subspaces U and W of X are incident (that is, $U * W$) if and only if either U is contained in W or W is contained in U.

Then, $\Gamma := (X, *, \text{type})$ is called a **projective geometry over** I.

4.20 Theorem. *Let P be a d-dimensional projective space. Then, P defines a thick geometry of rank d over $I := \{0, 1, \ldots, d - 1\}$.*

Proof. Let $\Gamma = (X, *, \text{type})$ be the projective geometry defined by P. In order to verify that Γ is a pregeometry, we consider two incident subspaces U and W of P

(w.l.o.g., let U be contained in W) of the same type. Since type(U) = dim U, it follows that dim U = dim W. By Theorem 4.13, we get $U = W$.

In order to show that Γ is a geometry, we consider a flag $F = \{U_1, \ldots, U_r\}$. Since F is a flag, w.l.o.g., we can assume that $U_1 \subseteq \ldots \subseteq U_r$. Repeated application of Theorem 4.12 yields a basis $B = \{x_0, \ldots, x_{d-1}\}$ of $\boldsymbol{P}$ such that for $i = 1, \ldots, r$, we have:

$$U_i = \langle x_0, \ldots, x_{\dim(U_i)} \rangle.$$

For $j = 0, 1, \ldots, d - 1$, let $W_j = \langle x_0, \ldots, x_j \rangle$. Then, $W := \{W_0, \ldots, W_{d-1}\}$ is a chamber of Γ containing F. Thus, Γ is a geometry.

In order to verify that Γ is thick, we consider a co-maximal flag F of type $I \setminus \{i\}$. Let $F = \{U_j \mid j \in I \setminus \{i\}\}$ with dim $U_j = j$.

1st case: Let $i = 0$: Then, F consists of a chain of subspaces of the form

$$U_1 \subseteq \ldots \subseteq U_{d-1}.$$

Since every line is incident with at least three points, there exist three points x, y and z on the line U_1. Hence, F is contained in the chambers $\{x\} \cup F$, $\{y\} \cup F$ and $\{z\} \cup F$.

2nd case: Let $0 < i < d - 1$. Then, F consists of a chain of subspaces of the form

$$U_1 \subseteq \ldots \subseteq U_{i-1} \subseteq U_{i+1} \subseteq \ldots \subseteq U_{d-1}.$$

By Theorem 4.19, there exists a line g in U_{i+1} such that g and U_{i-1} are complementary subspaces.[4] Let x, y, and z be three points on g, and let $U_x := \langle U_{i-1}, x \rangle$, $U_y := \langle U_{i-1}, y \rangle$ and $U_z := \langle U_{i-1}, z \rangle$. Then, F is contained in the chambers $\{U_x\} \cup F$, $\{U_y\} \cup F$ and $\{U_z\} \cup F$.

3rd case: Let $i = d - 1$. Then, F consists of a chain of subspaces of the form

$$U_1 \subseteq \ldots \subseteq U_{d-2}.$$

By Theorem 4.19, there exists a line g in $\boldsymbol{P}$ such that g and U_{d-2} are complementary subspaces. From now on, the proof is analogous to Case 2.   □

4.21 Theorem. *Let Γ be a projective geometry. Then, Γ is a set geometry.*

Proof. The proof is obvious since the subspaces of Γ are defined as subsets of the point set of the underlying projective space $\boldsymbol{P}$.   □

---

[4]Theorem 4.19 is applied as follows: Set $\boldsymbol{P} := U_{i+1}$, $U := U_i$ and $W := \emptyset$. Then, W' is the line g.

5 Affine Spaces

Affine and projective spaces are related in the same way as affine and projective planes (see Sect. 3). In the present section, we shall introduce affine spaces and we shall show that an affine space A can be constructed from a projective space P by deleting all points and lines of a hyperplane of P (Theorem 5.4).

Conversely, a projective space can be obtained from an affine space by adding **points and lines at infinity**. This process will be explained in Theorem 5.5.

Due to this correlation between affine and projective spaces, the notions basis, dimension, etc. can be transferred from projective to affine spaces.

Finally, we will introduce the notion of parallelism for the affine subspaces of an affine space and we will show that for any t-dimensional affine subspace U and any point x of an affine space, there is exactly one t-dimensional affine subspace through x parallel to U (Theorem 5.8).

Definition. Let L be a linear space.

(a) A **parallelism** of L is an equivalence relation $||$ on the set of lines of L such that for every line g and every point x of L, there is exactly one line h through x with $g || h$.[5]
(b) If g and h are two lines of L with $g || h$, the lines g and h are called **parallel**.
(c) If g is a line, the set of the lines parallel to g is called a **parallel class**.
(d) A subspace U of L is called **closed with respect to** $||$ if for every line g of U and every point x of U, the following holds: The (uniquely determined) line h of L through x with $g || h$ is contained in U.
(e) A subspace U of L is called an **affine subspace of L** if U is closed with respect to $||$.

5.1 Theorem. *Let L be a linear space with parallelism $||$. Then, for any two parallel lines g and h, we have either $g = h$, or g and h do not have a point in common.*

Proof. Let g and h be two distinct parallel lines. Assume that g and h intersect in a point x. By assumption, h is a line parallel to g through x. Since $||$ is reflexive, it follows that $g || g$. Hence, g and h are two distinct lines through x parallel to g, a contradiction. □

The converse of Theorem 5.1 is not true. In general, there exist disjoint lines which are not parallel. Two disjoint lines, that are not parallel, are called **skew**.

5.2 Theorem. *Let L be a linear space with parallelism $||$. The intersection of an arbitrary family of affine subspaces is also an affine subspace of L.*

Proof. Let $(U_i)_{i \in I}$ be a family of affine subspaces of L, and let x and g be a point and a line contained in U_i for all i of I. Then, the line through x parallel to g is contained in U_i for all of I. Hence, $\bigcap_{i \in I} U_i$ is closed with respect to $||$. □

---

[5]Since $||$ is an equivalence relation, $||$ is in particular reflexive, that is, $g || g$. If the point x is incident with g, it follows that $h = g$.

Definition. Let L be a linear space with parallelism $\|$, and let M be a set of points of L. Furthermore, let

$$\langle M \rangle_\| := \cap \, U \, | U \text{ is an affine subspace of } L \text{ with } M \subseteq U.$$

Then, $\langle M \rangle_\|$ is the smallest affine subspace of L containing M. $\langle M \rangle_\|$ is called the **affine subspace generated by** M.

Definition. Let A be a geometry of rank 2 over the type set {point, line}.

(a) A is called an **affine space** if A fulfils the following conditions:

   (AS_1) Every two points of A are incident with exactly one line.
   (AS_2) A admits a parallelism $\|$.
   (AS_3) Every affine subspace E of A generated by three non-collinear points is an affine plane. In particular, two lines g and h of E are parallel if and only if $g \cap h = \emptyset$.
   (AS_4) Every line of A is incident with at least two points. There are at least two lines.

(b) Let A be an affine space. An affine subspace of A generated by three non-collinear points of A is called an **affine (sub-)plane** of A.

5.3 Theorem. *Let A be an affine space.*

(a) *If every line of A is incident with at least three points, every subspace of A is an affine subspace, that is, it is closed with respect to $\|$.*
(b) *If every line of A is incident with exactly two points, every subset of A is a subspace of A. There are subspaces of A which are not closed with respect to $\|$.*[6]

Proof. (a) Let U be a subspace of A. Let g be a line of U, and let x be a point of U not incident with g. Furthermore, let h be the line through x parallel to g. We have to show that h is contained in U.

For, let y be an arbitrary point on g, and let z be a further point on the line xy different from x and y. Furthermore, let y_1 and y_2 be two points on g different from y.

Let E be the affine subspace of A generated by g and x. By Axiom (AS_3), E is an affine plane containing the line h. Since g and h are parallel, the lines zy_1 and zy_2 intersect the line h in two points x_1 and x_2.[7] Since z, y_1 and y_2 are contained in U, the points x_1 and x_2 are also contained in U.

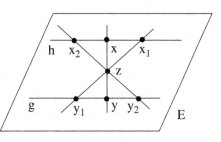

Thus, $h = x_1 x_2$ is contained in U, that is, U is closed with respect to $\|$.

---

[6] We shall see in Corollary 8.3 that any two lines of an affine space have the same cardinality.

[7] Otherwise, g and zy_1 would be two lines through y_1 parallel to h, a contradiction.

(b) Since every line is incident with exactly two points, every subset of A is a subspace of A.

Let $U := \{x, y, z\}$ be a set of three points. Then, U is a subspace of A. By assumption, there exists a line h through z parallel to the line $g := xy$. By Theorem 5.1, the lines g and h do not have a point in common. Let u be the point on h different from z. Since u is not contained in U, the line h is not contained in U. Hence, U is not closed with respect to $\|$, that is, U is not an affine subspace of A.

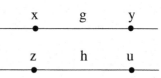

□

5.4 Theorem. *Let P be a projective space, and let H be a hyperplane of P. We define the geometry $A := P \backslash H$ over the type set $\{point, line\}$ as follows:*

The points and lines of A are the points and lines of P not contained in H.
A point and a line of A are incident if and only if they are incident in P.
Then, A is an affine space.

Proof. Verification of (AS_1): Since any two points of P are incident with exactly one line of P, any two points of A are incident with exactly one line of A.

Verification of (AS_2): We first define a parallelism on the lines of A: For any two lines g and h, let $g \| h$ if $g \cap H = h \cap H$.[8] Obviously, $\|$ is reflexive and symmetric. If $g \| h$ and $h \| l$, it follows that $g \cap H = h \cap H = l \cap H$, hence, $g \| l$. Therefore, $\|$ is transitive and, consequently, an equivalence relation.

Let x be a point and let g be a line of A. Let z be the intersection point of g and H, and let h be the line joining x and z. Then,

$$h \cap H = z = g \cap H,$$

that is, $g \| h$, and h is the only line with this property.

Verification of (AS_3): Let x, y and z be three non-collinear points of A, and let E_P and E_A be the projective plane in P and the affine plane in A generated by x, y and z. By Theorem 4.18, $E_P \cap H$ is a line l, and it follows that $E_A = E_P \backslash l$. By Theorem 3.7, E_A is an affine plane and for any two lines g and h of E_A, we have $g \| h$ if and only if $g \cap h = \emptyset$.

Verification of (AS_4): The assertion follows from the fact that every line of P has at least three points and that there are at least two lines in P which are not contained in H.

□

5.5 Theorem. *Let A be an affine space. Then, there exists a projective space P and a hyperplane H of P such that $A = P \backslash H$.*

---

[8]Note that, in view of Theorem 4.18, every line of P meets the hyperplane H in a point.

Proof. In order to define the geometry P, we first construct the hyperplane H: Let H be the set of all parallel classes of A. The point set of P is the (disjoint) union of the point set of A and the set H.

We will define the lines of P as subsets of the point set of P. The incidence will be defined by the set-theoretical inclusion.

For a line g of A, let $\pi = \pi_g$ be the parallel class with respect to $\|$ containing g. Then, π_g is an element of H. π_g is called the **point at infinity** of g. The set $g \cup \{\pi_g\}$ is called the **projective closure** of g. It will be denoted by g_P.

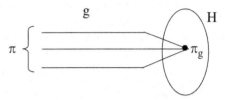

Let E be an affine plane of A, and let E_∞ be the set of the points at infinity of E, that is,

$$E_\infty := \{\pi \mid \exists \text{ line } g \text{ of } E \text{ with } g \in \pi\}.$$

The set E_∞ is called the **line at infinity** of E.

The line set of P is defined as follows: The lines of P are the projective closures of the lines of A and the lines at infinity defined by the affine planes of A.

The incidence between points and lines is defined as follows:

	Projective closure of a line g of A	Line at infinity E_∞ of an affine plane E of A
Point x of A	x and g are incident in A	Never
Point at infinity π	g is contained in π	π is contained in E_∞, that is, there is a line g of E such that g is contained in π.

We shall show that P is a projective space.

Verification of (PS_1): Let x and y be two points of P.

If x and y are points of A, the projective closure of the line of A joining x and y is the (unique) line of P incident with x and y.

Let x be a point of A, and let $y = \pi$ be a point at infinity. By Axiom (AS_2), there exists a unique line g of π through x. The projective closure of g is the unique line through x and $y = \pi$.

Let $x = \pi_x$ and $y = \pi_y$ be two points at infinity. First, we will show the **existence** of a line through x and y:

Let p be an arbitrary point of A. By Axiom (AS_2), there exist two lines g of π_x and h of π_y through p. The lines g and h generate an affine plane E of A. The line at infinity

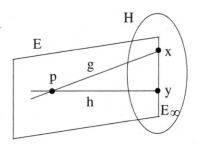

$$E_\infty = \{\pi \mid \exists \text{ line } l \text{ of } E \text{ with } l \in \pi\}$$

contains the points $x = \pi_x$ and $y = \pi_y$.

In order to verify the **uniqueness** of the line through x and y, let E'_∞ be a second line through x and y. Let E' be an affine plane of A such that E'_∞ is the line at infinity of E'. Let p' be a point of E', and let g' and h' be the lines of π_x and π_y, respectively, through p'. We have to show that $E_\infty = E'_\infty$. In view of symmetry, it suffices to verify that E_∞ is contained in E'_∞.

For, let $z = \pi_z$ be a point of E_∞. By definition of E_∞, there exists a line l of E such that z is the point at infinity of l.

W.l.o.g., we can assume p is not incident with l, since, otherwise, we can replace l by a line of E parallel to l.

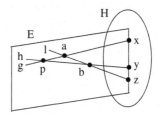

Since E is an affine plane and since l is neither parallel to g nor parallel to h,[9] the line l intersects the lines g and h in two points a and b.

Next, we will construct a line $l' = a'b'$ in E' parallel to the line $l = ab$.

For, consider the affine subplane $\langle p, p', a\rangle$ of A. Since g' and $g = pa$ admit the point x as their point at infinity, we have $g'\|g = pa$. Since the affine plane $\langle p, p', a\rangle$ is closed with respect to $\|$, the line g' is contained in the plane $\langle p, p', a\rangle$.

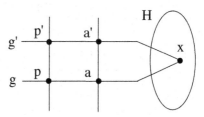

Hence, the line through a parallel to pp' intersects the line g' in a point a'.

Analogously, the line h' is contained in the affine plane $\langle p, p', b\rangle$ generated by p, p' and b, and the line of $\langle p, p', b\rangle$ through b parallel to pp' meets h' in a point b'.

Let $l' := a'b'$ be the line joining a' and b'. Since a', b' are contained in E', the line l' is also contained in E'. We will show that l and l' are parallel.

For, we first will show that l and l' are contained in a common affine plane: Since aa' and bb' both are parallel to pp', it follows that aa' and bb' are parallel, hence, there exists an affine subplane F such that aa' and bb' are contained in F. Since $l = ab$ and $l' = a'b'$, the lines l and l' are also contained in the plane F.

---

[9]Otherwise, we would have $z = x$ or $z = y$.

Assume that l and l' are not parallel. Then, l and l' have a point s in common. It follows that s is contained in $l \cap l' \subseteq E \cap E'$.

Let m_x and m_y be the lines through s with x and y as points at infinity. Then, m_x is parallel to g and to g'. Since E and E' both are closed with respect to $\|$, it follows that m_x is contained in $E \cap E'$. In the same way, it follows that m_y is contained in $E \cap E'$.

In summary, we have $E = \langle m_x, m_y \rangle = E'$, a contradiction.

Hence, the lines l and l' are parallel, that is, l' admits z as point at infinity. This means that z is contained in E'_∞, and it follows that E_∞ is contained in E'_∞. In view of the symmetry of E_∞ and E'_∞, it follows that $E_\infty = E'_\infty$. Hence, E_∞ is the unique line joining x and y.

Verification of (PS_2): We start the verification of the axiom of Veblen–Young with the proof of the following assertion (note that we have already seen that P is a linear space):

(A) Let $E_P := \langle x, y, z \rangle$ be a plane of P such that x is contained in A. Then, E_P is a projective plane.

Since every line of P either is the projective closure of a line of A or is contained in H, the lines xy and xz are projective closures of lines of A. Hence, there is a point a of A on xy different from x and a point b of A on xz different from x. The points x, a and b generate an affine subplane E of A. By construction of P, E_P is the projective closure of E. By Theorem 3.8, E_P is a projective plane.

In order to verify the axiom of Veblen–Young, we consider five points p, x, y, a and b such that the points p, x, y and p, a, b are collinear.

Let E_P be the plane of P generated by p, x and a. Note that y, b are contained in E_P.

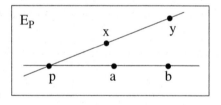

1st case: Outside of $E_P := \langle p, x, a \rangle$, there exists no point of A.

Then, A is an affine plane, and $P = E_P$ is the projective closure of A. By Theorem 3.8, E_P is a projective plane. In particular, the lines xa and yb intersect in a point.

2nd case: There is a point z of A outside of E_P.

Let p', x' and a' be three points of A on the lines zp, zx and za different from z.

Let $F_P := \langle z, p, x \rangle$ be the plane of P generated by z, p and x. By (A), F_P is a projective plane.

Since the points z, p, x, p', x' and y are contained in F_P, the lines $p'x'$ and zy intersect in a point y'.

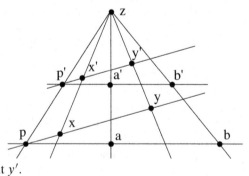

In the same way, it follows that the lines $p'a'$ and zb meet in a point b'.

Next, we consider the plane E'_P of P generated by p', x' and a'. Again, it follows from assertion (A) that E'_P is a projective plane. Hence, the lines $x'a'$ and $y'b'$ meet in a point s'.

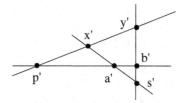

Furthermore, the points z, x, a, x', a' and s' are contained in the plane $G_P :=$ $\langle z, x, a \rangle$ generated by z, x and a. By assertion (A), G_P is a projective plane. Hence, the lines zs' and xa intersect in a point s_1. Analogously, the lines zs' and yb intersect in a point s_2.

We have $s_1 = s_2$, since, otherwise, z is contained in $s_1s_2 \subseteq E_P = \langle p, x, a \rangle$, a contradiction. Hence, the lines xa and yb intersect in the point $s_1 = s_2$, and the axiom of Veblen–Young is verified.

Verification of Axiom (PS_3): By assumption, there exist at least two lines. Since on every line of A there are at least two points, there are at least three points on every line of P not contained in H. Since the lines of H are the lines at infinity of the affine subplanes of A, it follows from Theorem 3.8 that the lines of H are incident with at least three points, too. $\square$

5.6 Theorem. *Let A be an affine space, and let $P = P(A)$ be the projective closure of A. Furthermore, let U be an affine subspace of A and let U_P be the projective closure of U.*

(a) U_P consists of the points of U and the points at infinity of the lines in U.
(b) We have $U = U_P \backslash (H \cap U_P)$ where H is the hyperplane at infinity of A.

Proof. (a) and (b) follow from of the proof of Theorem 5.5. $\square$

Theorems 5.5 and 5.6 allow to transfer the notions basis and dimension to affine spaces.

Definition. Let A be an affine space, and let P be a projective space with a hyperplane H such that $A = P \backslash H$.

(a) P is called the projective closure of A. It is denoted by $P(A)$. The hyperplane H is called the **hyperplane at infinity** of A.
(b) If U is an affine subspace of A, there exists a subspace U_P of P such that

$$U = U_P \backslash (U_P \cap H).$$

The subspace U_P is called the **projective closure** of U.

Definition. Let A be an affine space, and let U be an affine subspace of A. Furthermore, let B be a set of points of U.

(a) B is called a point set **generating** U if $\langle B \rangle = U$.
(b) B is called **independent** if for any point x of B, it holds that $\langle B \backslash \{x\} \rangle \neq \langle B \rangle$.
(c) B is called a **basis** of U if B is independent and if B generates U.

(d) If B is a basis of U with $d + 1$ elements, d is called the **dimension** of U and is denoted by dim U.

(e) A d-**dimensional** affine space is an affine space of **finite** dimension d.

5.7 Theorem. *Let A be an affine space, and let $P = P(A)$ be the projective closure of A. Furthermore, let U be an affine subspace of A and let U_P be the projective closure of U. Finally, let B be a set of points of U.*

(a) B generates U if and only if B generates U_P.

(b) B is independent in A if and only if B is independent in P.

(c) B is a basis of U if and only if B is a basis of U_P. There exists a basis of U_P which is completely contained in U.

(d) We have dim U = dim U_P.

Proof. (a) follows from the fact that in P the relation $\langle U \rangle = \langle U_P \rangle$ is valid.

(b) follows from (a).

(c) follows from (a) and (b). Note that since $\langle U \rangle = \langle U_P \rangle$, by Theorem 4.7, there exists a basis of U_P contained in U.

(d) follows from (c).   □

Definition. Let A be an affine space, let $P = P(A)$ be the projective closure of A, and let H be the hyperplane at infinity.

(a) Two t-dimensional affine subspaces U and W of A are called **parallel** if $U \cap H = W \cap H$.

(b) Two arbitrary affine subspaces U and W of A are called **parallel** if U is parallel to an affine subspace of W or if W is parallel to an affine subspace of U.

For an affine subspace X of A, we occasionally use the notation X_A to emphasize that X is an affine subspace. For the projective closure of X, we occasionally use the notation X_P.

5.8 Theorem. *Let A be an affine space. If U is a t-dimensional affine subspace of A and if x is a point of A, there exists a unique t-dimensional affine subspace W of A parallel to U such that x and W are incident.*

Proof. Let P be the projective closure of A, and let H be the hyperplane of P such that $A = P \backslash H$. For an affine subspace X of A, we shall use the notation X_A and X_P, respectively, to indicate whether X is considered as an (affine) subspace of A or as a subspace of P.

Let $Z := U_P \cap H$ (in P). Since U_A is an affine subspace of A, the subspace U_P is not contained in H. Hence, dim $Z = t - 1$.

Let $W_P := \langle x, Z \rangle_P$. Since x is not contained in H, it follows that $W_P \cap H = Z = U_P \cap H$. Hence, W_A is a t-dimensional affine subspace of A through x parallel to U_A.[10]

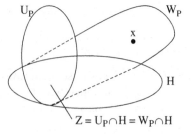

$Z = U_P \cap H = W_P \cap H$

---

[10]Note that it may occur that $U = W$.

On the other hand, for any t-dimensional affine subspace W'_A of A parallel to U_A, we have $W'_P \cap H = Z$, hence, $W'_P = \langle x, Z \rangle_P = W_P$. In summary, there is one and only one t-dimensional affine subspace through x parallel to U.   □

5.9 Theorem. *Let $A = P \backslash H$ be an affine space. If g and L are a line and a hyperplane of A, then g and L are parallel, or they intersect in a point.*

Proof. Let g and L be non-parallel. Then, g is not contained in L. We have to show that g and L meet in a point of A.

Since g (as a line of A) is not contained in H, by Theorem 4.18, there exist the intersection points $x := H \cap g$ and $y := L \cap g$ (in P).

It remains to show that y is a point of A. Assume on the contrary that y is not contained in A. Then, y is contained in H.

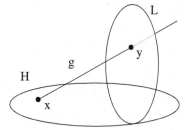

If $x \neq y$, the points x and y both are contained in H, hence, g is contained in H, a contradiction. If $x = y$, then $g \cap H$ is contained in $L \cap H$, hence, g is parallel to L contradicting the assumption that g and L are not parallel.   □

At the end of this section, we shall see that the subspaces of a d-dimensional affine space define a firm geometry of rank d over the type set $\{0, 1, \ldots, d-1\}$.

Definition. Let A be a d-dimensional affine space. Let X be the set of the subspaces of A with $X \neq \emptyset$ and $X \neq A$. Furthermore, let $I := \{0, 1, \ldots, d-1\}$. The type function type : $X \rightarrow I$ is defined by type$(U) := \dim U$. Two subspaces U and W of X are incident (that is, $U * W$) if and only if $U \subseteq W$ or $W \subseteq U$.

Then, $\Gamma := (X, *, \text{type})$ is called an **affine geometry over** I.

As for projective geometries, we will often use the notions "affine space" and "affine geometry" synonymously.

5.10 Theorem. *Let A be a d-dimensional affine space. Then, A defines a firm geometry of rank d over $I := \{0, 1, \ldots, d-1\}$.*

Proof. The proof of Theorem 5.10 is similar to the proof of Theorem 4.20 for projective geometries. The fact that A defines a firm and not necessarily a thick geometry is due to the fact that for an affine space, we only assume that the lines are incident with at least two points.   □

6   A Characterization of Affine Spaces

In Theorem 4.17 we have seen that a linear space P is a projective space if and only if every plane of P is a projective plane. The analogous result for affine spaces is true if one assumes that every line is incident with at least four points. However, the proof of this result due to F. Buekenhout [12] is much more complex. It is the main subject of the present section. Readers who do not want to invest too much time at this point are recommended to skip its proof at a first reading.

6.1 Theorem (**Buekenhout**). *Let A be a linear space, fulfilling the following conditions:*

   (i) Each line of A is incident with at least four points.
  (ii) Each plane of A is an affine plane.
 (iii) A contains at least two lines.

Then, A is an affine space.

Proof. We shall define a relation $\|$ on the lines of A of which we shall show that it is a parallelism.

For two lines g and h of A, we define $g \| h$ if g and h are contained in a common plane of A and if they are parallel in this (affine) plane. Two lines fulfilling the relation $\|$ are called **parallel**.

Verification of (AS_1): Axiom (AS_1) is fulfilled, since A is a linear space.

Verification of (AS_3): Axiom (AS_3) would immediately follow from Property (ii) and Axiom (AS_2). However, we need Axiom (AS_3) to verify Axiom (AS_2).

We need to show that if E is a plane of A and if g is a line of E, then every line of A parallel to g and intersecting E in at least one point is contained in E. We first will prove the following assertion (A):

(A) *Let g be a line of A, and let x be a point of A outside of g. Then, there exists exactly one line h through x parallel to g:*

Existence: Let $E := \langle x, g \rangle$ be the plane generated by x and g. Since E is an affine plane, there exists in E a line through x parallel to g.

Uniqueness: Let h_1 and h_2 be two lines through x parallel to g. By definition of the relation $\|$, the lines g and h_1 are contained in a plane E_1, and the lines g and h_2 are contained in a plane E_2. By Theorem 5.3 (a), it follows that

$$E_1 = \langle g, h_1 \rangle = \langle g, x \rangle = \langle g, h_2 \rangle = E_2.$$

Hence, h_1 and h_2 are two lines through x parallel to g in the affine plane E_1. It follows that $h_1 = h_2$. The assertion (A) is proved.

In order to verify Axiom (AS_3), we consider a plane E of A, a line g and a point x of E. Since E is an affine plane, there exists in E a line h through x parallel to g. By assertion (A), the line h is the only line through x parallel to g. Hence, every line through x parallel to g is contained in E, that is, E is closed with respect to $\|$.

Verification of (AS_4): The validity of Axiom (AS_4) follows from Assumptions (i) and (iii).

Verification of (AS_2): Obviously, the relation $\|$ is reflexive and symmetric. The verification of the transitivity is rather elaborate. The main part of the proof is to verify the following assertion (B):

(B) Let E be a plane of A, and let g be a line of A, intersecting the plane E in a point a. Furthermore, let U be the set of all points x of A such that x is either incident with g, or the planes $\langle x, g \rangle$ and E have a line in common.
  Then, U is a subspace of A.[11]
  We will prove Assertion (B) in several steps.

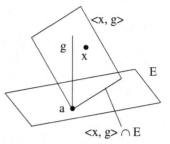

Step 1. Definition of a projection: Let u be a point of U not on g. By assumption, the planes $\langle g, u \rangle$ and E intersect in a line h.

Let u_g be the intersection point of g and the line through u parallel to h.[12]
  For any point p on g different from u_g, let $p(u)$ be the intersection point of the lines pu and h. Since h is contained in E, the point $p(u)$ is contained in E.
  If p is different from a, it follows that $p(u) \neq a$ since the point u is not on g.

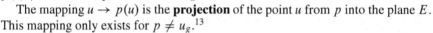

The mapping $u \to p(u)$ is the **projection** of the point u from p into the plane E. This mapping only exists for $p \neq u_g$.[13]
  In the proof of Assertion (B), we will repeatedly project points and lines into the plane E and use the fact that E is an affine plane.
  In order to verify Assertion (B), we will consider two points x and y of U, and we will show that each point on the line $m := xy$ is contained in U. W.l.o.g., we may assume that $m \neq g$.
  Step 2. If m and g are contained in a common plane, m is contained in U:
  For, let $F := \langle g, m \rangle$ be the plane generated by m and g, and let z be an arbitrary point on m. If z is incident with g, z is contained in U. Let z be a point not on g. W.l.o.g., the point x is also not incident with g.
  Since x is a point of U, the plane $\langle g, z \rangle = F = \langle g, m \rangle = \langle g, x \rangle$ intersects the plane E in a line. Hence, z is contained in U. It follows that m is also contained in U.
  In the following, we will assume that the lines m and g are not contained in a common plane. In particular, $E_p := \langle p, m \rangle$ is a plane for all points p on g.
  Step 3. Let p be a point on g outside of $\{a, x_g, y_g\}$. The point p exists since, by assumption, all lines of A have at least four points. Then, the planes E and $E_p = \langle p, m \rangle$ intersect in a line h_p:

---

[11]In fact, it will turn out that U is the 3-dimensional affine subspace of A generated by E and g.
[12]Since the plane $\langle g, u \rangle$ is closed with respect to $||$, the line l through u parallel to h is contained in $\langle g, u \rangle$. Since g and h meet in a point, l and g also meet in a point u_g.
[13]We will study projections in some more detail in Sect. 3 of Chap. 2.

For, we first will show that $p(x) \neq$
$p(y)$. Assume that $p(x) = p(y)$. Then, the
points x and y both lie on the line $pp(x)$.
Hence, the lines g and $m = xy$ intersect in
the point p, contradicting the assumption of
Step 2 that m and g are not contained in a
common plane.

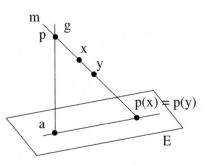

Since both $p(x)$ and $p(y)$ are contained
$E_p \cap E$, it follows that $h_p = E_p \cap E$ is the
line $p(x)p(y)$.

Step 4. Let p be a point on g outside of $\{a, x_g, y_g\}$. If the lines m and h_p are
parallel, the line m is contained in U.

Let z be an arbitrary point on m
different from x and y. Since, by
assumption, the lines m and h_p are
parallel lines in the plane $E_p =$
$\langle p, m \rangle$, the lines pz and h_p meet in a
point $p(z)$. Hence, the planes $\langle g, z \rangle$
and E intersect in the line $ap(z)$. It

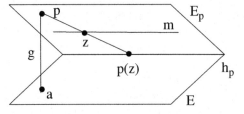

follows that z is contained in U, and, consequently, m is contained in U.

In view of Step 4, we may assume in the following that the lines m and h_p meet
in a point b.[14]

Since h_p is contained in E, the point b is contained in E. It follows that
$b = m \cap E$.

Step 5. We have $a \neq b$:

Assume on the contrary that $a = b$. Then, the lines g and m intersect in the point
$a = b$, in contradiction to the assumption of Step 2 saying that m and g are not
contained in a common plane.

Step 6. Let p be a point on g outside of $\{a, x_g, y_g\}$. Let c_p be the unique point
on $m = xy$ such that the line pc_p is parallel to the line h_p. Then, every point of m
different from c_p is contained in U.

In the plane $E_p = \langle p, m \rangle$ generated by p and m, there exists exactly
one line through p parallel to h_p.
This line intersects m in the point c_p.
For each point z of m different from
c_p, the lines pz and h_p intersect in a
point $p(z)$ contained in E. Since the
point p is different from a, we have
$p(z) \neq a$ (see Step 1). Hence, the

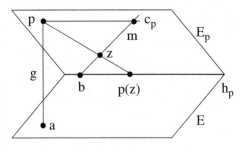

planes $\langle g, z \rangle$ and E intersect in the line $ap(z)$. It follows that z is contained in U.

---

[14]Note that in view of Step 3, both m and h_p are contained in the (affine) subplane E_p.

In Steps 7 and 8, we will investigate the case that $x_g = y_g$.

Step 7. If $x_g = y_g$, the planes $\langle x, y, x_g \rangle$ and E meet in the point b.

Assume that the planes $\langle x, y, x_g \rangle$ and E
intersect in a line h. In the plane $\langle x, y, x_g \rangle$,
at most one of the two lines xx_g and yx_g
is parallel to h. W.l.o.g., we assume that the
lines xx_g and h meet in a point r on h. Note
that the line h is contained in E.

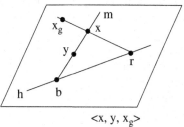

On the other hand, the line xx_g is parallel
to the line $\langle g, x \rangle \cap E = ra$.[15] Since r is on
two parallel lines ra and xx_g, it follows that $ra = xx_g$. Thus, the point x on ra is
contained in E, in contradiction to the choice of x.

Step 8. If $x_g = y_g$, m is contained in U.

By Step 7, the planes $\langle x, y, x_g \rangle$ and E intersect in the point b. Since every line
of A is incident with at least four points, there exist two points x' and x'' on the line
xx_g different from x and x_g.

The line $x'y$ is contained in U:
Otherwise, it follows from Step 4
that the line $x'y$ and the plane E
intersect in a point b'. It follows that

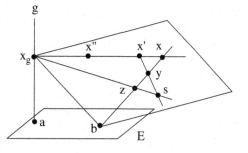

$$b' = x'y \cap E \subseteq \langle x, y, x_g \rangle \cap E = b,$$

thus, $b = b'$. Hence, the point x'
lies on the line $b'y = by = xy$, a
contradiction.

In the same way, it follows that the line $x''y$ is contained in U.

Let z be a point on m different from x, y and b. Then, at most one of the lines
$x'y$ or $x''y$ is parallel to the line $x_g z$.

W.l.o.g., we assume that $x_g z$ and $x'y$ meet in a point s. Since the line $x'y$ is
contained in U, the point s is contained in U as well, that is, the planes $\langle g, s \rangle$ and
E intersect in a line. Since $\langle g, s \rangle = \langle g, z \rangle$, the point z is contained in U. It follows
that m is contained in U.

Step 9. It follows from Step 8 that we can assume for all points r, s of $m \cap U$
that $r_g \neq s_g$.

Step 10. For every point p on g different from a, the planes $E_p := \langle p, m \rangle$ and
E intersect in a line h_p:

---

[15]Note that the line xx_g is contained in the plane $\langle g, u \rangle$ (see Step 1), hence the point r is contained
in $\langle g, u \rangle$. Therefore, the points a and r are contained in both planes $\langle g, u \rangle$ and E.

By Step 6, there exists at most one point c_p on m which is not contained in U. Since m is incident with at least four points, there are at least two points u, v of $m \cap U$ different from b and c_p. By Step 9, we may assume that $u_g \neq v_g$. W.l.o.g., let $u_g \neq p$. By Step 1, the line pu and the plane E intersect in a point $p(u)$.

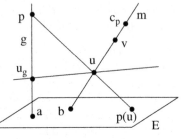

Since the lines pu and m are distinct, it follows that $p(u) \neq b$. Hence, $h_p = E_p \cap E$ is the line $p(u)b$.

Step 11 (Definition of two auxiliary lines l_1 and l_2). Let p be a point on g different from a. Consider the line of $E_p = \langle p, m \rangle$ through p parallel to m. This line intersects the line $h_p = E_p \cap E$ in a point p'.

Let l_1 be the line ap', and let l_2 be the line through a parallel to h_p.[16]

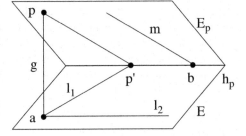

Step 12. Let F be a plane through g meeting the plane E in a line l. If F and m are disjoint, then $l = l_1$ or $l = l_2$:

In particular, there exist at most two planes through g intersecting the plane E in a line and being disjoint to m, namely the planes $\langle g, l_1 \rangle$ and $\langle g, l_2 \rangle$.

Since $g \cap E = a$, the point a is incident with the line $l = F \cap E$. Let $l \neq l_1$ and $l \neq l_2$. Then, l meets the line h_p in a point r different from p'.

Consider the plane $E_p = \langle m, p \rangle$. Since the line pp' is the line through p parallel to m (Step 11), the lines pr and m meet in a point s. It follows that

$$F \cap m \supseteq pr \cap m = s.$$

Hence, F and m are not disjoint.

Step 13. For a point $p \neq a$ of g, let $F(p) := \langle g, l_2 \rangle$ be the plane generated by g and $l_2 = l_2(p)$. Then, either m is contained in U or $F(p) \cap m = \emptyset$:

---

[16]Note that the lines l_1 and l_2 depend on the choice of the point p. More precisely, the lines should be denoted by $l_1(p)$ and $l_2(p)$.

Let s be a point of $F(p) \cap m$. We first will show that $s = c_p$ (see Step 6).

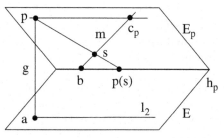

Assume that $s \neq c_p$. By definition, c_p is the unique point on m such that the line pc_p is parallel to the line h_p.

Since $s \neq c_p$, the lines ps and h_p intersect in a point $p(s)$. We have

$$p(s) \in F(p) \cap E = l_2.$$

Since l_2 and h_p are parallel and since $p(s)$ is contained in $l_2 \cap h_p$, it follows that $l_2 = h_p$.

Thus, the point a is contained in $l_2 = h_p = E \cap \langle m, p \rangle$, hence, the line $g = pa$ is contained in the plane $E_p = \langle m, p \rangle$. It follows that g and m are co-planar, a contradiction (see Step 2).

It follows that $c_p = s$ is contained in $F(p)$. It follows from $F = \langle g, l_2 \rangle$ that c_p is contained in U. By Step 6, the whole line m is contained in U.

Step 14. Assume that the line m is not contained in U.

This assumption will lead to a contradiction, finalizing the proof of Assertion (B).

By Step 12, there exist at most two planes through g disjoint to m and intersecting E in a line.

Let p, p' and p'' be three points on g different from a. By Step 13, we can assume that the planes $F(p)$, $F(p')$ and $F(p'')$ are disjoint to m. (Otherwise, m is contained in U, and the assertion (B) is proven.) Since any of the planes $F(p)$, $F(p')$ and $F(p'')$ intersects the plane E in a line, at least two of the three planes $F(p)$, $F(p')$ and $F(p'')$ must be identical. W.l.o.g., let $F(p) = F(p')$. It follows that $l_2(p) = l_2(p')$.[17] As before, let $h_p := \langle p, m \rangle \cap E$ and $h_{p'} := \langle p', m \rangle \cap E$.

Since h_p is parallel to $l_2(p)$, since $h_{p'}$ is parallel to $l_2(p')$ and since $l_2(p) = l_2(p')$, it follows that h_p is parallel to $h_{p'}$ ($\|$ is an equivalence relation in E).

Since b is contained in $h_p \cap h_{p'}$, it follows that $h_p = h_{p'}$. Hence, the points p and p' are contained in the plane $\langle m, h_p \rangle$.[18] It follows from $g = pp'$ that m and g are co-planar, a contradiction.

The assertion (B) is proven.

In order to verify the transitivity of $\|$, let g_1, g_2 and g_3 be three lines of A such that g_1 is parallel to g_2 and such that g_2 is parallel to g_3. We have to show that g_1 is parallel to g_3.

If the lines g_1, g_2 and g_3 are contained in a common plane, it follows that g_1 and g_3 are parallel since every plane of A is an affine plane. In the following, we will assume that the lines g_1, g_2 and g_3 are not contained in a common plane.

---

[17]By, assumption, the planes $F(p)$ and $F(p')$ are disjoint to m. By definition, we have $F(p) = \langle g, l_2(p) \rangle$ and $F(p') = \langle g, l_2(p') \rangle$. Since $l_2(p)$ and $l_2(p')$ are lines contained in E, we have $l_2(p) = E \cap F(p) = E \cap F(p') = l_2(p')$.

[18]Note that in view of Step 10, the line h_p is contained in the plane $E_p = \langle p, m \rangle$. Hence, the point p is contained in the plane $E_p = \langle p, m \rangle = \langle m, h_p \rangle$.

Let E be the plane generated by the
lines g_1 and g_2, and let x and y be two
points on the lines g_2 and g_3, respectively.
Furthermore, let $m := xy$ be the line
through x and y, and let U be the subspace
generated by E and m. Finally, let h be the
line through y parallel to g_1 (this line exists
in view of assertion (A)). We shall prove the
parallelism of g_1 and g_3 in several steps:

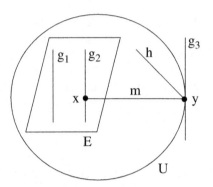

Step 1. The line h is contained in U:

Since U is a subspace of A, the plane $\langle g_1, y\rangle$ is contained in U. Since the planes
of A are closed with respect to $\|$, it follows that h is contained in $\langle g_1, y\rangle$, hence h
is contained in U.

Step 2. We have $h \cap E = \varnothing$:

Otherwise, the line h would be a line through $h \cap E$ parallel to g_1. Since the
plane E is closed with respect to $\|$, it follows that h is contained in E. In particular,
the point y is contained in E. Hence, g_3 is a line through the point y contained in
E which is parallel to g_2. Again, since E is closed with respect to $\|$, it follows that
g_3 is contained in E, in contradiction to the assumption that the lines g_1, g_2 and g_3
are not contained in a common plane.

Step 3. Let l be the line of A through x parallel to h. Furthermore, let $F :=$
$\langle m, h\rangle$ be the plane generated by m and h. Then, $E \cap F = l$:

Since U is a subspace of A, F is contained in U. In view of Assertion (B), the
planes F and E intersect in a line l'.[19] Since $h \cap E = \varnothing$ (Step 2), it follows that

$$h \cap l' \subseteq h \cap E = \varnothing.$$

Hence, l' is a line through x in F parallel to h, that is, $l' = l$.

Step 4. We have $l = g_2$:

Assume that $l \neq g_2$. Then, l and g_2 meet
in the point x. Since g_1 and g_2 are parallel
and since there is exactly one line through
x parallel to g_1, the lines g_1 and l are not
parallel. On the other hand, the lines g_1 and
l both are contained in the plane E, hence,
they meet in a point z.

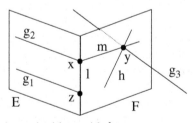

Since $E \cap h = \varnothing$ (Step 2), it follows that z is not incident with h.

---

[19]Let s be an arbitrary point on h different from $y = h \cap m$. By Assertion (B), the subspace U
consists of all points t of A such that t is either incident with m, or the planes $\langle t, m\rangle$ and E have a
line in common. Since the line h is contained in U, the point s is contained in U as well. Since the
point s is not incident with m, it follows that the planes $\langle s, m\rangle = \langle h, m\rangle = F$ and E have a line
in common. This line is denoted by l'.

Since g_1 and h are parallel, they generate a common plane. It follows that $\langle g_1, h \rangle = \langle z, h \rangle = \langle h, l \rangle$ (since the lines h and l are parallel, they generate a plane).

In particular, it follows that h is contained in $\langle g_1, l \rangle = E$, in contradiction to the fact that $E \cap h = \varnothing$.

Step 5. We have $g_1 \parallel g_3$:

By Step 4, we have $l = g_2$. By construction (Step 3), the line h is parallel to l, and the line g_3 is parallel to g_2. Since $l = g_2$, it follows that h and g_3 are lines through y parallel to l. Hence, $h = g_3$. Since h is parallel to g_1, it follows that g_1 and g_3 are parallel, and the proof is completed.   □

7 Residues and Diagrams

"J. Tits has achieved a far-reaching generalization of projective geometry, including a geometric interpretation of all simple groups of Lie type (...). One of the most fascinating features of this theory is that each of its geometries is essentially determined by a diagram (the Coxeter diagram). These nice and simple pictures with an enormous potential of information might very well appear as pieces of that universal language that some people want to elaborate in order to communicate with hypothetical extraterrestrial beings." [14]

These diagrams are the subject of the present section. They are ideograms of geometries containing essential information about the structure of the geometry. Strongly connected to diagrams are the so-called **residues** which are also introduced in this section.

The main results are as follows: The diagrams of the projective and affine spaces are determined (Theorem 7.6 and Theorem 7.7). Conversely, it will be seen that residually connected geometries with these diagrams are projective and affine spaces (Theorem 7.15 and Theorem 7.16).

Definition. Let $\Gamma = (X, *, \text{type})$ be a geometry over a type set I, and let F be a flag of Γ of type $I \setminus J \neq I$.[20]

(a) The **residue of F in Γ** is the pregeometry

$$\Gamma_F = (X_F, *|_F, \text{type}|_F)$$

over the type set $J = I \setminus \text{type}(F)$ where $X_F := \{x \in X \setminus F \mid x * F\}$ is the set of all elements x of $X \setminus F$ such that x is incident with every element of F.

(b) If Γ_F is a geometry of rank n, Γ_F is called a **rank-n-residue over J of Γ**.

(c) If the flag $F = \{x\}$ consists of one element, the residue $\Gamma_F = \Gamma_{\{x\}}$ is also denoted by Γ_x.

---

[20]For a residue Γ_F, we will from now on always assume that $\text{type}(F) \neq I$. Otherwise, X_F is the empty set. Conversely, F may be the empty set. In this case, we have $\Gamma_F = \Gamma$.

7.1 Example. Let Γ be the geometry of the cube with notations as in the figure below.

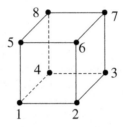

The residue Γ_1 consists of the edges 12, 14 and 15 and of the faces 1234, 1458 and 1256. Hence, the residue Γ_1 is the following triangle:

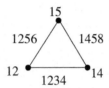

7.2 Theorem. *Let* $\Gamma = (X, *, \text{type})$ *be a geometry over the type set* I, *and let* F *be a flag of* Γ *of type* $I \setminus J$.

(a) *A subset* A *of* X *is a flag of* Γ_F *if and only if* $A \cap F = \emptyset$ *and if* $A \cup F$ *is a flag of* Γ.
(b) Γ_F *is a geometry.*
(c) *Let* A *be a flag of* Γ_F. *Then,* $(\Gamma_F)_A = \Gamma_{F \cup A}$.

Proof. The proof is obvious.   □

7.3 Theorem. *Let* $P = (X, *, \text{type})$ *be a d-dimensional projective geometry, and let* $d \geq 3$.

(a) *If* H *is a hyperplane of* P, Γ_H *is a* $(d-1)$-*dimensional projective space.*
(b) *If* x *is a point of* P, Γ_x *is a* $(d-1)$-*dimensional projective space. The points, lines, planes, etc. of* Γ_x *are the lines, planes, 3-dimensional subspaces, etc. of* Γ *through* x.

Proof. (a) By Theorem 4.1, Γ_H is a projective space. By Theorem 4.18, we have dim H $= d - 1$.
(b) We shall apply Theorem 4.17.
    Step 1. Γ_x is a linear space: Let g and h be two points of Γ_x, that is, two lines of P through x. The unique plane of P generated by g and h is the unique line joining g and h in Γ_x.

Step 2. Every line of Γ_x is incident with at least three points: Let E be a line of Γ_x. Then, E is a plane of $\boldsymbol{P}$ through x. Let l be a line of E, not containing x, and let a, b, c be three points on l. Then, the lines xa, xb and xc are three distinct lines of E through x, that is, three distinct points of Γ_x on the line E.

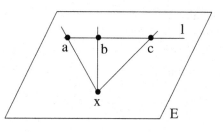

Step 3. Every plane of Γ_x is a projective plane: Let E and F be two lines of Γ_x which are contained in a common plane U. Then, E and F are planes of $\boldsymbol{P}$ through x which are contained in the 3-dimensional subspace U of $\boldsymbol{P}$. By the dimension formula (Theorem 4.15), it follows that

$$\dim (E \cap F) = \dim E + \dim F - \dim \langle E, \, F \rangle = 1.$$

Hence, the planes E and F intersect in a line g (through x). This means that in Γ_x, the lines E and F meet in the point g.

In order to show that Γ_x contains at least two lines, consider a line g of $\boldsymbol{P}$ through x. By Theorem 4.20, $\boldsymbol{P}$ is a thick geometry, hence, there are at least three planes through g (dim $\boldsymbol{P} \geq 3$). In Γ_x, these three planes are lines, it follows that there exist at least three lines in Γ_x. $\quad\square$

7.4 Theorem. *Let $A = \Gamma = (X, \, *, \text{ type})$ be a d-dimensional affine geometry, and let $d \geq 3$.*

(a) If H is a hyperplane of A, Γ_H is a $(d - 1)$-dimensional affine space.
(b) If x is a point of A, Γ_x is a $(d - 1)$-dimensional projective space. The points, lines, planes, etc. of Γ_x are the lines, planes, 3-dimensional subspaces, etc. of A through x.

Proof. Let $\boldsymbol{P} = \boldsymbol{P}(A)$ be the projective closure of A with hyperplane H_∞ at infinity.

(a) For a subspace U of A, we denote by U_P the corresponding subspace in $\boldsymbol{P}$. Then, H_P is a $(d - 1)$-dimensional subspace of $\boldsymbol{P}$, it follows that H_P is a projective space that we denote by $\boldsymbol{H}$.

Let $H'_\infty := H_P \cap H_\infty$. Then, H'_∞ is a hyperplane of $\boldsymbol{H}$. Since Γ_H consists of the subspaces of A that are contained in H, we have $\Gamma_H = \boldsymbol{H} \backslash H'_\infty$. Hence, Γ_H is a $(d - 1)$-dimensional affine space.

(b) Since x is a point of A, x is not contained in H_∞. Since every subspace of Γ_x contains the point x, the subspaces of Γ_x are the subspaces of $\boldsymbol{P}$ through x. By Theorem 7.3, Γ_x is a $(d - 1)$-dimensional projective space. $\quad\square$

As mentioned in the introduction of this section many geometries can be described by a small ideogram, the so-called diagram of the geometry. These ideograms are not only graphic representations of the geometries, but also they often contain a lot of information about the structure of the geometries. Before introducing diagrams, we shall introduce the notion of a generalized digon.

Definition. Let Γ be a geometry of rank 2 over the type set $\{0, 1\}$. Γ is called a **generalized digon** if every element of type 0 is incident with every element of type 1.

Definition. Let Γ be a geometry of rank 2 over the type set $\{0, 1\}$.

(a) Γ has the diagram $\overset{\mathrm{X}}{\underset{0 \qquad 1}{\circ\!\!-\!\!\!-\!\!\circ}}$

(b) If Γ is a generalized digon, Γ has the diagram $\underset{0}{\circ} \quad \underset{1}{\circ}$

(c) If Γ is a linear space, Γ has the diagram $\overset{\mathrm{L}}{\underset{0 \qquad 1}{\circ\!\!-\!\!\!-\!\!\circ}}$

(d) If Γ is an affine plane, Γ has the diagram $\overset{\mathrm{Af}}{\underset{0 \qquad 1}{\circ\!\!-\!\!\!-\!\!\circ}}$

(e) If Γ is a generalized projective plane, Γ has the diagram $\underset{0 \qquad\qquad 1}{\circ\!\!-\!\!\!-\!\!\!-\!\!\!-\!\!\circ}$

If the type set of Γ is endowed with an ordering, the diagram is often used without an explicit labelling. The diagram $\overset{\mathrm{X}}{\underset{0 \qquad 1}{\circ\!\!-\!\!\!-\!\!\circ}}$ is then abbreviated by

$\overset{\mathrm{X}}{\circ\!\!-\!\!\!-\!\!\circ}$

The definition that every rank-2-geometry has the diagram $\overset{\mathrm{X}}{\circ\!\!-\!\!\!-\!\!\circ}$ guarantees that every rank-2-geometry has at least one diagram. Often, a geometry has more than one diagram. For example, a projective plane has the following diagrams:

So far, we have assigned diagrams only to a small class of geometries. During the rest of this book we shall introduce further diagrams.

Definition. (a) Let I be a non-empty set. A **diagram** is a weighted graph D with I as set of vertices. Every edge of D is of the form

$$\overset{\mathrm{X}}{\underset{i \qquad j}{\circ\!\!-\!\!\!-\!\!\circ}}$$

where i and j are different elements of I and where X is an arbitrary sequence of numbers or letters that can be empty.

(b) Let Γ be a geometry over the type set I, and let D be a diagram over I. The geometry Γ **belongs to the diagram** D if for any two elements i and j of I and for every flag F of type $I \setminus \{i, \ j\}$, the following holds: If i and j are joined by the edge

in D (this includes the case that i and j are not joined by an edge), the geometry Γ_F has the diagram

If the set I is endowed with an ordering, the diagram often is used without an explicit labelling.

7.5 Theorem. *Let* $\Gamma = (X, *, \text{type})$ *be a geometry over a type set* I *with the diagram* D. *Furthermore, let* Γ_F *be a residue of* Γ *over the type set* $J \subseteq I$.
   Then, Γ_F *has a diagram* D_J. *The diagram* D_J *arises from* D *by deleting all vertices of* $I \setminus J$ *and all edges through these vertices.*

Proof. The proof is obvious.   □

   Theorem 7.5 is useful to determine the diagram of a geometry of higher rank. We shall illustrate this technique with the example of projective spaces.

7.6 Theorem. *Let* P *be a* d-*dimensional projective space. Then,* P *has the following diagram:*

Proof. We shall prove the assertion by induction on d. If $d = 2$, P is a projective plane and belongs to the diagram

   Let $d \geq 3$. If x is a point of P, by Theorem 7.3, the residue P_x is a projective space of dimension $d - 1$. By induction, P_x has the diagram

with $d - 1$ vertices.
   If H is a hyperplane of P, the residue P_H, again by Theorem 7.3, is a projective space of dimension $d - 1$ and has the diagram

Finally, let $F = \{U_1, \ldots, U_{d-2}\}$ be a flag of type $\{1, \ldots, d-2\} = I \backslash \{0, d-1\}$. Then, (with a suitable numbering) the elements $U_1, \ldots, U_{d-2}$ are subspaces of $\boldsymbol{P}$ with dim $U_i = i$ and $U_1 \subset U_2 \subset \ldots \subset U_{d-2}$. The residue $\boldsymbol{P}_F$ consists of the points contained in U_1 and the hyperplanes of $\boldsymbol{P}$ through U_{d-2}. It follows that $\boldsymbol{P}_F$ is a generalized digon. From the diagrams

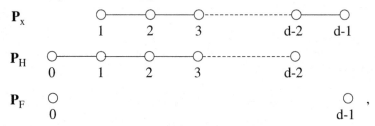

it follows together with Theorem 7.5 that $\boldsymbol{P}$ has the diagram

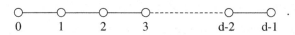

□

7.7 Theorem. *Let $\boldsymbol{P}$ be a d-dimensional affine space. Then, $\boldsymbol{P}$ has the following diagram:*

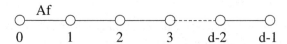

Proof. The proof is similar to the proof of Theorem 7.6.   □

At the end of this section, we shall deal with the question whether the converse of Theorems 7.6 and 7.7 is true, that is, whether every geometry Γ with the diagram of a projective space (respectively an affine space) is a projective space (respectively an affine space).

In order to answer this question, we need the notion of a **residually connected geometry**.

Definition. Let G be a graph.

(a) Let $x_0, \ldots, x_r$ be a sequence of vertices such that for $i = 1, \ldots, r$, the vertices x_{i-1} and x_i are joined by an edge and such that $x_{i-2} \neq x_i$ for $i = 2, \ldots, r$. If $x_0 \neq x_r$, the sequence $x_0, \ldots, x_r$ is called a **path of length r from x_0 to x_r**.
(b) The graph G is called **connected** if for any two vertices x and y of G, there exists a path from x to y.

Definition. Let Γ be a pregeometry over a type set I. The **incidence graph** $I(\Gamma)$ of Γ is defined as follows: The vertices of $I(\Gamma)$ are the elements of Γ. Two distinct vertices x and y of $I(\Gamma)$ are joined by an edge if and only if x and y are incident.

7.8 Example. Let Γ be a 3-gon, and let $I(\Gamma)$ be the incidence graph of Γ. Then, Γ and $I(\Gamma)$ are of the following form:

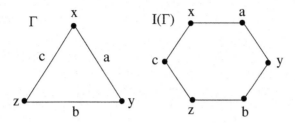

Definition. Let Γ be a pregeometry of rank d over a type set I.

(a) Γ is called **connected** if the incidence graph $I(\Gamma)$ of Γ is connected.
(b) Γ is called **residually connected** if all rank-m-residues ($2 \leq m \leq d$) of Γ are connected.

If Γ is a pregeometry and if F is the empty flag of Γ, we have $\Gamma_F = \Gamma$. It follows that residually connected pregeometries are connected. For pregeometries of rank 2, the notions connected and residually connected are identical.

7.9 Theorem. *(a) Let P be a d-dimensional projective space. Then, P is a residually connected geometry.*
(b) Let A be a d-dimensional affine space. Then, A is a residually connected geometry.

Proof. (a) We will prove the assertion by induction on d.

Let $d = 2$. We have to show that P is connected.

Let X and Y be two elements of P. If X and Y are two points, there exists a line g through X and Y. In the incidence graph $I(P)$, the sequence $X - g - Y$ is a path from X to Y. If X and Y are two lines and if p is the intersection point of X and Y, $X - p - Y$ is a path from X to Y.

Finally, let X be a point, and let Y be a line. If X is on Y, X and Y are joined by an edge in $I(P)$. If X is not on Y, let p be an arbitrary point on Y, and let g be the line joining X and p. Then, $X - g - p - Y$ is a path from X to Y.

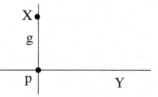

Let $d > 2$. Let F be a flag of P of corank at least 2, and let X and Y be two elements of the residue P_F. With a suitable numbering we have $F = \{U_1, \ldots, U_r, U_{r+1}, \ldots, U_{r+s}, U_{r+s+1}, \ldots, U_{r+s+t}\}$ with

$$U_1 \subset \ldots \subset U_r \subset X \subset U_{r+1} \subset \ldots \subset U_{r+s} \subset U_{r+s+1} \subset \ldots \subset U_{r+s+t} \quad \text{and}$$
$$U_1 \subset \ldots \subset U_r \subset U_{r+1} \subset \ldots \subset U_{r+s} \subset Y \subset U_{r+s+1} \subset \ldots \subset U_{r+s+t}.$$

1st case. Let $s \neq 0$. Then, $X \subset U_{r+1} \subset \ldots \subset U_{r+s} \subset Y$. It follows that X is contained in Y, that is, X and Y are incident. In the incidence graph $I(P)$, the elements X and Y are joined by an edge.

2nd case. Let $s = 0$. W.l.o.g., let dim $X \leq$ dim Y. We distinguish three cases:

(i) Let $r \geq 1$. In this case, the subspace U_1 of F is contained in X and in Y. If p is a point of U_1, the flag F and the subspaces X and Y are contained in the residue P_p. By Theorem 7.3, P_p is a $(d-1)$-dimensional projective space. By induction, P_p is residually connected. It follows that there exists a path from X to Y in $(P_p)_F = P_{F \cup \{p\}}$. Obviously, this path is also contained in P_F.

(ii) Let $t \geq 1$. In this case, the subspaces X and Y are contained in the subspace U_{r+s+t}. Let H be a hyperplane of P through U_{r+s+t}. The flag F and the subspaces X and Y are contained in the residue P_H. As in Part (i), we use the fact that P_H is a $(d-1)$-dimensional projective space which is, by induction, residually connected.

(iii) Let $r = 0$ and $t = 0$. Since $r = s = t = 0$, the flag F is the empty flag. Let x be a point of X, let y be a point of Y, and let $g := xy$ be the line joining x and y. Then, $X - x - g - y - Y$ is a path from X to Y in $P = P_F$.

(b) The proof is similar to the proof of Part (a). Instead of Theorem 7.3, we apply Theorem 7.4.   □

7.10 Theorem (Tits). *Let $\Gamma = (X, *, \text{type})$ be a residually connected geometry of rank r over a type set I. Let x and y be two elements of Γ, and let i and j be two types of I.*

There exists a path from x to y such that all elements of this path different from x and y are of type i or j (for short: there is an $\{i, j\}$-path from x to y).

Proof. We shall prove the assertion by induction on r. If $r = 2$, Γ is a connected geometry over a type set $\{a, b\}$. It follows that $\{i, j\} = \{a, b\}$, and there exists an $\{i, j\}$-path from x to y.

Let $r \geq 3$. Since Γ is connected, there exists a path $x = x_0, x_1, \ldots, x_n = y$ from x to y. If every element $x_1, \ldots, x_{n-1}$ is of type i or j, there is nothing to show. Otherwise, let a be the smallest index of $\{1, \ldots, n-1\}$ such that x_a is not of type i or j. Let k be the type of x_a. The residue Γ_{x_a} is a residually connected geometry over $I \backslash \{k\}$ of rank $r - 1$.

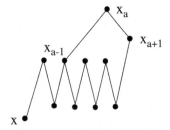

By induction, in Γ_{x_a}, there exists an $\{i, j\}$-path from x_{a-1} to x_{a+1}.

We replace the path $x_{a-1} - x_a - x_{a+1}$ by this $\{i, j\}$-path. Successive application of this procedure on all elements x_b of the path $x = x_0, x_1, \ldots, x_n = y$ which are not of type i or j yields an $\{i, j\}$-path from x to y.   □

7.11 Theorem (**Tits**). *Let Γ be a residually connected geometry of rank r over a type set I with a diagram D which is not connected.*

If i and j are two types of I contained in different connected components of D, every element of type i is incident with every element of type j.

Proof. We prove the assertion by induction on r.

Let $r = 2$. Since the diagram consists of two vertices and since the diagram is not connected, D is the diagram of a generalized digon. It follows that every element of type i is incident with every element of type j.

Let $r > 2$, and let x and y be two elements of type i and j, respectively. We have to show that x and y are incident. By Theorem 7.10, there exists an $\{i, j\}$-path

$$x = x_0, \ x_1, \ldots, x_n = y$$

from x to y. Let $x = x_0, \ x_1, \ldots, x_n = y$ be such a path of minimal length. If $n = 1$, x and y are incident, and the assertion is proven.

Assume that $n > 1$. Let k be a type of I different from i and j. W.l.o.g., let k be contained in another connected component of D than i. Furthermore, let

$$x_1 = z_0, \ z_1, \ldots, z_m = x_3$$

be a $\{j, k\}$-path from x_1 to x_3.

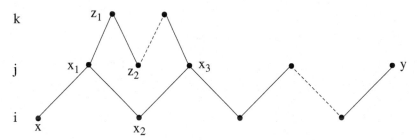

By Theorem 7.5, the geometry Γ_{x_1} has a diagram D_1. Since the types i and k are in different connected components of D, the types i and k are in different connected components of D_1. By induction, it follows that x and z_1 are incident in Γ_{x_1}. It follows that x and z_1 are incident in Γ.

Hence, x and z_2 are elements of the residue Γ_{z_1}, and by induction, x and z_2 are incident. By successive application of this argument, it follows that x and $z_m = x_3$ are incident. Therefore,

$$x = x_0, \ x_3, \ldots, x_n = y$$

is an $\{i, j\}$-path of length $n - 2$ from x to y, in contradiction to the fact that n is minimal. It follows that x and y are incident.   □

Definition. Let Γ be a geometry over the type set $\{0, 1, \ldots, d-1\}$ with the diagram

$$\underset{0}{\circ} \overset{\gamma_1}{\rule{1cm}{0.4pt}} \underset{1}{\circ} \overset{\gamma_2}{\rule{1cm}{0.4pt}} \underset{2}{\circ} \overset{\gamma_3}{\rule{1cm}{0.4pt}} \underset{3}{\circ} \text{-----} \underset{d-2}{\circ} \overset{\gamma_{d-1}}{\rule{1cm}{0.4pt}} \underset{d-1}{\circ}$$

The geometry Γ and the diagram of Γ are called **linear**.

7.12 Theorem. *Let* $\Gamma = (X, *, \text{type})$ *be a residually connected geometry over the type set* $\{0, 1, \ldots, d-1\}$ *with the following linear diagram:*

$$\underset{0}{\circ}\;\overset{\gamma_1}{\underline{\hspace{1.3cm}}}\;\underset{1}{\circ}\;\overset{\gamma_2}{\underline{\hspace{1.3cm}}}\;\underset{2}{\circ}\;\overset{\gamma_3}{\underline{\hspace{1.3cm}}}\;\underset{3}{\circ}\text{-------}\underset{d\text{-}2}{\circ}\;\overset{\gamma_{d\text{-}1}}{\underline{\hspace{1.3cm}}}\;\underset{d\text{-}1}{\circ}$$

(a) *Let* x, y *and* z *be three elements of* Γ *with* $\text{type}(x) < \text{type}(y) < \text{type}(z)$. *If* x *and* y *and* y *and* z *are incident,* x *and* z *are also incident.*[21]
(b) *Let* U *be an element of* Γ *of type* i, *and let* Γ_1 *and* Γ_2 *be the following two pregeometries over* $I_1 := \{0, 1, \ldots, i-1\}$ *and* $I_2 := \{i+1, i+2, \ldots, d-1\}$, *respectively:*

   The elements of Γ_1 *are the elements of* Γ *of type* $0, 1, \ldots$ *or* $i-1$ *which are incident with* U.

   The elements of Γ_2 *are the elements of* Γ *of type* $i+1, i+2, \ldots$ *or* $d-1$ *which are incident with* U.

   The incidence relation and the type function of Γ_1 *and* Γ_2 *are induced by* Γ.

   If $i > 1$, Γ_1 *is a geometry with the diagram*

$$\underset{0}{\circ}\;\overset{\gamma_1}{\underline{\hspace{1.3cm}}}\;\underset{1}{\circ}\;\overset{\gamma_2}{\underline{\hspace{1.3cm}}}\;\underset{2}{\circ}\text{-------}\underset{i\text{-}2}{\circ}\;\overset{\gamma_{i\text{-}1}}{\underline{\hspace{1.3cm}}}\;\underset{i\text{-}1}{\circ}$$

   If $i < d-2$, Γ_2 *is a geometry with the diagram*

$$\underset{i+1}{\circ}\;\overset{\gamma_{i+2}}{\underline{\hspace{1.3cm}}}\;\underset{i+2}{\circ}\;\overset{\gamma_{i+3}}{\underline{\hspace{1.3cm}}}\;\underset{i+3}{\circ}\text{-------}\underset{d\text{-}2}{\circ}\;\overset{\gamma_{d\text{-}1}}{\underline{\hspace{1.3cm}}}\;\underset{d\text{-}1}{\circ}$$

Proof. (a) Let $i := \text{type}(y)$. Then, the residue Γ_y has the diagram

$$\underset{0}{\circ}\;\overset{\gamma_1}{\underline{\hspace{1.3cm}}}\;\underset{1}{\circ}\text{-------}\underset{i\text{-}1}{\circ}\quad\underset{i+1}{\circ}\;\overset{\gamma_{i+2}}{\underline{\hspace{1.3cm}}}\;\underset{i+2}{\circ}\text{-------}\underset{d\text{-}1}{\circ}$$

By Theorem 7.11, every element of Γ_y of type $0, 1, \ldots, i-1$ is incident with every element of Γ of type $i+1, \ldots, d-1$. In particular, x and z are incident.

(b) Let $i > 0$, and let $F = \{U, U_{i+1}, \ldots, U_{d-1}\}$ be a flag of Γ of type $\{i, \ldots, d-1\}$ containing U. By (a), the elements of Γ of type $0, 1, \ldots, i-1$ being incident with U are exactly the elements of Γ being incident with $U, U_{i+1}, \ldots, U_{d-1}$. Since Γ_F has the diagram

---

[21] This means that one can imagine incident elements of Γ as subspaces, where one subspace is contained in the other one. If the point x lies on the line y and if the line y lies in the plane z, x lies in z. For geometries with non-linear diagrams, this idea normally is false.

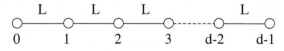

the assertion follows. The second assertion of (b) follows in the same way. □

There is the following relation between set geometries and linear geometries:

7.13 Theorem. *(a) Every set geometry over the type set $I = \{0, 1, \ldots, d-1\}$ is a linear geometry.*

(b) Let Γ be a linear geometry over the type set $I = \{0, 1, \ldots, d-1\}$ fulfilling the following two conditions:

  (i) For any two non-incident elements U and W, there exists a point (element of type 0) of U non-incident with W and a point of W non-incident with U.

  (ii) For an element X of Γ, denote by X_0 the set of points incident with X. Suppose that $U_0 \neq W_0$, whenever $U \neq W$.

Then, Γ is a set geometry.

Proof. (a) Let Γ be a set geometry, and let V be an element of Γ with $0 < \text{type}(V) < d-1$. We have to show that every two elements U and W of Γ_V with $\text{type}(U) < \text{type}(V) < \text{type}(W)$ are incident. Since Γ is a set geometry, it follows from $U * V$ that U_0 is contained in V_0. In the same way, it follows that V_0 is contained in W_0. Hence, U_0 is contained in W_0. Since Γ is a set geometry, it follows that U_0 and W_0 are incident.

(b) Let U and W be two subspaces of Γ with $0 \leq \text{type}(U) < \text{type}(W)$.

  Step 1. If U and W are incident, U_0 is contained in W_0: Let x be a point of U_0. Since x and W are incident with U, it follows from the linearity of Γ that x and W are incident. Hence, U_0 is contained in W_0.

  Step 2. If U_0 is contained in W_0, U and W are incident: Assume that U and V are not incident. By assumption (i), there exists a point x of U_0 not contained in W_0, in contradiction to the assumption that U_0 is contained in W_0.

  Step 3. We have $U_0 \neq W_0$, whenever $U \neq W$: This follows from assumption (ii). □

7.14 Theorem. *Let Γ be a residually connected geometry over the type set $I = \{0, 1, \ldots, d-1\}$ with the diagram*

$$\underset{0}{\circ} \overset{L}{-\!\!-} \underset{1}{\circ} \overset{L}{-\!\!-} \underset{2}{\circ} \overset{L}{-\!\!-} \underset{3}{\circ} -\!-\!-\!-\!- \underset{d\text{-}2}{\circ} \overset{L}{-\!\!-} \underset{d\text{-}1}{\circ}$$

where the elements of type 0, 1, 2 are called points, lines and planes, respectively. Then, any two points of Γ are joined by exactly one line.

Proof. We will prove the assertion by induction on d.

If $d = 2$, Γ is a linear space, and the assertion is obvious.

Let $d > 2$, and let x and y be two points of Γ. We first shall prove the existence of a line through x and y. By Theorem 7.10, there exists a $\{0, 1\}$-path

$$x = x_0, x_1, \ldots, x_n = y$$

from x to y in Γ. Let $x = x_0, x_1, \ldots, x_n = y$ be such a path of minimal length.

If $n = 1$, $x - x_1 - y$ is a $\{0, 1\}$-path from x to y, that is, x and y are incident with the line x_1.

Assume that $n > 1$. We consider the path $x = x_0 - x_1 - x_2 - x_3 - x_4$. Then, x, $r := x_2$, $s := x_4$ are points, and $g := x_1$ and $h := x_3$ are lines of Γ.

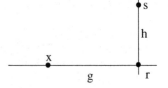

Since Γ_r is residually connected, since Γ_r has the diagram

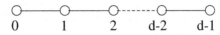

and since g and h are points of Γ_r, by induction, there is a line E of Γ_r through g and h. In Γ, E is a plane containing the lines g and h. By Theorem 7.12 (b), the points and lines of E form a linear space, that is, in E there exists a line l through x and s. It follows that

$$x = x_0, 1, s = x_4, \ldots, x_n = y$$

is a path of length $n - 2$ from x to y, in contradiction to the minimality of n.

Next, we will show the uniqueness of the line through x and y: Assume that there are two lines g and h through the points x and y. Considering the residue Γ_x it follows by induction that there is a plane E of Γ_x through g and h. Since y is incident with g (and h), by Theorem 7.12 (a), y is incident with E. It follows from Theorem 7.12 (b) that the points and lines of E form a linear space, in contradiction to the fact that there are two lines g and h of E incident with x and y.   □

7.15 Theorem (Tits). *Let Γ be a residually connected geometry over the type set $I = \{0, 1, \ldots, d - 1\}$ $(d \geq 2)$ with the diagram*

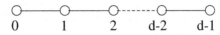

If every line of Γ contains at least three points, Γ is a d-dimensional projective geometry.

Proof. Step 1. Let L be the geometry of rank 2 whose point set is the point set of Γ and whose line set is the line set of Γ. The incidence in L shall be induced by the incidence in Γ. By Theorem 7.14, there is exactly one line through any two points of Γ. Hence, L is a linear space. By assumption, every line of Γ (and of L) is incident with at least three points.

Step 2. The planes of L (that is, the subspaces of L generated by three non-collinear points) are exactly the point sets of the elements of type 2 of Γ:

(i) Let E be an element of Γ of type 2, and let F be the point set of E. Then, F is a plane of L: For, let x and y be two points of F. By definition, x and y are two points incident with E. By Step 1, there is exactly one line g of Γ through x and y. Since Γ_E is a projective plane, the line g must belong to Γ_E, that is, g is incident with E. Since Γ is a residually connected linear geometry, it follows that the points of g are incident with E. Hence, every point of g is contained in F, that is, F is a linear subspace of L. Since Γ_E is a projective plane, it follows that the points and lines of F form a projective plane. In particular, the subspace F of L is generated by any three non-collinear points of F.
(ii) Conversely, let F be a subspace of L generated by three non-collinear points x, y and z. Consider the lines $g = xy$ and $h = xz$. In the residue Γ_x, g and h are two points. By Theorem 7.14, there is exactly one line E through g and h, that is, there is an element E of type 2 through g and h.

Since Γ is a residually connected linear geometry, it follows from Theorem 7.12 that the points x, y and z are incident with E. By Step (i), it follows that the point set of E is a subspace of L generated by x, y and z. It follows that the point set of E equals F.

Step 3. It follows from Steps 1 and 2 that L is a linear space with the property that every line of L is incident with at least three points and that every plane of L is a projective plane. By Theorem 4.17, L is a projective space.

Step 4. For $i = 0, \ldots, d - 1$, the point sets of the elements of Γ of type i are the i-dimensional subspaces of L:

We shall prove the assertion by induction on i.

$i = 0, 1$: The assertion follows from the definition of L.
$i = 2$: The assertion follows from Step 2.
$i - 1 \rightarrow i$:

(i) Let U be an element of Γ of type i. By induction, Γ_U is an i-dimensional projective geometry. As in Step 2, Part (i), it follows that the point set W of U is a subspace of L. Since Γ_U and W are projective spaces defined over the same point and line sets, it follows that W is an i-dimensional subspace of L.
(ii) Conversely, let W be an i-dimensional subspace of L. Let W' be an $(i - 1)$-dimensional subspace of W, let x be a point of W', and let g be a line of W through x not contained in W'. By induction, there is an element U' of Γ of type $i - 1$ such that W' is the point set of U'. By induction, Γ_x is a $(d - 1)$-dimensional projective space, hence, there exists an element U of type i through U' and g. By Part (i), the point set of U is an i-dimensional subspace of L containing W' and g. Hence, W is the point set of U. □

7.16 Theorem. *Let* Γ *be a residually connected geometry over the type set* $I = \{0, 1, \ldots, d - 1\}$ $(d \geq 2)$ *with the diagram*

$$\underset{0}{\circ}\!\!\overset{\text{Af}}{\underline{\qquad}}\!\!\underset{1}{\circ}\!\!\underline{\qquad}\!\!\underset{2}{\circ}\!\!-\!-\!-\!-\!-\!-\!\underset{d\text{-}2}{\circ}\!\!\underline{\qquad}\!\!\underset{d\text{-}1}{\circ}$$

If every line of Γ *is incident with at least four points,* Γ *is an affine geometry.*

Proof. The proof is similar to the proof of Theorem 7.15. Instead of Theorem 4.17, we make use of Theorem 6.1.   □

8 Finite Geometries

In the present section, we shall consider finite geometries, that is, geometries consisting of finitely many elements. The main results are as follows:

- Principle of Double Counting (Theorem 8.1)
- Introduction of the order of a projective or an affine space (Theorem 8.2 and Corollary 8.3)
- Determination of the number of subspaces of a projective or an affine space (Theorem 8.6).

Definition. A pregeometry $\Gamma = (X, *, \text{type})$ is called **finite** if the set X is finite.

8.1 Theorem (Principle of Double Counting). *Let* $\Gamma = (X, *, \text{type})$ *be a finite pregeometry of rank 2 over the type set* $\{0, 1\}$. *For an element x of X, let $[x]$ be the number of elements of $X \setminus \{x\}$ incident with x. Let A and B be the sets of elements of type 0 and type 1 of X, respectively. Then,*

$$\sum_{a \in A} [a] = \sum_{b \in B} [b].$$

Proof. Let $M := \{(a, b) | a \in A, b \in B$ such that a and b are incident$\}$. Then,

$$\sum_{a \in A} [a] = |M| = \sum_{b \in B} [b].$$

□

8.2 Theorem. *Let* P *be a projective space of dimension at least 2, and let g and h be two lines of P. If P_g and P_h are the sets of points incident with g and h, respectively, the sets P_g and P_h have the same cardinality.*

Proof. Step 1. If g and h have a point in common, the sets P_g and P_h have the same cardinality:

For, let g and h be two lines of P intersecting in a point z, and let E be the plane generated by g and h. Furthermore, let p be a point of E which is neither incident with g nor with h.

Let the mapping $\alpha : P_g \to P_h$ be defined by $\alpha(x) := px \cap h$. Note that $\alpha(z) = z$. Then, α is a bijective mapping of the point set of g onto the point set of h.

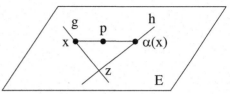

Step 2. The sets P_g and P_h have the same cardinality: Let g and h be two arbitrary lines of P. If g and h meet in a point, the assertion follows from Step 1. Otherwise, choose a point x on g and a point y on h and consider the line l through x and y. Let P_l be the set of points incident with l. Since g and l meet in a point, the sets P_g and P_l have the same cardinality. Similarly, P_h and P_l have the same cardinality. It follows that P_g and P_h have the same cardinality.   □

8.3 Corollary. *Let A be an affine space of dimension at least 2, and let g and h be two lines of A. If P_g and P_h are the sets of points incident with g and h, respectively, the sets P_g and P_h have the same cardinality.*

Proof. Let P be the projective closure of A. The projective closure of any line of A consists of the points of A and the point at infinity. Thus, the assertion follows from Theorem 8.2.   □

Definition. (a) Let P be a finite projective space such that every line of P is incident with exactly $q + 1$ points. The number q is called the **order** of P.
(b) Let A be a finite affine space such that every line of A is incident with exactly q points. The number q is called the **order** of A.

   If A is a finite affine space of order q and if P is the projective closure of A, P is a projective space of order q.

8.4 Theorem. *Let P be a finite projective space of order q, and let U be a subspace. Furthermore, let x be a point of P.*

(a) U is a finite projective space of order q.
(b) If $\dim P \geq 3$, the residue P_x is a finite projective space of order q.

Proof. (a) is obvious.
(b) By Theorem 7.3, P_x is a projective space. Let E be a plane of P through x. Let h be a line of E, not containing the point x. Since every line of E through x intersects the line h, the number of lines of E through x equals the number of points on h. In E, there are $q + 1$ lines incident with x. Thus, every line of P_x contains $q + 1$ points, that is, P_x is of order q.   □

8.5 Theorem. *Let A be a finite affine space of order q, and let U be a subspace. Furthermore, let x be a point of A.*

(a) U is a finite affine space of order q.
(b) If $\dim A \geq 3$, the residue A_x is a finite projective space of order q.

Proof. (a) is obvious.

(b) Let P be the projective closure of A. Then, P is a projective space of order q. Since $A_x = P_x$, the assertion follows from Theorem 8.4.   □

8.6 Theorem. *Let P be a finite d-dimensional projective space of order q, and let U be a t-dimensional subspace of P.*

(a) *U has $q^t + \ldots + q + 1$ points. In particular, P has exactly $q^d + \ldots + q + 1$ points.*

(b) *In U, there are $q^t + \ldots + q + 1$ subspaces of dimension $t - 1$. In particular, P contains exactly $q^d + \ldots + q + 1$ hyperplanes.*

(c) *Let $t := \dim U \geq 1$, and let W be a $(t - 2)$-dimensional subspace of U. Then, U contains exactly $q + 1$ subspaces of dimension $t - 1$ through W. In particular, there are exactly $q + 1$ hyperplanes of P through a $(d-2)$-dimensional subspace of P.*

(d) *Any point of U is incident with exactly $q^{t-1} + \ldots + q + 1$ lines of U. In particular, any point of P is incident with exactly $q^{d-1} + \ldots + q + 1$ lines.*

(e) *The number of lines of U is*

$$\frac{\left(q^t + q^{t-1} + \cdots + q + 1\right)\left(q^{t-1} + \cdots + q + 1\right)}{q + 1}.$$

Proof. (a) We will prove the assertion by induction on t. For $t = 0$, the subspace U consists of a point, and the assertion is obvious.

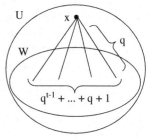

Let $t > 0$. Let x be a point of U, and let W be a $(t - 1)$-dimensional subspace of U not containing x. By induction, W contains exactly $q^{t-1} + \ldots + q + 1$ points. Since every line of U through x intersects the subspace W in exactly one point, there are exactly $q^{t-1} + \ldots + q + 1$ lines of U through x. Any of these lines is incident with q further points of U besides x. It follows that

$$|U| = 1 + (q^{t-1} + \ldots + q + 1) \cdot q = q^t + \ldots + q + 1.$$

(b) Again, we will prove the assertion by induction on t. For $t = 1$, U is a line. Hence, U is incident with exactly $q + 1$ points.

Let $t > 1$. Let x be a point of U. The residue U_x is a $(t - 1)$-dimensional projective space of order q (Theorem 8.4). By induction, U_x contains exactly $q^{t-1} + \ldots + q + 1$ hyperplanes, that is, U contains exactly $q^{t-1} + \ldots + q + 1$ subspaces of dimension $t - 1$ through x.

Let T be the set of the $(t - 1)$-dimensional subspaces of U. We shall apply the principle of double counting (Theorem 8.1) on the set

$$M := \{(x, W) \mid x \in U, W \in T, x \text{ is contained in } W\}.$$

We get

$$|M| = \sum_{x \in U} [x] = \sum_{x \in U} q^{t-1} + \cdots + q + 1 = |U| \left(q^{t-1} + \cdots + q + 1 \right)$$

$$|M| = \sum_{W \in T} [W] = \sum_{W \in T} q^{t-1} + \cdots + q + 1 = |T| \left(q^{t-1} + \cdots + q + 1 \right)$$

By (a), it follows that $|T| = |U| = q^t + \ldots + q + 1$.

(c) Let W be a $(t-2)$-dimensional subspace of U, and let g be a line of U disjoint to W. Then, every $(t-1)$-dimensional subspace of U through W intersects the line g in exactly one point. It follows that the number of the $(t-1)$-dimensional subspaces of U through W equals the number of points on g. The number of points on g is $q + 1$.

(d) The number of lines of U through x equals the number of points of the residue U_x. By Theorem 8.4 and by (a), this number equals $q^{t-1} + \ldots + q + 1$.

(e) Let G be the set of lines of U. We shall apply the principle of double counting on the set

$$M := \{(x, g) \mid x \in U, g \in G, x \in g\}.$$

We get

$$|M| = \sum_{x \in U} [x] = \sum_{x \in U} q^{t-1} + \cdots + q + 1 = |U| \left(q^{t-1} + \cdots + q + 1 \right)$$

$$= \left(q^t + q^{t-1} + \cdots + q + 1 \right) \left(q^{t-1} + \cdots + q + 1 \right)$$

$$|M| = \sum_{g \in G} [g] = \sum_{g \in G} q + 1 = |G| \, (q + 1).$$

It follows that

$$|G| = \frac{\left(q^t + q^{t-1} + \cdots + q + 1 \right) \left(q^{t-1} + \cdots + q + 1 \right)}{q + 1}.$$

$\square$

Summarizing Theorem 8.6, we get the following parameters for a d-dimensional projective space of order q:

Cardinality	Value
Number of points of P	$q^d + \ldots + q + 1$
Number of lines of P	$\frac{\left(q^d + q^{d-1} + \cdots + q + 1 \right)\left(q^{d-1} + \cdots + q + 1 \right)}{q+1}$
Number of points on a line of P	$q + 1$
Number of lines through a point of P	$q^{d-1} + \ldots + q + 1$
Number of hyperplanes of P	$q^d + \ldots + q + 1$
Number of hyperplanes of P through a $(d-2)$-dimensional subspace of P	$q + 1$

At the end of this section, we shall study the projective planes of order 2.

8.7 Theorem. *Every finite projective plane of order 2 is of the following form:*

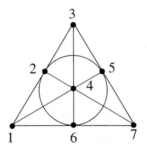

*In other words, the projective plane **P** has seven points and seven lines, and–with a suitable numbering 1, 2, 3, 4, 5, 6, 7 of the points–the lines of **P** are the points sets* {1, 2, 3}, {1, 4, 5}, {1, 6, 7}, {2, 4, 7}, {2, 5, 6}, {3, 4, 6} *and* {3, 5, 7}.

Proof. Let {1, 2, 3, 4, 5, 6, 7} be the set of points of **P**.[22] There are exactly three lines through the point 1. W.l.o.g., we may assume that these three line are the lines {1, 2, 3}, {1, 4, 5} and {1, 6, 7}.

The line through the points 3 and 5 intersects the line {1, 6, 7}. W.l.o.g., we may assume that these two lines intersect in the point 7. Then, the line through the points 3 and 4 must intersect the line {1, 6, 7} in the point 6. Otherwise, there would be two lines through the points 3 and 7 or 3 and 1.

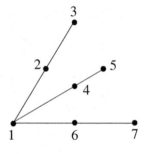

With the same argument, the line through the points 2 and 4 meets the line {3, 5, 7} in the point 7, and the line through the points 2 and 5 meets the line {1, 6, 7} in the point 6.   □

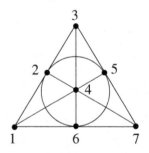

----

[22]In view of Theorem 8.6, **P** has $q^2 + q + 1 = 7$ points.

Chapter 2
Isomorphisms and Collineations

1 Introduction

Isomorphisms and collineations are transformations of one geometry into another preserving the geometrical structure. In Sect. 2, we shall introduce these transformations. In Sect. 3, we shall deal with projections. The concept of projections has been implicitly used in the classification of affine spaces in Sect. 6 of Chap. 1. In Sect. 4, we will introduce collineations. Collineations are isomorphisms of projective or affine spaces.

A particularly important class of collineations are the central collineations. They are investigated in Sect. 5. Finally, in Sect. 6, it is shown that every projective space fulfilling the Theorem of Desargues admits as many central collineations as possible. Furthermore, it is shown that every projective space of dimension at least 3 fulfils the Theorem of Desargues.

2 Morphisms

Morphisms are mappings between two geometries which preserve the geometrical structure. They are used to define isomorphisms between geometries and to "measure" the symmetry of a geometry. Often, the knowledge about the automorphism group of a geometry provides useful information about the geometry itself.

In the present section, we shall introduce different types of morphisms (homomorphisms, isomorphisms, correlations, polarities).

Definition. Let $\Gamma = (X, *, \text{type})$ and $\Gamma' = (X', *', \text{type}')$ be two geometries over the type sets I and I', respectively, and let α be a mapping from X to X'. α is called a **morphism** if α fulfils the following conditions:

(i) For any two incident elements x and y of Γ, the elements $\alpha(x)$ and $\alpha(y)$ are incident in Γ'.

J. Ueberberg, *Foundations of Incidence Geometry*, Springer Monographs in Mathematics, 57
DOI 10.1007/978-3-642-20972-7_2, © Springer-Verlag Berlin Heidelberg 2011

(ii) Two elements x and y of Γ are of the same type in Γ if and only if $\alpha(x)$ and $\alpha(y)$ are of the same type in Γ'.

Note that there are different definitions of morphisms in the literature depending on the context where they are used. For example, there exist definitions of morphisms without Condition (ii).

2.1 Example. Let Γ_3 and Γ_4 be a 3-gon and a 4-gon with the following notations for the vertices and edges:

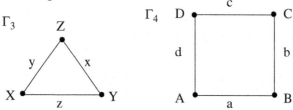

Let $\alpha : \Gamma_3 \to \Gamma_3$ and β, γ, $\delta : \Gamma_4 \to \Gamma_4$ be defined as follows:

$\alpha : X \to x, Y \to y, Z \to z$
$\quad\; x \to X, y \to Y, z \to Z$

$\beta : A \to B, B \to C, C \to D, D \to A$ ("rotation through 90°")
$\quad\; a \to b, \; b \to c, \; c \to d, \; d \to a$

$\gamma : A \to a, B \to b, C \to c, D \to d$ ("rotation through 45°")
$\quad\; a \to B, b \to C, c \to D, d \to A$

$\delta : A \to A, B \to A, C \to D, D \to D$ ("folding of the right side onto
$\quad\; a \to a, \; b \to d, \; c \to c, \; d \to d$ the left side").

The mapping α is a morphism of Γ_3 into itself, and the mappings β, γ, δ are morphisms of Γ_4 into itself.

Definition. Let $\Gamma = (X, \, ^*, \, \text{type})$ and $\Gamma' = (X', \, ^{*\prime}, \, \text{type}')$ be two geometries over the same type set I.

(a) If α is a morphism of Γ into Γ', such that $\text{type}(\alpha(x)) = \text{type}(x)$ for all elements x of X, α is called a **homomorphism**.
(b) If $\alpha : X \to X'$ is a bijective homomorphism, such that $\alpha^{-1} : X' \to X$ is a homomorphism as well, α is called an **isomorphism**.
(c) An isomorphism of Γ into Γ is called an **automorphism**. The group of automorphisms of Γ is denoted by Aut (Γ). Every subgroup of Aut (Γ) is called an **automorphism group** of Γ.

The mappings β and δ defined in Example 2.1 are homomorphisms of the 4-gon, β is also an automorphism.

Definition. Let $\Gamma = (X, \, ^*, \, \text{type})$ and $\Gamma' = (X', \, ^{*\prime}, \, \text{type}')$ be two geometries over the type sets I and I', respectively.

(a) If $\alpha : X \to X'$ is a bijective morphism, such that $\alpha^{-1} : X' \to X$ is a morphism as well, α is called a **correlation**.
(b) A correlation of Γ into Γ is called an **autocorrelation**. The group of the autocorrelations of Γ is denoted by $\text{Cor}(\Gamma)$.
(c) Let n be a finite number. An autocorrelation α of Γ is called of **order** n if $\alpha^n = id$ (identity) and if $\alpha^m \neq id$ for all $m \in \{1, \ldots, n - 1\}$. If $\alpha^m \neq id$ for all natural numbers m, the autocorrelation α is called of infinite **order**.

The mappings α, β and γ defined in Example 2.1 are autocorrelations of order 2, 4 and 8, respectively.

2.2 Theorem. *Let Γ be a geometry over the type set I, and let α be an autocorrelation of Γ.*

(a) The autocorrelation α induces a permutation α_I on I.
(b) The autocorrelation α is an automorphism of Γ if α induces the identity on I.

Proof. (a) Let i be a type of I. By definition, the morphism α maps all elements of type i on elements of a type j. Let $\alpha_I(i) := j$. Then, α_I is a permutation on I.
(b) follows from the definition of an automorphism.   □

Definition. Let Γ be a geometry over the type set I, and let α be an autocorrelation of Γ.

(a) If the mapping α induces a permutation α_I of order 2 on I, α is called a **duality**.
(b) If both α and α_I are of order 2, α is called a **polarity**.

The mapping γ defined in Example 2.1 is a duality of the 4-gon. The mapping α defined in Example 2.1 is a polarity of the 3-gon.

2.3 Theorem. *Let Γ be a geometry over the type set I with diagram D, and let α be an autocorrelation of Γ. The permutation α_I induced by α maps the diagram D onto itself.*

Proof. The proof follows from Theorem 2.2.   □

3 Projections

In the present section, we will introduce projections. They form an important tool for the investigation of projective spaces. Whereas in projective spaces, a point can always be projected into a hyperplane (Theorem 3.1), this is not the case in affine spaces. The exact situation is described in Theorem 3.2.

Definition. Let P be a projective space, let H be a hyperplane of P, and let p be a point of P outside of H. For each point x different from p, let $\pi(x)$ be the intersection point of the line px with H. Then, π is called the **projection** of $P \backslash \{p\}$ on H.[1]

---

[1] If x is a point of H, $\pi(x) = x$.

3.1 Theorem. *Let P be a d-dimensional projective space, let H be a hyperplane of P, and let p be a point of P outside of H. Let U be a subspace of P not containing the point p, and let π be the projection of $P \setminus \{p\}$ on H.*

Then, $\pi(U)$ is a subspace of H, and $\pi : U \to \pi(U)$ is a collineation from U onto $\pi(U)$.

Proof. Step 1. π maps the point set of U bijectively on the point set of $\pi(U)$:

Obviously, π is surjective. In order to verify the injectivity of π, let $\pi(x) = \pi(y)$ for two points x and y of U. Then, the lines px and py intersect the hyperplane H in the same point $\pi(x) = \pi(y)$. It follows that the points x and y are incident with the line $p\pi(x)$. Assuming $x \neq y$, the line xy would be contained in U, and it follows that p is contained in U, in contradiction to the assumption.

Step 2. π maps subspaces of U on subspaces. In particular, $\pi(U)$ is a subspace of H.

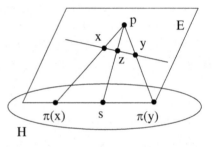

Let W be a subspace of U, and let $\pi(x)$ and $\pi(y)$ be two points of $\pi(W)$ with $x, y \in W$. Let s be a point on the line $\pi(x)\pi(y)$.

The points p, $\pi(x)$, $\pi(y)$ generate a plane E intersecting H in the line $\pi(x) \pi(y)$. Obviously, s is contained in E. Since p, s, x, y are contained in E, the lines xy and ps intersect in a point z. Since W is a subspace of U, the point z is contained in W. Furthermore,

$$s = pz \cap \pi(x)\,\pi(y) = pz \cap H = \pi(z).$$

It follows that s is contained in $\pi(W)$, that is, $\pi(W)$ is a subspace of $\pi(U)$.

Step 3. π^{-1} maps subspaces of $\pi(U)$ on subspaces of U.
The proof is analogous to the proof of Step 2.

Step 4. $\pi : U \to \pi(U)$ is a collineation:
For every point x and every line g of U, we have:

$$x \text{ is incident with } g \Leftrightarrow \pi(x) \text{ is incident with } \pi(g).$$

Hence, $\pi : U \to \pi(U)$ is a collineation.   □

Next, we will consider projections of affine spaces:

3.2 Theorem. *Let A be an affine space, let H be a hyperplane of A, and let p be a point of A outside of H.*

Let H' be the hyperplane through p parallel to H. For each point $x \neq p$ of A, the following holds:

(a) If x is not contained in H', px and H meet in a point $\pi(x)$.
(b) If x is contained in H', px and H do not intersect, that is, the point $\pi(x)$ is not defined.

Proof. Let $P = P(A)$ be the projective closure of A and let H_∞ be the hyperplane at infinity.

(a) Let H_P be the subspace of P defined by H, and let $z := px \cap H_P$. Assume that the point z is contained in the hyperplane H_∞. Then,

$$z \in H_P \cap H_\infty = H'_P \cap H_\infty, \text{ that is, } z \in H'_P.$$

Since the point p is contained in H'_P, the line pz is contained in H' as well. Since x lies on pz, it follows that x is contained in H', in contradiction to the choice of x.

(b) If x is contained in H', we have the following situation in P:

$$px \cap H_P \in H'_P \cap H_P \subseteq H_\infty.$$

It follows that px and H do not intersect in A. $\qquad\qquad\qquad\qquad\square$

By extending an affine space to its corresponding projective space, it is possible to project any point from a given point p to a given hyperplane H. This is the reason why projective spaces are called *projective*.

4 Collineations of Projective and Affine Spaces

In the present section, we will investigate collineations of projective and affine spaces. Compared to isomorphisms, collineations have the advantage that they are only defined on the point set of the projective or affine space. Therefore, it is often easier to verify whether a certain mapping is a collineation than to verify whether a certain mapping is an isomorphism.

At the beginning of this section, we will give a formal definition of collineations and parallelism-preserving collineations. We shall see that a collineation between two affine spaces is a parallelism-preserving collineation if the lines of the affine spaces are incident with at least three points (Theorem 4.4).

Furthermore, we will show that each parallelism-preserving collineation can be uniquely extended to a projective collineation (Theorem 4.6).

Finally, we will see that each collineation between two projective spaces and each parallelism-preserving collineation between two affine spaces defines an isomorphism (Theorems 4.8 and 4.9).

Definition. Let L and L' be two linear spaces, and let P and P' be the point sets of L and L', respectively.

A mapping $\alpha : P \to P'$ is called a **collineation** if α is bijective and if any three points x, y and z of L are collinear if and only if $\alpha(x)$, $\alpha(y)$ and $\alpha(z)$ are collinear.

4.1 Theorem. *Let L and L' be two linear spaces, and let $\alpha : L \to L'$ be a collineation.*

(a) $\alpha^{-1} : L' \to L$ is a collineation from L' onto L.
(b) α maps the set of lines of L bijectively onto the set of lines of L'.
(c) If U is a subspace of L, $\alpha(U)$ is a subspace of L'.
(d) If U' is a subspace of L', $\alpha^{-1}(U)$ is a subspace of L.

Proof. (a) follows from the definition of a collineation.
(b) Let g be a line of L, and let x and y be two points on g. Let g' be the line through $\alpha(x)$ and $\alpha(y)$. Then, we have for each point z of L:

$$z \text{ lies on } g$$
$$\Leftrightarrow x,\ y \text{ and } z \text{ are collinear}$$
$$\Leftrightarrow \alpha(x),\ \alpha(y) \text{ and } \alpha(z) \text{ are collinear}$$
$$\Leftrightarrow \alpha(z) \text{ lies on } g'.$$

It follows that $\alpha(g) = g'$, that is, α maps lines on lines. It follows from (a) that α^{-1} also maps lines on lines. Hence, α maps the lines of L bijectively onto the lines of L'.
(c) Let U be a subspace of L. Let $\alpha(x)$ and $\alpha(y)$ be two points of $\alpha(U) = \{\alpha(x) \mid x \in U\}$, and let g' be the line through $\alpha(x)$ and $\alpha(y)$ in L'. Let z' be a point on g'. We need to show that z' is contained in $\alpha(U)$.

Since α is bijective, there exists a point z of L such that $\alpha(z) = z'$. Since $\alpha(x)$, $\alpha(y)$ and $z' = \alpha(z)$ lie on g', the points x, y and z are collinear, that is, the point z is incident with the line xy which is contained in U.

It follows that $z' = \alpha(z)$ is contained in $\alpha(U)$.
(d) Since, by (a), $\alpha^{-1} : L' \to L$ is a collineation, the assertion follows from (c). □

4.2 Theorem. *Let P and P' be two projective spaces, and let $\alpha : P \to P'$ be a collineation.*

(a) Let U be a subspace of P, and let B be a basis of U. Then, $\alpha(B)$ is a basis of $\alpha(U)$.
(b) We have $\dim \alpha(U) = \dim U$ for all subspaces U of P.
(c) We have $\dim P = \dim P'$.

Proof. (a) Step 1. Let $\{x_1, \ldots, x_r\}$ be a finite subset of B. Then, $\alpha(\langle x_1, \ldots, x_r\rangle) = \langle \alpha(x_1), \ldots, \alpha(x_r)\rangle$: We shall prove the assertion by induction on r:
For $r = 1$, we have $\alpha(\langle x_1 \rangle) = \alpha(x_1) = \langle \alpha(x_1)\rangle$.
$r - 1 \to r$: By induction, we have $\alpha(\langle x_1, \ldots, x_{r-1}\rangle) = \langle \alpha(x_1), \ldots, \alpha(x_{r-1})\rangle$.
Let x be a point of $\langle x_1, \ldots, x_r\rangle$ not contained in $\langle x_1, \ldots, x_{r-1}\rangle$. By Theorem 4.2 of Chap. 1, there exists a point p of $\langle x_1, \ldots, x_{r-1}\rangle$ such that x is incident with the line px_r. Since α is a collineation, it follows that $\alpha(x)$ is incident with the line $\alpha(p)\alpha(x_r)$. Again, by Theorem 4.2 of Chap. I, the point $\alpha(x)$ is contained in the subspace $\langle \alpha(\langle x_1, \ldots, x_{r-1}\rangle),\ \alpha(x_r)\rangle = \langle \alpha(x_1), \ldots, \alpha(x_r)\rangle$. Hence, $\alpha(\langle x_1, \ldots, x_r\rangle)$ is contained in $\langle \alpha(x_1), \ldots, \alpha(x_r)\rangle$.

Conversely, let y be a point of $\langle \alpha(x_1), \ldots, \alpha(x_r)\rangle$ not contained in $\langle \alpha(x_1), \ldots, \alpha(x_{r-1})\rangle$. There exists a point b of $\langle \alpha(x_1), \ldots, \alpha(x_{r-1})\rangle$ such that y is incident with the line $b\alpha(x_r)$. By induction, there is a point a of $\langle x_1, \ldots, x_{r-1}\rangle$ such that

$\alpha(a) = b$. Since α is a collineation, there is a point x on the line ax_r such that $\alpha(x) = y$. Hence, $\langle \alpha(x_1), \ldots, \alpha(x_r) \rangle$ is contained in $\alpha(\langle x_1, \ldots, x_r \rangle)$.

Step 2. $\alpha(B)$ generates $\alpha(U)$: For, let y be a point of $\alpha(U)$, and let x be a point of U such that $\alpha(x) = y$. By Theorem 4.8 of Chap. 1, there exists a finite subset $\{x_1, \ldots, x_r\}$ of B such that x is contained in $\langle x_1, \ldots, x_r \rangle$. It follows from Step 1 that

$$y = \alpha(x) \in \alpha(\langle x_1, \ldots, x_r \rangle) = \langle \alpha(x_1), \ldots, \alpha(x_r) \rangle \subseteq \langle \alpha(B) \rangle.$$

Step 3. $\alpha(B)$ is linearly independent: Assume that $\alpha(B)$ is linearly dependent. Then, there exists an element y of $\alpha(B)$ such that y is contained in $\langle \alpha(B) \backslash \{y\} \rangle$. By Theorem 4.8 of Chap. 1, there exists a finite subset $\{y_1, \ldots, y_r\}$ of $\alpha(B) \backslash \{y\}$ such that y is contained in $\langle y_1, \ldots, y_r \rangle$. Let $x_1, \ldots, x_r$ and x be the points of B such that $\alpha(x_1) = y_1, \ldots, \alpha(x_r) = y_r$ and $\alpha(x) = y$. By Step 1, it follows that

$$\alpha(x) = y \in \langle y_1, \ldots, y_r \rangle = \langle \alpha(x_1), \ldots, \alpha(x_r) \rangle = \alpha(\langle x_1, \ldots, x_r \rangle).$$

Hence, the point x is contained in $\langle x_1, \ldots, x_r \rangle$. Since $\alpha(x) = y$ and $y \neq y_1, \ldots, y_r$, we have $x \neq x_1, \ldots, x_r$, contradicting the fact that B is a basis of U.

(b) and (c) follow from (a). $\qquad \Box$

Definition. Let A and A' be two affine spaces, and let $\alpha : A \to A'$ be a collineation. α is called a **parallelism-preserving collineation** if any two lines g and h are parallel in A if and only if $\alpha(g)$ and $\alpha(h)$ are parallel in A'.

4.3 Theorem. *Let A and A' be two affine spaces, and let $\alpha : A \to A'$ be a parallelism-preserving collineation.*

(a) $\alpha^{-1} : A' \to A$ is a parallelism-preserving collineation as well.
(b) α maps the set of affine planes of A bijectively onto the set of affine planes of A'.

Proof. (a) follows from the definition.
(b) Let E be an affine plane of A, and let x, y and z be three non-collinear points of E. Let E' be the affine plane of A' generated by $\alpha(x)$, $\alpha(y)$ and $\alpha(z)$.

Let p be a point of E. If p lies on the line xy or on the line xz, we have

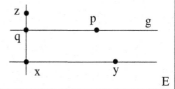

$$\alpha(p) \in \alpha(x)\alpha(y) \subseteq E' \text{ or}$$
$$\alpha(p) \in \alpha(x)\alpha(z) \subseteq E'.$$

If p neither lies on xy nor on xz, there exists a line g through p in E parallel to xy meeting the line xz in a point q.

Since α maps parallel lines on parallel lines, the point $\alpha(p)$ is incident with the line $g' := \alpha(g)$ which is parallel to $\alpha(x)\alpha(y)$ and which meets the line $\alpha(x)\alpha(z)$ in the point $\alpha(q)$. Since E' is closed with respect to $\|$, it follows that g' is contained in E', hence, the point $\alpha(p)$ is also contained in E'.

It follows that $\alpha(E) \subseteq E'$. In the same way, it follows from (a) that $\alpha^{-1}(E') \subseteq E$, hence, $\alpha(E) = E'$. Altogether, α maps the affine planes of A bijectively on the affine planes of A'.   □

4.4 Theorem. *Let A and A' be two affine spaces such that every line of A is incident with at least three points.*

Then, every collineation $\alpha : A \to A'$ is a parallelism-preserving collineation.

Proof. Let g and h be two parallel lines of A.

If $g = h$, $\alpha(g) = \alpha(h)$, and $\alpha(g)$ and $\alpha(h)$ are parallel.

Let $g \neq h$. Then, $g \cap h = \varnothing$. It follows that $\alpha(g) \cap \alpha(h) = \varnothing$. In order to show that $\alpha(g)$ and $\alpha(h)$ are parallel lines, we need to show that $\alpha(g)$ and $\alpha(h)$ are contained in an affine subplane of A'.

Since g and h are parallel, g and h generate a plane E of A. Let x_1 be a point on g, and let y_1 be a point on h. Since there are at least three points on any line of A, there exists a point z on the line $x_1 y_1$ different from x_1 and y_1.

Let x_2 be a point on g different from x_1. Since g and h are parallel lines, the lines $x_2 z$ and h meet in a point y_2 different from y_1.

Let E' be the plane of A' generated by the points $\alpha(x_1)$, $\alpha(x_2)$ and $\alpha(z)$. By Theorem 5.3 of Chap. I, the plane E' is an affine subplane of A'.

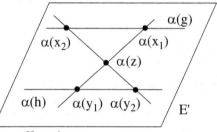

By definition of an affine space, E' is an affine plane.

Since α is a collineation, the points $\alpha(x_1)$, $\alpha(z)$ and $\alpha(y_1)$, and $\alpha(x_2)$, $\alpha(z)$ and $\alpha(y_2)$ are collinear, hence, the points $\alpha(y_1)$ and $\alpha(y_2)$ are contained in E'. It follows that $\alpha(h) = \alpha(y_1)\alpha(y_2)$ is contained in E'. It follows that $\alpha(g)$ and $\alpha(h)$ are parallel lines.   □

Remark. If A and A' are two affine spaces such that every line is incident with exactly two points, every bijective mapping from the point set of A onto the point set of A' is a collineation, but only a few of them preserve parallelism.

4.5 Theorem. *Let P and P' be two projective spaces, and let H and H' be two hyperplanes of P and P', respectively. Furthermore, let $A := P \setminus H$ and $A' := P' \setminus H'$ be the affine spaces defined by P and H and P' and H', respectively.*

If $\alpha : P \to P'$ is a collineation with $\alpha(H) = H'$, α induces a parallelism-preserving collineation from A to A'.

Proof. Obviously, the mapping $\alpha|_A$ maps the point set of A bijectively on the point set of A'. Furthermore, three points x, y and z are collinear in A if and only if $\alpha(x)$, $\alpha(y)$ and $\alpha(z)$ are collinear in A'.

For a line l of A or A', we denote by l_P the projective closure of l in P or P', respectively. Then, for any two lines g and h of A, we have:

g and h are parallel in A

$\Leftrightarrow g_P$ and h_P meet in a point x of H

$\Leftrightarrow \alpha(g_P)$ and $\alpha(h_P)$ meet in the point $\alpha(x)$ of H'

$\Leftrightarrow \alpha(g)$ and $\alpha(h)$ are parallel in A'.

It follows that α is a parallelism-preserving collineation.   □

Definition. Let A and A' be two affine spaces with projective closures P and P', and let $\alpha : A \rightarrow A'$ be a collineation. If $\alpha^* : P \rightarrow P'$ is a collineation such that $\alpha^*|_A = \alpha$, the collineation α^* is called the **projective closure** of α.

4.6 Theorem. *Let A and A' be two affine spaces, and let $\alpha : A \rightarrow A'$ be a parallelism-preserving collineation.*

If P and P' are the projective closures of A and A', respectively, there exists exactly one projective closure of α.

Proof. Let H and H' be the hyperplanes of P and P' such that $A = P \backslash H$ and $A' = P' \backslash H'$. For an affine line g of A or A', we denote by g_P the projective closure of g in P or in P'.

We first will prove the existence of the mapping α^*.

Step 1. Definition of the mapping α^*:

If x is a point of A, set $\alpha^*(x) := \alpha(x)$.

If x is a point of H, there exists a line g of A such that $x = g_P \cap H$. Set

$$\alpha^*(x) := \alpha(g)_P \cap H'.$$

We have to show that α^* is well-defined: For, let g and h be two lines of A such that

$$x = g_P \cap H = h_P \cap H.$$

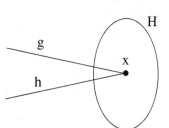

Then, g and h are parallel in A. Since α is a parallelism-preserving collineation, the lines $\alpha(g)$ and $\alpha(h)$ are also parallel in A', that is,

$$\alpha(g)_P \cap H' = \alpha(h)_P \cap H'.$$

It follows that α^* is well-defined.

Step 2. Let P and P' be the point sets of P and P', respectively. Then, the mapping

$$\alpha^* : P \rightarrow P'$$

is bijective:

Since $\alpha : A \rightarrow A'$ is a collineation, we only need to show that $\alpha^*|_H : H \rightarrow H'$ is bijective.

The mapping α^* is surjective: Let x' be a point of H', and let g' be a line of A' such that $x' = g'_P \cap H'$. Since α is a collineation, there exists a line g of A with $\alpha(g) = g'$. For $x := g_P \cap H$, we have

$$\alpha^*(x) = \alpha(g)_P \cap H' = g'_P \cap H' = x'.$$

The mapping α is injective: Let x and y be two points of H such that $\alpha^*(x) = \alpha^*(y)$. By definition of α^*, there exist two lines g and h of A such that $x = g_P \cap H$ and $y = h_P \cap H$. It follows from $\alpha^*(x) = \alpha^*(y)$ that $\alpha(g)_P \cap H' = \alpha(h)_P \cap H'$.

It follows that the lines $\alpha(g)$ and $\alpha(h)$ are parallel in H'. Since α is a parallelism-preserving collineation, the lines g and h are parallel as well. It follows that $x = g_P \cap H = h_P \cap H = y$.

Step 3. Let x, y and z be three collinear points of P. Then, the points $\alpha^*(x)$, $\alpha^*(y)$ and $\alpha^*(z)$ are collinear as well:

First case. There is a line g of A such that x, y and z are incident with g_P. By definition of α^*, the points $\alpha^*(x)$, $\alpha^*(y)$ and $\alpha^*(z)$ are incident with the line $\alpha(g)_P$.

Second case. The points x, y and z are contained in H. Since x, y and z are collinear, there exists an affine plane E of A such that x, y and z are incident with the line at infinity of E. By Theorem 4.3, $\alpha(E)$ is also an affine plane. It follows that the points $\alpha^*(x)$, $\alpha^*(y)$ and $\alpha^*(z)$ are incident with the line at infinity of $\alpha(E)$. In particular, they are collinear.

Step 4. If $\alpha^*(x)$, $\alpha^*(y)$ and $\alpha^*(z)$ are three collinear points of P', x, y and z are collinear as well: Since α^* is bijective, the assertion follows as in Step 3 if one replaces α^* by α.

It follows from Steps 2 to 4 that α^* is a collineation from P to P'.

Let β^* be a further projective closure of α. Obviously, $\beta^*(x) = \alpha(x) = \alpha^*(x)$ for all points x of A. For a point x of H, it follows that $\beta^*(x) = \alpha^*(x)$, since α^* and β^* map lines onto lines. It follows that $\beta^* = \alpha^*$ which shows the uniqueness of α^*. □

4.7 Theorem. *Let A and A' be two affine spaces, and let $\alpha : A \to A'$ be a parallelism-preserving collineation.*

(a) *Let U be a subspace of A, and let B be a basis of U. Then, $\alpha(B)$ is a basis of $\alpha(U)$.*

(b) *We have $\dim \alpha(U) = \dim U$ for all subspaces U of A.*

(c) *We have $\dim A = \dim A'$.*

Proof. (a) Let P and P' be the projective closures of A and A', respectively. By Theorem 4.6, α admits a unique closure $\alpha : P \to P'$. For a subspace U of A, let U_P be the subspace of P defined by U. It follows from Theorem 5.7 of Chap. 1 that B is a basis of U_P. By Theorem 4.2, $\alpha(B)$ is a basis of $\alpha(U_P) = \alpha(U)_P$. Again, by Theorem 5.7 of Chap. 1, $\alpha(B)$ is a basis of $\alpha(U)$.

(b) and (c) follow from (a).   □

4.8 Theorem. *Let **P** and **P′** be two d-dimensional projective spaces, and let* α :
***P** → **P′** be a collineation. Then,* α *induces an isomorphism of the projective geometry defined by **P** onto the projective geometry defined by **P′**.*

Proof. Let $\Gamma = (X, *, \text{type})$ and $\Gamma' = (X', *', \text{type}')$ be the projective geometries defined by **P** and **P′**. Then, X is the set of subspaces of **P** different from $\varnothing$ and **P**.

By Theorem 4.1, α maps the set X bijectively onto the set X'. Furthermore, by Theorem 4.2, we have

$$\text{type}'\, \alpha(U) = \dim\, \alpha(U) = \dim\, U = \text{type}\, U \text{ for all subspaces } U \text{ of } \boldsymbol{P}.$$

It follows that α is an isomorphism.   □

4.9 Theorem. *Let **A** and **A′** be two d-dimensional affine spaces, and let* α : **A** → **A′** *be a parallelism-preserving collineation. Then,* α *induces an isomorphism from the affine geometry defined by **A** onto the affine geometry defined by **A′**.*

Proof. The proof is similar to the proof of Theorem 4.8. Instead of Theorem 4.2, we make use of Theorem 4.7.   □

Definition. (a) Two projective spaces **P** and **P′** are called **isomorphic** if there exists a collineation α : **P** → **P′** from **P** onto **P′**.

(b) Two affine spaces **A** and **A′** are called **isomorphic** if there exists a parallelism-preserving collineation α : **A** → **A′** from **A** onto **A′**.

5 Central Collineations

In the present section, we shall introduce central collineations. They are the main tool in the proof of the fundamental theorem of projective geometry classifying all projective (and affine) spaces of dimension at least 3 (see Chap. 3).

A central collineation is defined by the fact that it has a centre and an axis (for the definition, see below). In Theorems 5.3 and 5.6, we will see that any collineation admitting an axis or a centre is already a central collineation.

Theorem 5.8 gives a necessary condition for the existence of a central collineation. In fact, it will turn out in Section 6 that for projective spaces of dimension at least 3, this condition is also sufficient.

Finally, in Theorem 5.9, the results about central collineations in projective spaces are transferred to affine spaces.

Definition. Let $\Gamma = (X, *, \text{type})$ be a geometry, and let α : $X \to X$ be a morphism.

(a) An element x of X is called a **fixed element** (with respect to α) if $\alpha(x) = x$.

(b) If Γ is a linear space, a point x is called a **fixed point** if $\alpha(x) = x$, and a line g is called a **fixed line** if $\alpha(g) = g$.[2]

5.1 Lemma. *Let L be a linear space, and let $\alpha : L \to L$ be a collineation.*

(a) If x and y are two points of L, we have:

$$\alpha(xy) = \alpha(x)\alpha(y).$$

In particular, any two fixed points are joined by a fixed line.
(b) If g and h are two lines of L, we have:

$$\alpha(g \cap h) = \alpha(g) \cap \alpha(h).$$

In particular, if two fixed lines g and h meet in a point x, x is a fixed point.

Proof. The assertions follow directly from the definition of a collineation.   □

Definition. Let P be a projective space, and let $\alpha : P \to P$ be a collineation.

(a) A point z is called a **centre** of α if every line through z is fixed by α. A hyperplane H is called an **axis** of α if every point of H is fixed by α.
(b) α is called a **central collineation** if α has a centre and an axis.
(c) A central collineation τ is called an **elation** if the centre of τ and the axis of τ are incident.
(d) A central collineation δ is called a **homology** if the centre of δ and the axis of δ are not incident.

Every central collineation is either an elation or a homology.

As we shall see in Theorems 5.3 and 5.6, the existence of a centre implies the existence of an axis and vice versa.

5.2 Theorem. *Let P be a projective space, and let z and H be a point and a hyperplane of P, respectively. Then, the set $Z(p, H)$ of all central collineations of P with centre z and axis H forms a group.*

Proof. The proof is obvious.   □

5.3 Theorem. *Let P be a projective space, and let $\alpha : P \to P$ be a collineation. If α has an axis H, α is a central collineation.*

Proof. First case. There is a fixed point z of α which is not contained in H. Then, z is a centre of α, and α is a homology:

Let g be a line through z. By Theorem 4.18 of Chap. 1, g and H meet in a point. By assumption, this point is a fixed point, hence, g is a fixed line. It follows that z is a centre of α.

Second case. Outside of H, there are no fixed points of α. Then, α is an elation:

---

[2]This means that the line g is fixed as a line. In particular, we have $\alpha(x) \in g$ for all $x \in g$. The points of g may be fixed, but, in general, this is not the case.

Let x be a point of P outside of H.
Furthermore, let $g := x\alpha(x)$ be the line
through x and $\alpha(x)$. Then, g and H meet in
a fixed point z. It follows that

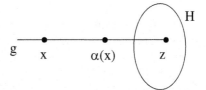

$$\alpha(g) = \alpha(xz) = \alpha(x)z = g.$$

It follows that g is a fixed line.
We shall show that z is a centre of α:

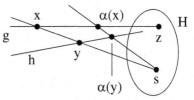

For, let y be a point outside of H and g,
and let $h := y\alpha(y)$ be the line through y and
$\alpha(y)$. As above, it follows that h is a fixed
line. Let $s := xy \cap H$. Then, s is a fixed
point. Since the line xy passes through s, the
line $\alpha(xy) = \alpha(x)\alpha(y)$ passes through s as
well. It follows that the lines g and h are contained in the plane $E := \langle s, g \rangle$ and
therefore meet in a point a.

Since g and h are fixed lines, the point a is a fixed point, and it follows from the
assumption that a is contained in H. Hence, a is contained in $g \cap H = z$, that is,
$a = z$. For every point y of P, the line yz is a fixed line, that is, z is a centre of α.   □

5.4 Theorem. *Let P be a projective space, and let $\alpha : P \to P$ be a collineation
with a centre z.*

(a) α fixes every subspace through z.
(b) If g is a fixed line of P not containing z, every point of g is fixed by α.
(c) Every line of P has a fixed point.
*(d) Let g be a line of P such that every point of g is fixed by α, and let x be a fixed
point outside of g different from z. Then, every point of the plane $\langle x, g \rangle$ is fixed.*
*(e) Let g be a line through z. If there are at least two further fixed points (besides
z) on g, every point of g is fixed by α.*

Proof. (a) Let U be a subspace of P through z, and let x be a point of U. Then, the
line $g := zx$ is contained in U, and it follows:

$$\alpha(x) \in \alpha(g) = g \subseteq U.$$

(b) Let g be a fixed line of P not containing z,
and let x be a point on g. Then, $h := zx$ is
a fixed line different from g. Hence, x is
the intersection point of two fixed lines, in
particular, x is a fixed point.

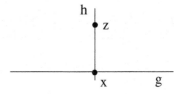

(c) Let g be a line of P. W.l.o.g., let z
and g be non-incident. By (b), we
may assume that g is not a fixed
line. Let $h := \alpha(g) \neq g$. By (a), α
fixes the plane $E := \langle z, g \rangle$. It follows that $h = \alpha(g)$ is contained in
E, hence, the lines g and h meet in

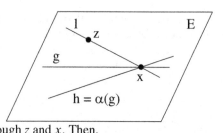

a point x. Let $l := zx$ be the line through z and x. Then,

$$\alpha(x) = \alpha(g \cap l) = \alpha(g) \cap \alpha(l) = h \cap l = x.$$

(d) Let g be a line of P such that every point of g is fixed by α, and let x be a fixed
point outside of g different from z. Let $E := \langle x, g \rangle$ be the plane generated by
z and g.
If z is not contained in E, by (b), the points of all lines of E through x are fixed.
Hence, all points of the plane E are fixed.
If z is contained in E, again by (b), the points of all lines through x different
from xz are fixed. Hence, all points of E are fixed.

(e) Let g be a line through z, and let a and b be two fixed points on g different
from z. Let c be a further point on g.
   Assume that c is not a fixed point.
Let h be an arbitrary line through c
different from g. By (c), h contains
a fixed point $s \neq c, z$. Since the line
as does not contain the point z, any
point of the line as is fixed by α in
view of (b). The plane $E := \langle z, as \rangle$

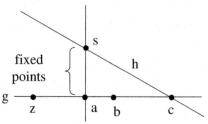

contains the fixed point b. By (d),
every point of the plane E is fixed. In particular, we have $\alpha(c) = c$, a
contradiction.   □

5.5 Theorem. *Let P be a projective space, and let $\alpha : P \to P$ be a collineation.
If α has a centre z, one of the following possibilities occurs:*

(i) Each line through z has exactly two fixed points (z being one of them).
*(ii) If g is a line through z, either all points on g are fixed, or g contains exactly
one fixed point, namely z.*

Proof. Step 1. If there is a line g through z with exactly two fixed points a and z,
every line through z has at least two fixed points:

For, let h be a line through z different from g. Let l be a line, intersecting the lines g and h in two points different from z and a. By Theorem 5.4, 1 contains a fixed point b. If $b = l \cap h$, h contains the fixed point b different from z. If $b \neq l \cap h$, the line ab is a fixed line in $\langle g, h \rangle$, intersecting h in a fixed point s different from z.

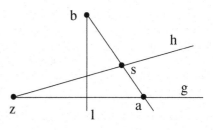

Step 2. If there is a line g through z with exactly two fixed points, every line through z has exactly two fixed points:

By Step 1, every line through z has at least two fixed points. Assume that there exists a line l through z with at least three fixed points.

By Theorem 5.4 (e), every point on l is fixed. Let x be an arbitrary point on g outside of l, and let $E := \langle x, l \rangle$ be the plane generated by x and l. By assumption, the line $g := zx$ contains a fixed point y. By Theorem 5.4(b) (d), all points of E are fixed. In particular, x is a fixed point, in contradiction to the assumption that g admits exactly two fixed points.

Step 3. Suppose that no line through z admits exactly two fixed points. Then, by Theorem 5.4 (e), every line g through z contains either exactly one fixed point, namely z, or all points on g are fixed. □

5.6 Theorem. *Let* P *be a projective space, and let* $\alpha : P \to P$ *be a collineation. If* α *has a centre,* α *is a central collineation.*

Proof. By Theorem 5.5, one of the following cases occurs:

(i) Each line through z contains exactly two fixed points (z being one of them).
(ii) If g is a line through z, either every point on g is fixed, or g contains exactly one fixed point, namely z.

Case (i): Let $H := \{x \in P \mid x$ is a fixed point of $\alpha, x \neq z\}$. We will show that α is a central collineation with centre z and axis H:

H is a subspace: For, let x and y be two points of H, and let $g := xy$ be the line through x and y. Since x and y are fixed points, the line g is fixed by α. Since $x, y \neq z$ and since, by assumption, the lines through z contain exactly one further fixed point besides z, it follows that z is not incident with g. By Theorem 5.4(b), every point on g is fixed. It follows that g is contained in H.

H is a hyperplane: By assumption and by Theorem 5.4, every line contains a fixed point different from z. Hence, every line contains a point of H. It follows from Theorem 4.18 of Chap. 1 that H is a hyperplane.

Case (ii): Let H be the set of fixed points of α (including z). Then, α is an elation with axis H.

H is a subspace: Let x and y be two points of H, and let $g := xy$ be the line through x and y. By assumption and by Theorem 5.4, every point on g is fixed by α. It follows that g is contained in H.

H is a hyperplane: The lines through z intersect H at least in the point z. By Theorem 5.4, all other lines contain at least a fixed point and therefore a point of H. By Theorem 4.18 of Chap. 1, H is a hyperplane. $\square$

5.7 Theorem. *Let P be a projective space, and let $\alpha : P \to P$ be a central collineation with centre z and axis H.*

(a) Let x be a point outside of $H \cup \{z\}$. If x is a fixed point of α, we have $\alpha = id$.
(b) Let x be a point outside of $H \cup \{z\}$. Then, the point $\alpha(x)$ is incident with the line zx.

Proof. (a) follows from Theorem 5.3.
(b) follows from the fact that the line zx is a fixed line. $\square$

5.8 Theorem. *Let P be a projective space. Let H be a hyperplane of P, and let z be a point of P. Furthermore, let g be a line outside of H through z. Finally, let x and y be two points on g outside of H and different from z.*

Then, there exists at most one central collineation α with centre z and axis H such that $\alpha(x) = y$.

Proof. Let α and β be two central collineations with centre z and axis H such that $\alpha(x) = \beta(x) = y$. Then, $\beta^{-1} \circ \alpha$ is also a central collineation with centre z and axis H. Furthermore

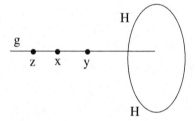

$$(\beta^{-1} \circ \alpha)(x) = \beta^{-1}(\alpha(x)) = \beta^{-1}(y) = x.$$

By Theorem 5.7, it follows that $\beta^{-1} \circ \alpha = id$, that is, $\alpha = \beta$. $\square$

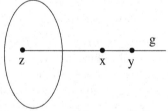

Definition. Let A and A' be two affine spaces, and let $\alpha : A \to A'$ be a parallelism-preserving collineation.

(a) α is called a **translation** if there exists a parallel class π such that α fixes every line of π and if either $\alpha = id$ or α has no fixed points.
(b) α is called a **homology** if there exists a point z such that α fixes every line through z. The point z is called the **centre** of α.
(c) If α is a translation or a homology, α is called a **central collineation**.

5.9 Theorem. *Let **P** be a projective space, let H be a hyperplane of **P**, and let* $A := P \backslash H$ *be the affine space defined by **P** and H.*

(a) If $\alpha : P \to P$ *is an elation with axis H, α induces a translation in **A**.*

(b) If $\alpha : P \to P$ *is a homology with axis H and centre z, α induces a homology with centre z in **A**.*

(c) If $\alpha : A \to A$ *is a translation, the projective closure of α is an elation with axis H. The centre of α is the point of H defined by the parallel class which is element-wise fixed by α.*

(d) If $\alpha : A \to A$ *is a homology with centre z, the projective closure of α is a homology with axis H and centre z.*

Proof. (a) Let z be the centre of α. Note that z and H are incident since α is an elation. Then, the lines through z not contained in H form a parallel class of A. Any of these lines is fixed by $\alpha|_A$. If $\alpha \neq id$, by Theorem 5.7, there exist no fixed points outside of H. It follows that $\alpha|_A$ is a translation of A.

(b) The proof is obvious.

(c) Let π be the parallel class consisting of the lines fixed by α, and let z be the projective closure of π.

We will show that the point z is the centre of the projective closure α^* of α: For, by Theorem 5.6, α^* is a central collineation with centre z and axis G. Since α (for $\alpha \neq id$) does not have a fixed point in A, it follows that $G = H$.

(d) Since α fixes every line through z, the projective closure α^* of α fixes every point of H. It follows that α^* is a central collineation with centre z and axis H.

$\square$

6 The Theorem of Desargues

In Sect. 5, we have seen that given a projective space P, a point z, a hyperplane H and two points x and y outside of H and collinear with z, there exists at most one central collineation α with centre z and axis H such that $\alpha(x) = y$. In the present section, we shall see that for every projective space fulfilling the so-called Theorem of Desargues, such a central collineation exists (Theorem 6.2). Furthermore, we shall see that every projective space of dimension at least 3 fulfils the Theorem of Desargues (Theorem 6.1).

Definition. Let P be a projective space.

(a) Let p, a_1, a_2, a_3, b_1, b_2, b_3 be seven points such that the lines $a_1 b_1$, $a_2 b_2$ and $a_3 b_3$ intersect in the point p and such that the points a_1, a_2, a_3 and b_1, b_2, b_3 are not collinear. Then, the seven points p, a_1, a_2, a_3, b_1, b_2, b_3 are called a **Desargues configuration**.

(b) Let p, a_1, a_2, a_3, b_1, b_2, b_3 be a Desargues configuration, and let the intersection points s_{12}, s_{13} and s_{23} be defined as follows:

$$s_{12} := a_1 a_2 \cap b_1 b_2, \; s_{13} := a_1 a_3 \cap b_1 b_3 \text{ and } s_{23} := a_2 a_3 \cap b_2 b_3.$$

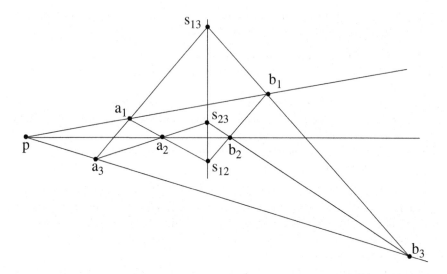

In P the **Theorem of Desargues** holds if the points s_{12}, s_{13} and s_{23} are collinear.

(c) The projective space P is called **Desarguesian** if in P the Theorem of Desargues holds.

Definition. An affine space A is called **Desarguesian** if the projective closure $P(A)$ of A is Desarguesian.

6.1 Theorem. *Let P be a projective space. If $d \geq 3$, P is Desarguesian.*

Proof. Let p, a_1, a_2, a_3, b_1, b_2, b_3 be seven points of a Desargues configuration, that is, the lines a_1b_1, a_2b_2 and a_3b_3 meet in the point p, and the points a_1, a_2, a_3 and b_1, b_2, b_3 are not collinear.

Let the intersection points s_{12}, s_{13} and s_{23} be defined as follows:

$$s_{12} := a_1a_2 \cap b_1b_2, \; s_{13} := a_1a_3 \cap b_1b_3 \text{ and } s_{23} := a_2a_3 \cap b_2b_3.$$

We need to show that the points s_{12}, s_{13} and s_{23} are collinear.

First case: The planes $E := \langle a_1, a_2, a_3 \rangle$ and $F := \langle b_1, b_2, b_3 \rangle$ are distinct.

Then, the point p is not contained in E, since, otherwise, F would be contained in $\langle p, E \rangle = E$, that is, $F = E$, a contradiction.

Obviously, the points $s_{12} = a_1a_2 \cap b_1b_2$, $s_{13} = a_1a_3 \cap b_1b_3$ and $s_{23} = a_2a_3 \cap b_2b_3$ are contained in $E \cap F$. Since, by assumption, $E \neq F$, the set $E \cap F$ is a line. Hence, the points s_{12}, s_{13} and s_{23} are collinear.

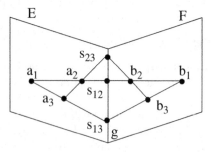

Second case: The points a_1, a_2, a_3 and b_1, b_2, b_3 are contained in a common plane E.

Obviously, the point p is contained in E.

Step 1. We first will construct three auxiliary points r_1, r_2 and r_3:

For, let g be a line of $\boldsymbol{P}$ intersecting the plane E in the point p. Let x and y be two points on g different from p. Let $F_1 := \langle g, a_1 b_1 \rangle$ be the plane generated by the lines g and $a_1 b_1$.

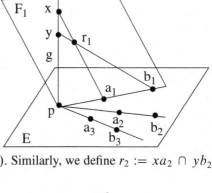

Set $r_1 := xa_1 \cap yb_1$ (the lines xa_1 and yb_1 have a common point since the points x, y, a_1, b_1 are all contained in F_1). Similarly, we define $r_2 := xa_2 \cap yb_2$ and $r_3 := xa_3 \cap yb_3$.

Step 2. The points r_1, r_2 and r_3 are not collinear:

Otherwise, $F := \langle x, r_1, r_2, r_3 \rangle$ would be a plane containing the points a_1, a_2 and a_3. It follows that a_1, a_2, a_3 are contained in $E \cap F$. This contradicts the assumption that the points a_1, a_2 and a_3 are not collinear.

Step 3. The points s_{12}, s_{13} and s_{23} are collinear:

By Step 2, the subspace $G :=$

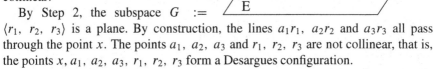

$\langle r_1, r_2, r_3 \rangle$ is a plane. By construction, the lines $a_1 r_1$, $a_2 r_2$ and $a_3 r_3$ all pass through the point x. The points a_1, a_2, a_3 and r_1, r_2, r_3 are not collinear, that is, the points $x, a_1, a_2, a_3, r_1, r_2, r_3$ form a Desargues configuration.

By construction, the points r_1, r_2, r_3 are not contained in $E = \langle a_1, a_2, a_3 \rangle$, that is, the planes $G = \langle r_1, r_2, r_3 \rangle$ and E are distinct. By Case 1, the Theorem of Desargues holds for this configuration, hence, the points

$$u_{12} := r_1 r_2 \cap a_1 a_2, \ u_{13} := r_1 r_3 \cap a_1 a_3 \text{ and } u_{23} := r_2 r_3 \cap a_2 a_3$$

are incident with a common line l. Since $u_{ij} = r_i r_j \cap a_i a_j$ is contained in $G \cap E$, it follows that $l = G \cap E$.

Similarly, it follows that the points

$$v_{12} := r_1 r_2 \cap b_1 b_2, \ v_{13} := r_1 r_3 \cap b_1 b_3 \text{ and } v_{23} := r_2 r_3 \cap b_2 b_3$$

are incident with the line $l = G \cap E$.

Since the point u_{12} is incident with the lines l and $r_1 r_2$, it follows that $u_{12} = r_1 r_2 \cap l$. In the same way, it follows that $v_{12} = r_1 r_2 \cap l$, hence, $u_{12} = v_{12}$. Similarly, it follows that $u_{13} = v_{13}$ and $u_{23} = v_{23}$.

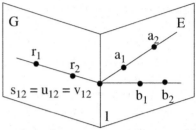

The point $u_{12} = v_{12}$ is incident with the lines $a_1 a_2$ and $b_1 b_2$, therefore, $u_{12} = a_1 a_2 \cap b_1 b_2 = s_{12}$. In the same way, it follows that $u_{13} = s_{13}$ and $u_{23} = s_{23}$. Thus, the points s_{12}, s_{13} and s_{23} are all incident with the line l. □

Theorem 6.1 cannot be generalized to projective planes. There are (many) examples of projective planes which are not Desarguesian. The interested reader is referred to [23] and [31].

6.2 Theorem. *Let P be a projective space. Let H be a hyperplane of P, and let z be a point of P. Furthermore, let g be a line through z not contained in H, and let p and p' be two points on g different from z such that p and p' are not contained in H.*

If P is Desarguesian, there exists exactly one central collineation α with centre z and axis H such that $\alpha(p) = p'$.

We first sketch the idea of the proof: Given a central collineation γ with centre z and axis H, consider a point $x \neq z$ outside of H, and let $x' := \gamma(x)$. Since z is the centre of γ, the point $x' = \gamma(x)$ is incident with the line zx. If y is a point outside of H not incident with the line zx, the point $y' := \gamma(y)$ can be constructed as follows:

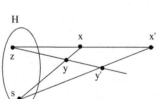

Let $s := xy \cap H$ be the intersection point of the line xy and the hyperplane H. It follows that

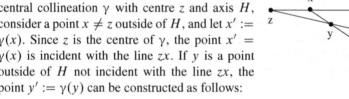

$$y' = \gamma(y) = \gamma(zy \cap sx) = \gamma(zy) \cap \gamma(sx) = zy \cap sx'.$$

Proof. The uniqueness of α has already been shown in Theorem 5.8. For the definition of α, we will make use of auxiliary functions $\alpha_{a,a'}$:

Step 1. Definition of the mapping $\alpha_{a,a'}$:

Let a and a' be two points of $P \backslash H$ which are collinear with z. For a point x of $P \backslash H$ non-incident with the line za, we define $x' := \alpha_{a,a'}(x)$ as follows:

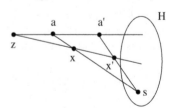

Let $s := ax \cap H$. Set $\alpha_{a,a'}(x) := zx \cap a's$.
(The point $\alpha_{a,a'}(x)$ exists since the lines
zx and $a's$ are contained in the plane
$\langle z, a, x \rangle$.)

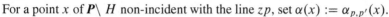

Step 2. Definition of α:
Set $\alpha(z) := z$ and $\alpha(x) := x$ for all points
x of H.

For a point x of $P \setminus H$ non-incident with the line zp, set $\alpha(x) := \alpha_{p,p'}(x)$.

Let x be a point of $P \setminus H$ incident with the line zp. For the definition of $\alpha(x)$,
we consider a point q of $P \setminus H$ such that q is not incident with zp. Let $q' = \alpha(q) = \alpha_{p,p'}(q)$, and set $\alpha(x) := x' := \alpha_{q,q'}(x)$.

We have to show that $\alpha(p) = \alpha_{q,q'}(p) = p'$:
Since $q' = \alpha_{p,p'}(q)$, it follows for $s := pq \cap H$
that $q' = zq \cap p's$. Thus, $\alpha_{q,q'}(p) = zp \cap q's = p'$.

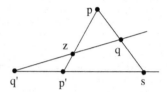

Altogether, $\alpha(x)$ is defined for all points x of P,
and we have $\alpha(p) = p'$.

Step 3. α is bijective:
Let β be defined as in Step 2 under the assumption that $\beta(p') = p$ and that q is
replaced by q'. We will show that $\beta = \alpha^{-1}$. For, let x be an arbitrary point of P.

First case. If $x = z$ or if x is contained in H, it
follows that $\beta(\alpha(x)) = \beta(x) = x$.

Second case. Let x be a point outside of H non-
incident with the line zp. Then, we have for $s :=$
$px \cap H$:

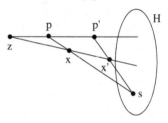

$$x' := \alpha(x) = zx \cap sp'.$$

It follows that $p'x' \cap H = s$, and, therefore,
$\beta(x') = zx' \cap sp = x$. In particular, it follows
that $\beta(q') = q$ for the point q defined in
Step 2.

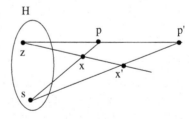

Third case. Let $x \neq z$ be a point outside of H incident with the line zp. The
assertion $\beta(\alpha(x)) = x$ follows as in Case 2 using the relation $\beta(q') = q$. So,
$\beta = \alpha^{-1}$.

Step 4. Let a and a' be two points on a line through z both different from z such that a, a' are not contained in H. Let b be a point of $P\setminus H$ non-incident with the line za, and let $b' := \alpha_{a,a'}(b)$. Finally, let c be a point of $P\setminus H$ such that c is neither incident with the line za nor with the line zb. Then, $\alpha_{a,a'}(c) = \alpha_{b,b'}(c)$:

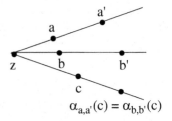

$$\alpha_{a,a'}(c) = \alpha_{b,b'}(c)$$

By definition of $\alpha_{a,a'}$ (Step 1), for $s := ab \cap H$, $t := ac \cap H$ and $u := bc \cap H$, the following hold:

$$b' = \alpha_{a,a'}(b) = zb \cap a's$$
$$\alpha_{a,a'}(c) = zc \cap a't$$
$$\alpha_{b,b'}(c) = zc \cap b'u.$$

First case. Suppose that the points a, b and c are collinear.

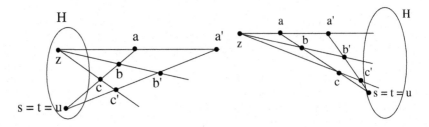

Then

$$\alpha_{a,a'}(c) = zc \cap a't = c'$$
$$\alpha_{b,b'}(c) = zc \cap b'u = c'.$$

Hence, $\alpha_{a,a'}(c) = \alpha_{b,b'}(c)$.

Second case. Suppose that the points a, b and c are non-collinear. First, observe that the plane generated by the points a, b and c intersects the hyperplane H in a line. It follows that the points s, t and u are collinear and that

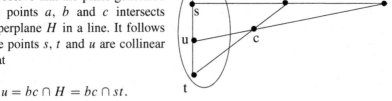

$$u = bc \cap H = bc \cap st.$$

We shall apply the Theorem of Desargues: By construction, the point triples $(a,$
$a', z)$, (a, b, s) and (a, c, t) are collinear, that is, the lines za', bs and ct meet in the
point a. Furthermore, the points z, b, c are not collinear since, by assumption, c is
not incident with zb. The points a', s, and t are not collinear since the points s and
t are contained in H, whereas the point a' is not contained in H. It follows that the
points a, z, b, c, a', s, t form a Desargues configuration.

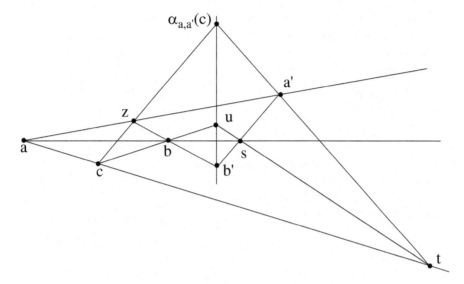

By the Theorem of Desargues, the points $b' = zb \cap a's$, $u = bc \cap H = bc \cap st$
and $\alpha_{a,a'}(c) = zc \cap a't$ are collinear.

In particular, the point $\alpha_{a,a'}(c) = zc \cap a't$ lies on the lines $a't$, zc and $b'u$. It
follows that

$$\alpha_{a,a'}(c) = b'u \cap zc = \alpha_{b,b'}(c).$$

Step 5. Let a be a point of $P \backslash H$ different from z, and let $a' := \alpha(a)$. Then, for
any point x of $P \backslash H$ non-incident with the line za, we have $\alpha(x) = \alpha_{a,a'}(x)$:

We shall apply Step 4 repeatedly:

(i) Let q be the point outside of H and non-
incident with the line zp as chosen in Step 2.
Then, for any point r of $P \backslash H$ such that r is
neither incident with zp nor with zq, we have

$$\alpha(r) = \alpha_{p,p'}(r) = \alpha_{q,q'}(r):$$

By definition of α (Step 2), we have $\alpha(r) = \alpha_{p,p'}(r)$. By Step 4, we have
$\alpha_{p,p'}(r) = \alpha_{q,q'}(r)$. Altogether, we have

$$\alpha(r) = \alpha_{p,p'}(r) = \alpha_{q,q'}(r).$$

(ii) We have $\alpha(x) = \alpha_{a,a'}(x)$ for all x of $P\backslash H$ non-incident with the line za.

First case. Suppose that p is not incident with
za and not incident with zx.

By definition of α, we have $\alpha(x) = \alpha_{p,p'}(x)$.
It follows from Step 3 that

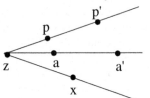

$$\alpha(x) = \alpha_{p,p'}(x) = \alpha_{a,a'}(x).$$

Second case. Suppose that p is incident with
za and that q is not incident with zx.

By definition of α, we have $\alpha(a) = \alpha_{q,q'}(a)$.
By (i), we have $\alpha_{p,p'}(x) = \alpha_{q,q'}(x)$. It follows
from Step 3 that

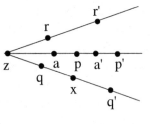

$$\alpha(x) = \alpha_{p,p'}(x) = \alpha_{q,q'}(x) = \alpha_{a,a'}(x).$$

Third case. Suppose that p is incident with za
and that q is incident with zx.

Let r be a point of $P\backslash H$ such that r is not
incident with za and neither with zx. Furthermore, let $r' := \alpha(r) = \alpha_{p,p'}(r)$.

By (i), we have $r' = \alpha_{q,q'}(r)$. Thus, it follows
from Step 3 that

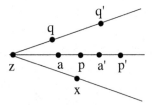

$$\alpha(a) = \alpha_{q,q'}(a) = \alpha_{r,r'}(a).$$

Again by Step 3, it follows that $\alpha_{r,r'}(x) = \alpha_{a,a'}(x)$.

In view of $r' = \alpha_{p,p'}(r)$, it follows from Step 3 that $\alpha(x) = \alpha_{p,p'}(x) = \alpha_{r,r'}(x)$. Altogether, we have $\alpha(x) = \alpha_{p,p'}(x) = \alpha_{r,r'}(x) = \alpha_{a,a'}(x)$.

Fourth case. Suppose that p is incident with
zx and that q is not incident with za.

By (i), we have $a' := \alpha(a) = \alpha_{p,p'}(a) = \alpha_{q,q'}(a)$. It follows from Step 3 that

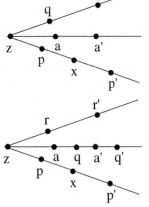

$$\alpha(x) = \alpha_{q,q'}(x) = \alpha_{a,a'}(x).$$

Fifth case. Suppose that p is incident with zx
and that q is incident with za.

Let r be a point of $P\backslash H$ such that r is not
incident with za and neither with zx. Furthermore, let $r' := \alpha(r) = \alpha_{p,p'}(r)$.

By (i), we have $r' = \alpha_{q,q'}(r)$. Thus, it follows
from Step 3 that

$$\alpha(x) = \alpha_{q,q'}(x) = \alpha_{r,r'}(x).$$

Again by Step 3, it follows that $\alpha_{r,r'}(x) = \alpha_{a,a'}(x)$, hence $\alpha(x) = \alpha_{a,a'}(x)$.

Step 6. α is a collineation:

Let a, b, c be three collinear points of P, and let g be the line through a, b and c.

First case. Suppose that g is contained in H. Then, it follows that $\alpha(a) = a$, $\alpha(b) = b$ and $\alpha(c) = c$. In particular, the points $\alpha(a)$, $\alpha(b)$ and $\alpha(c)$ are collinear.

Second case. Suppose that z is incident with g. By construction, the points $\alpha(a)$, $\alpha(b)$ and $\alpha(c)$ are on the line g, thus, they are collinear.

Third case. Suppose that g is not contained in H and that z is not incident with g. Let $s := g \cap H$, and let $a' := \alpha(a)$. Then, by Step 5, α maps any point x of g ($x \neq a$, s) on the line a's.[3] In particular, the points $a' = \alpha(a)$, $\alpha(b)$ and $\alpha(c)$ are on the line a's.

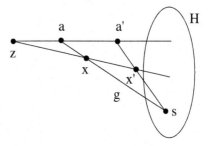

Similarly, the mapping α^{-1} (Step 3) maps any three collinear points on three collinear points.

Thus, α is a collineation.   □

---

[3] More precisely, in view of Step 4, we have $\alpha(x) = \alpha_{a,a'}(x) = zx \cap a's$.

Chapter 3
Projective Geometry Over a Vector Space

1 Introduction

In this chapter we will construct projective and affine geometries by means of vector spaces (Sects. 2, 3 and 5). For these projective and affine spaces defined by vector spaces, we will investigate their automorphism groups (Sects. 4 and 6).

The First Fundamental Theorem (Sect. 7) states that every projective or affine space fulfilling the Theorem of Desargues, in particular, every projective or affine space of dimension at least 3, stems from a vector space. This theorem provides a complete classification of projective and affine spaces fulfilling the Theorem of Desargues.

The Second Fundamental Theorem (Sect. 8) determines the automorphism groups of projective and affine spaces stemming from a vector space.

2 The Projective Space *P(V)*

So far, projective spaces have been introduced by their defining axioms. In the present section, we shall present an algebraic construction of projective spaces based on vector spaces (see Theorem 2.2). The relation between the subspaces of the underlying vector space and the subspaces of the projective space is described in Theorem 2.3.

Finally, in Theorem 2.4, it is shown that the projective spaces introduced in this section are Desarguesian.

Definition. Let V be a left vector space of dimension at least 3 over a skew field K.

(a) The geometry $P = P(V) = (X, *, \text{type})$ over the type set {point, line} is defined as follows:

J. Ueberberg, *Foundations of Incidence Geometry*, Springer Monographs in Mathematics, DOI 10.1007/978-3-642-20972-7_3, © Springer-Verlag Berlin Heidelberg 2011

(i) The points of **P** are the 1-dimensional subspaces of **V**.

(ii) The lines of **P** are the 2-dimensional subspaces of **V**.

(iii) A point $p = \langle v \rangle$, $0 \neq v \in V$, and a line $g = \langle v_1, v_2 \rangle$ are called **incident** if the 1-dimensional subspace $\langle v \rangle$ is contained in the 2-dimensional subspace $\langle v_1, v_2 \rangle$.

The geometry $P = P(V)$ is called the **projective space over the vector space** V.[1]

For a subspace W of V, we denote by $P(W)$ the set of the 1-dimensional subspaces of W. Thus, $P(V)$ is the point set of **P**.

(b) If **V** is a $(d + 1)$-dimensional vector space over a skew field K, the projective space $P = P(V)$ is also denoted by $PG(d, K)$.

Often, a point p of $P = P(V)$ is denoted by $p = \langle v \rangle$. This means implicitly that v is a vector of **V** distinct from 0.

Remark. Throughout this chapter, vector spaces are left vector spaces. Of course, all results are also valid for right vector spaces. However, sometimes the results have to be slightly reformulated. At those points, we shall give a short hint.

2.1 Proposition. *Let $P = P(V)$ be a projective space over a left vector space V. Let $x = \langle v \rangle$ and $y = \langle w \rangle$ be two (distinct) points of P. The 2-dimensional subspace $\langle v, w \rangle$ of V is the unique line of P through x and y.*

Proof. Since x and y are different points of **P**, it follows that $\langle v \rangle \neq \langle w \rangle$, that is, $\langle v, w \rangle$ is a 2-dimensional subspace of **V** and therefore a line of **P** through x and y. The line $\langle v, w \rangle$ is unique since there exists exactly one 2-dimensional subspace of **V** through $\langle v \rangle$ and $\langle w \rangle$. □

2.2 Theorem. *Let $P = P(V)$ be a projective space over a left vector space V. Then, P is a projective space.*

Proof. Verification of (PS_1): The fact that any two points are incident with exactly one line has been shown in Proposition 2.1.

Verification of (PS_2); Axiom of Veblen-Young: Let $p = \langle v \rangle$, $x_1 = \langle w_1 \rangle$, $x_2 = \langle w_2 \rangle$, $y_1 = \langle u_1 \rangle$ and $y_2 = \langle u_2 \rangle$ be five points of **P** such that the points p, x_1, x_2 and p, y_1, y_2 are collinear.

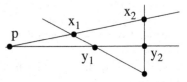

Let $U := \langle v, w_1, u_1 \rangle$ be the subspace of **V** generated by v, w_1 and u_1. Since $x_2 = \langle w_2 \rangle$ is incident with the line $px_1 = \langle v, w_1 \rangle$, it follows that w_2 is contained in U. Similarly, it follows that u_2 is contained in U. It follows from the dimension formula for vector spaces that

---

[1] We shall see in Theorem 2.2 that $P = P(V)$ is indeed a projective space.

$$\dim_V(\langle w_1, u_1 \rangle \cap \langle w_2, u_2 \rangle) = \dim_V \langle w_1, u_1 \rangle + \dim_V \langle w_2, u_2 \rangle - \dim_V \langle w_1, u_1, w_2, u_2 \rangle$$

$$= \dim_V \langle w_1, u_1 \rangle + \dim_V \langle w_2, u_2 \rangle - \dim_V U$$

$$= 2 + 2 - 3 = 1.$$

It follows that the lines $x_1 y_1 = \langle w_1, u_1 \rangle$ and $x_2 y_2 = \langle w_2, u_2 \rangle$ meet in the point $\langle w_1, u_1 \rangle \cap \langle w_2, u_2 \rangle$.

Verification of (PS_3): Let $g = \langle v, w \rangle$ be a line of P, that is, a 2-dimensional subspace of V. Then, $g = \langle v, w \rangle$ contains the pairwise different 1-dimensional subspaces $\langle v \rangle$, $\langle w \rangle$ and $\langle v + w \rangle$. Thus, g is incident with at least three points.

Since $\dim_V V \geq 3$, there are three linearly independent vectors v_1, v_2 and v_3 contained in V. It follows that P contains the three lines $\langle v_1, v_2 \rangle$, $\langle v_1, v_3 \rangle$ and $\langle v_2, v_3 \rangle$.   □

In what follows, we shall investigate the subspace structure of $P = P(V)$. Since P is a projective space, P has subspaces. We shall see that the subspaces of P correspond to the subspaces of V.

2.3 Theorem. *Let $P = P(V)$ be a projective space over a left vector space V.*

(a) *Let W be a subspace of V. Then, the set $P(W)$ of the 1-dimensional subspaces of W forms a subspace of P.*

(b) *Let A be a set of non-zero vectors of V, and let $W := \langle A \rangle$ and $U := \langle p := \langle v \rangle \mid v \in A \rangle$. Then, $U = P(W)$.*

(c) *Let U be a subspace of P. Then, there exists a subspace W of V such that the points of U are exactly the 1-dimensional subspaces of W, that is, $U = P(W)$.*

(d) *Let A be a set of non-zero vectors of V, and let $B := \{p := \langle v \rangle \mid v \in A\}$. Then, the set A is a linearly independent subset of V if and only if the set B is an independent subset of P.*

(e) *Let U be a subspace of P, and let W be a subspace of V with $U = P(W)$. Furthermore, let $\{p_i = \langle v_i \rangle \mid i \in I\}$ be a point set of U. Then, the set $\{p_i \mid i \in I\}$ is a basis of U if and only if $\{v_i \mid i \in I\}$ is a basis of W.*

(f) *Let U be a subspace of P, and let W be a subspace of V with $U = P(W)$. Then, $\dim U = \dim_V W - 1$.*

Proof. (a) Let $x = \langle v \rangle$ and $y = \langle w \rangle$ be two (distinct) points of $P(W)$. Then, by Proposition 2.1, for the line $g := xy$ joining x and y, we have:

$$g = xy = \langle v, w \rangle \subseteq W.$$

Since the points of g are the 1-dimensional subspaces of $\langle v, w \rangle$, it follows that g is contained in $P(W)$. Thus, $P(W)$ is a subspace of P.

(b) (i) The set U is contained in $P(W)$: For, let v be an element of A. By definition of $P(W)$, the point $p := \langle v \rangle$ is contained in $P(W)$. By Part (a), $P(W)$ is a subspace of P. Hence, we have

$$U = \langle p := \langle v \rangle \mid v \in A \rangle \subseteq P(W).$$

(ii) The set $P(W)$ is contained in U: For a natural number r, let

$$W_r := \{w \in W \mid \exists\, C \subseteq A, \; |C| = r, \; w \in \langle C \rangle\}$$

be the set of vectors of W which are contained in a subspace of W which is generated by at most r elements of A. We shall prove by induction on r that $P(W_r)$ is contained in U for all natural numbers r:

$r = 0$: Let $0 \neq w$ be an element of W_0. Then, there exists an element v of A such that $\langle w \rangle = \langle v \rangle$. It follows that $\langle w \rangle = \langle v \rangle$ is contained in U.

$r - 1 \rightarrow r$: Let w be an element of W_r. Then, there exists a subset $C = \{c_1, \ldots, c_r\}$ of A such that w is contained in $\langle C \rangle$. Let $W' := \langle C \rangle$, and let $W'' := \langle c_1, \ldots, c_{r-1} \rangle$. Let $U' := \langle p = \langle v \rangle \mid v \in W' \rangle$ and $U'' := \langle p = \langle v \rangle \mid v \in W'' \rangle$. By induction, we have $P(W'') = U''$.

The vector w is contained in $W' = \langle W'', c_r \rangle$. Hence, there exist two elements λ and μ of K such that $w = \lambda\, w'' + \mu\, c_r$.

It follows that the point $\langle w \rangle$ is on the line $\langle w'', c_r \rangle$. Since the points $\langle w'' \rangle$ and $\langle c_r \rangle$ both are contained in U, it follows that $\langle w \rangle$ is contained in U; hence $P(W_r)$ is contained in U.

Since $W = \bigcup_{r \in N} W_r$, it follows that $P(W)$ is contained in U.

(c) follows from (a).

(d) (i) Suppose that B is an independent subset of P. Assume that A is a linearly dependent subset of V. Then, there exists an element v of A such that

$$\langle A \setminus \{v\} \rangle = \langle A \rangle.$$

By (b), it follows that

$$\langle B \setminus \{\langle v \rangle\} \rangle = \langle B \rangle,$$

in contradiction to the independency of B.

(ii) Conversely, suppose that A is a linearly independent subset of V. Assume that B is a dependent subset of P. Then, there exists an element $p = \langle v \rangle$ of B such that

$$\langle B \setminus \{\langle v \rangle\} \rangle = \langle B \rangle.$$

By (b), it follows that

$$\langle A \setminus \{v\} \rangle = \langle A \rangle$$

in contradiction to the independency of A.

(e) follows from (b) and (d).

(f) follows from (e). $\square$

If $P = P(V)$ is a projective space over a vector space V, there is the following relation between the subspaces of P and the subspaces of V:

Subspaces of V	Subspaces of $P = P(V)$
0-dimensional subspace	Empty set
1-dimensional subspaces	Points
2-dimensional subspaces	Lines
3-dimensional subspaces	Planes
$(t+1)$-dimensional subspaces	t-dimensional subspaces
V	P

2.4 Theorem. *Let $P = P(V)$ be a projective space over a left vector space V. Then, P is Desarguesian.*

Proof. By Theorem 6.1 of Chap. II, every projective space of dimension $d \geq 3$ is Desarguesian. We therefore can assume that P is a projective plane.

Let $p, a_1, a_2, a_3, b_1, b_2, b_3$ be a Desargues configuration in P. Since the points $a_1 = \langle v_1 \rangle, a_2 = \langle v_2 \rangle$ and $a_3 = \langle v_3 \rangle$ are non-collinear, the vectors v_1, v_2 and v_3 are linearly independent. Since P is a plane, the vector v with $p = \langle v \rangle$ is contained in the vector space generated by v_1, v_2 and v_3. Since p is not incident with any of the lines a_1a_2, a_2a_3 or a_1a_3, there exist elements c_1, c_2 and c_3 of $K \setminus \{0\}$ such that $v = c_1 v_1 + c_2 v_2 + c_3 v_3$. Substituting v_i by $c_i v_i$, we get $a_1 = \langle v_1 \rangle, a_2 = \langle v_2 \rangle, a_3 = \langle v_3 \rangle$ and $p = \langle v_1 + v_2 + v_3 \rangle$.

Since the points p, a_i, b_i for $i = 1, 2, 3$ are collinear, there exist elements k_1, k_2, k_3 of K such that

$$b_1 = \langle v_1 + v_2 + v_3 + v_1 k_1 \rangle = \langle (1 + k_1) v_1 + v_2 + v_3 \rangle$$
$$b_2 = \langle v_1 + (1 + k_2) v_2 + v_3 \rangle$$
$$b_3 = \langle v_1 + v_2 + (1 + k_3) v_3 \rangle.$$

For the intersection point $s_{12} := a_1 a_2 \cap b_1 b_2$, it follows:

$$s_{12} = \langle v_1, v_2 \rangle \cap \langle (1 + k_1) v_1 + v_2 + v_3, v_1 + (1 + k_2) v_2 + v_3 \rangle$$
$$= \langle k_1 v_1 - k_2 v_2 \rangle.$$

The last equation follows from the fact that the vector $k_1 v_1 - k_2 v_2$ is contained in $\langle v_1, v_2 \rangle$ and in $\langle (1 + k_1) v_1 + v_2 + v_3, v_1 + (1 + k_2) v_2 + v_3 \rangle$.

Similarly, it follows for the intersection points $s_{13} := a_1 a_3 \cap b_1 b_3$ and $s_{23} := a_2 a_3 \cap b_2 b_3$ that

$$s_{13} = \langle k_1 v_1 - k_3 v_3 \rangle$$
$$s_{23} = \langle k_2 v_2 - k_3 v_3 \rangle.$$

Since the vectors $k_1 v_1 - k_2 v_2$, $k_1 v_1 - k_3 v_3$ and $k_2 v_2 - k_3 v_3$ are contained in the vector space $\langle k_1 v_1 - k_2 v_2, k_3 v_3 - k_1 v_1 \rangle$, it follows that the points s_{12}, s_{23} and s_{13} are on the line $\langle k_1 v_1 - k_2 v_2, k_3 v_3 - k_1 v_1 \rangle$. Thus, they are collinear. $\square$

3   Homogeneous Coordinates of Projective Spaces

Since the projective space $P = P(V)$ is defined by means of a vector space, it admits the definition of coordinates. We shall introduce coordinates only for finite-dimensional projective spaces.

Definition. Let $P = P(d, K)$ be a d-dimensional projective space over a left vector space V, and let $\{e_0, \ldots, e_d\}$ be a basis of V. If $x = \langle v \rangle$ is a point of P with

$$v = \lambda_0 e_0 + \ldots + \lambda_d e_d \text{ where } \lambda_0, \ldots, \lambda_d \in K,$$

the vector $(\lambda_0, \ldots, \lambda_d)$ is called the **homogeneous coordinates** of x with respect to the basis $\{e_0, \ldots, e_d\}$. The vector $(\lambda_0, \ldots, \lambda_d)$ is denoted by $\bar{x}$.

$\bar{x} = (\lambda_0, \ldots, \lambda_d)$ are the homogenous coordinates of v with respect to the basis $\{e_0, \ldots, e_d\}$. Often, we shall introduce homogeneous coordinates without mentioning the basis $\{e_0, \ldots, e_d\}$ of V explicitly.

Remark. If V is a right vector space, the homogenous coordinates of $x = \langle v \rangle$ are denoted by $(\lambda_0, \ldots, \lambda_d)^t$ and not as $(\lambda_0, \ldots, \lambda_d)$.

3.1 Theorem. *Let $P = PG(d, K)$ be a d-dimensional projective space over a left vector space V, and let $x = \langle v \rangle$ and $y = \langle w \rangle$ be two points of P with homogeneous coordinates $\bar{x} = (\lambda_0, \ldots, \lambda_d)$ and $\bar{y} = (\mu_0, \ldots, \mu_d)$. Then, $x = y$ if and only if there exists an element $0 \neq \lambda$ of K such that $\bar{x} = \lambda \bar{y}$.*

Proof. The proof is obvious.   □

Although the homogeneous coordinates of a point of $P = PG(d, K)$ are only defined up to a multiple of K, one often speaks of "the" homogeneous coordinates of a point.

Remark. One might think that the addition in V could be used to define an addition on the points of $P = PG(d, K)$ by setting $x + y := \langle v + w \rangle$ for any two points $x = \langle v \rangle$ and $y = \langle w \rangle$ of P. Unfortunately, this definition is **not** well-defined. If the field K has at least three elements and if λ is an element of K with $\lambda \neq 0, 1$, we have $y = \langle \lambda w \rangle$, but $x + y = \langle v + w \rangle \neq \langle v + \lambda w \rangle = x + y$, a contradiction.

3.2 Theorem. *Let $P = PG(d, K)$ be a d-dimensional projective space over a left vector space V, and let $A: V \to V$ be a bijective linear transformation with matrix representation*

$$M_A = \begin{pmatrix} a_{00} & \ldots & a_{0d} \\ \vdots & \ddots & \vdots \\ a_{d0} & \ldots & a_{dd} \end{pmatrix}.$$

Let α be the transformation[2] of P induced by A. If x is a point of P with homogenous coordinates $\bar{x} = (\lambda_0, \ldots, \lambda_d)$, the point $\alpha(x)$ has homogenous coordinates

$$\bar{y} = \bar{x}\, A = (\lambda_0, \ldots, \lambda_d)\, M_A = (\lambda_0, \ldots, \lambda_d) \begin{pmatrix} a_{00} & \cdots & a_{0d} \\ \vdots & \ddots & \vdots \\ a_{d0} & \cdots & a_{dd} \end{pmatrix}.$$

Proof. The proof is obvious.   □

Remark. If V is a right vector space, we have $\bar{x} = (\lambda_0, \ldots, \lambda_d)^t$ and

$$\bar{y} = A\,\bar{x} = M_A\,(\lambda_0, \ldots, \lambda_d)^t = \begin{pmatrix} a_{00} & \cdots & a_{0d} \\ \vdots & \ddots & \vdots \\ a_{d0} & \cdots & a_{dd} \end{pmatrix} \begin{pmatrix} \lambda_0 \\ \vdots \\ \lambda_d \end{pmatrix}.$$

3.3 Theorem. *Let $P = PG(d, K)$ be a d-dimensional projective space over a left vector space V, and let H be a hyperplane of P. Then, there exist $\mu_0, \ldots, \mu_d$ of K with the following property:*

A point x of P with homogeneous coordinates $\bar{x} = (\lambda_0, \ldots, \lambda_d)$ is contained in H if and only if $\sum_{i=0}^{d} \lambda_i \mu_i = 0$.

Proof. In V, the hyperplane H is a d-dimensional subspace. It follows that there exist $\mu_0, \ldots, \mu_d$ of K such that for any vector v of V with coordinates $(\lambda_0, \ldots, \lambda_d)$, we have:

$$v \in H \Leftrightarrow \sum_{i=0}^{d} \lambda_i \mu_i = 0. \qquad\qquad □$$

Definition. Let $P = PG(d, K)$ be a d-dimensional projective space, and let H be a hyperplane of P. Then, the elements $\mu_0, \ldots, \mu_d$ of Theorem 3.3 are called the **homogeneous coordinates** of H. They are denoted by $\bar{H} = [\mu_0, \ldots, \mu_d]^t$.

If x is a point of P with homogeneous coordinates $\bar{x} = (\lambda_0, \ldots, \lambda_d)$, set

$$\bar{x} * \bar{H} := \sum_{i=0}^{d} \lambda_i \mu_i.$$

For a point x and a hyperplane H, the point x is contained in H if and only if $\bar{x} * \bar{H} = 0$.

Alternatively, a hyperplane can be described by a linear equation: If $[\mu_0, \ldots, \mu_d]^t$ are the homogeneous coordinates of a hyperplane H of $P = PG(d, K)$, H is defined by the equation

$$x_0\, \mu_0 + x_1\, \mu_1 + \ldots + x_d\, \mu_d = 0.$$

---

[2]In Theorem 4.3, we shall see that α is a collineation.

This means that H consists of the points $(x_0, x_1, \ldots, x_d)$ for which the above equation is fulfilled.

Remark. If V is a right vector space, the hyperplane H has homogeneous coordinates $[\mu_0, \ldots, \mu_d]$. A point $(x_0, x_1, \ldots, x_d)^t$ of P is contained in H if and only if

$$\mu_0 \, x_0 + \mu_1 \, x_1 + \ldots + \mu_d \, x_d = 0.$$

3.4 Theorem. *Let* $P = PG(d, K)$ *be a d-dimensional projective space over a left vector space* V, *and let* H *and* L *be two hyperplanes of* P *with homogeneous coordinates* $\bar{H} = [\mu_0, \ldots, \mu_d]^t$ *and* $\bar{L} = [\nu_0, \ldots, \nu_d]^t$.
    Then, $H = L$ *if and only if there is an element* $0 \neq \lambda$ *of* K *such that* $\bar{H} = \bar{L}\lambda$, *that is,* $[\mu_0, \ldots, \mu_d]^t = [\nu_0 \lambda, \ldots, \nu_d \lambda]^t$.

Proof. The corresponding relation holds in V. $\square$

Remark. If V is a right vector space, we have $H = L$ if and only if there is an element $0 \neq \lambda$ of K such that $\bar{H} = \lambda \bar{L}$.

4   Automorphisms of *P(V)*

If $P = P(V)$ is a projective space over two vector space V, the semilinear and the linear transformations of V induce automorphisms of P. In the present section, we shall see that central collineations are induced by semilinear and linear transformations of V (see Theorems 4.7 and 4.8).

Definition. Let V and V' be two left vector spaces over two skew fields K and K', and let $\rho : K \to K'$ be an isomorphism from K onto K'. A transformation $A: V \to V'$ is called a **semilinear transformation with accompanying isomorphism** ρ if for all x, y of V and for all k of K, we have:

$$A(x + y) = A(x) + A(y) \text{ and}$$
$$A(k\ x) = \rho(k)\ A(x).$$

If $V = V'$, the semilinear transformation A is called the **the semilinear transformation with accompanying automorphism** ρ.

4.1 Theorem. *Let* V *and* V' *be two left vector spaces over two skew fields* K *and* K', *and let* $A: V \to V'$ *be a bijective semilinear transformation with accompanying isomorphism* ρ.

(a) *For any subspace* U *of* V, *the set* $A(U)$ *is a subspace of* V'.
(b) *If* U *is a subspace of* V, *and if* B *is a basis of* U, *then* $A(B) := \{A(v) \mid v \in B\}$ *is a basis of* $A(U)$. *In particular, we have* $\dim A(U) = dim\ U$.

Proof. (a) Let U be a subspace of V, and let x and y be two elements of U. Then, $A(x) + A(y) = A(x + y)$ is an element of $A(U)$. Furthermore, for an element k' of K', the element $k'A(x) = A(\rho^{-1}(k')x)$ is also an element of $A(U)$.

(b) Let $w = A(u)$ be an element of $A(U)$. Then, there exist elements $v_1, \ldots, v_r$ of B and $k_1, \ldots, k_r$ of K such that $u = k_1 u_1 + \ldots + k_r u_r$. It follows that $w = A(u) = \rho(k_1) A(u_1) + \ldots + \rho(k_r) A(u_r)$.

Let $v_1, \ldots, v_r$ be some elements of B, and let $k'_1, \ldots, k'_r$ be some elements of K', and let $0 = k'_1 A(u_1) + \ldots + k'_r A(u_r) = A(\rho^{-1}(k'_1) u_1 + \ldots + \rho^{-1}(k'_r) u_r)$. Since A is a bijective transformation with $A(0) = 0,$[3] it follows that $\rho^{-1}(k'_1) u_1 + \ldots + \rho^{-1}(k'_r) u_r = 0$. Since B is a basis, it follows that $\rho^{-1}(k'_1) = \ldots = \rho^{-1}(k'_r) = 0$, hence, $k'_1 = \ldots = k'_r = 0$. □

4.2 Theorem. *Let V and V' be two left vector spaces over two skew fields K and K', and let $A: V \to V'$ be a bijective semilinear transformation with accompanying isomorphism ρ.*

(a) *The transformation A^{-1} is a semilinear transformation with accompanying isomorphism ρ^{-1}.*

(b) *If $V = V'$, the set of bijective semilinear transformations of V forms a group.*

Proof. (a) Let x' and y' be two elements of V', and let k' be an element of K'. We have

$$A(A^{-1}(x' + y')) = x' + y' = A(A^{-1}(x')) + A(A^{-1}(y'))$$
$$= A(A^{-1}(x') + A^{-1}(y')), \text{ and}$$
$$A(A^{-1}(k' x')) = k' x' = k' A(A^{-1}(x'))$$
$$= \rho(\rho^{-1}(k') A(A^{-1}(x'))) = A(\rho^{-1}(k') A^{-1}(x')).$$

Since A is bijective, it follows that

$$A^{-1}(x' + y') = A^{-1}(x') + A^{-1}(y'), \text{ and } A^{-1}(k' x') = \rho^{-1}(k') A^{-1}(x').$$

(b) follows from (a) and the fact that the product of a semilinear transformation with accompanying automorphism ρ_1 and a semilinear transformation with accompanying automorphism ρ_2 is a semilinear transformation with accompanying automorphism $\rho_2 \rho_1$. □

4.3 Theorem. *Let $P = P(V)$ and $P' = P(V')$ be two projective spaces over the left vector spaces V and V', and let $A: V \to V'$ be a bijective semilinear transformation. Then, A induces a collineation α from P into P'.*

Proof. Since $A: V \to V'$ is bijective, the mapping $\alpha : P \to P'$ is a bijective mapping from the point set of P into itself. Furthermore, for three points $x = \langle v \rangle$, $y = \langle w \rangle$ and $z = \langle u \rangle$ of P, we have:

---

[3] We have $A(x) = A(x + 0) = A(x) + A(0)$, hence $A(0) = 0$.

The points $x = \langle v \rangle$, $y = \langle w \rangle$ and $z = \langle u \rangle$ are collinear.

$\Leftrightarrow$ The vectors v, w and u are contained in a 2-dimensional subspace of V.

$\Leftrightarrow$ The vectors $A(v)$, $A(w)$ and $A(u)$ are contained in a 2-dimensional subspace of V.

$\Leftrightarrow$ The points $\alpha(x) = \langle A(v) \rangle$, $\alpha(y) = \langle A(w) \rangle$ and $\alpha(z) = \langle A(u) \rangle$ are collinear.

Hence, $\alpha : P \to P'$ is a collineation. $\square$

Definition. Let $P = P(V)$ be a projective space over a left vector space V.

(a) The group of the collineations of P, which are induced by bijective semilinear transformations, is denoted by $P\Gamma L(V)$. If $P = PG(d, K)$ is a d-dimensional projective space, the group $P\Gamma L(V)$ is denoted by $P\Gamma L(d + 1, K)$.

(b) If $A: V \to V$ is a bijective *linear* transformation, the automorphism of P induced by A is called a **projective collineation** of P.

The group of the projective collineations of $P = P(V)$ is denoted by $PGL(V)$. If $P = PG(d, K)$ is a d-dimensional projective space, the group $PGL(V)$ is denoted by $PGL(d + 1, K)$.

The parameter d in $PG(d, K)$ denotes the dimension of P, whereas the parameter $d + 1$ in $PGL(d + 1, K)$ and in $P\Gamma L(d + 1, K)$ denotes the dimension of V. The reason for this somewhat confusing terminology is that the parameter $d + 1$ denotes the size of the matrices of the corresponding linear transformations. In Sect. 8, we shall see that the group $P\Gamma L(V)$ is the whole automorphism group of the projective space $P = P(V)$.

4.4 Theorem. *Let $P = P(V)$ and $P' = P(V')$ be two projective spaces over two left vector spaces V and V', and let A and B be two bijective linear transformations from V onto V'. Let α and β be the projective collineations induced by A and B. Finally, let $Z(K')$ be the centre of K'. Then, we have $\alpha = \beta$ if and only if there exists an element λ of $Z(K') \setminus \{0\}$ with $A = \lambda B$.*[4]

Proof. (i) Let $A = \lambda B$ for some λ of $Z(K') \setminus \{0\}$. Then, for any point $x = \langle v \rangle$ with $0 \neq v \in V$, we have

$$\alpha(\langle v \rangle) = A(\langle v \rangle) = \langle Av \rangle = \langle \lambda Bv \rangle = \langle Bv \rangle = B(\langle v \rangle) = \beta(\langle v \rangle).$$

It follows that $\alpha = \beta$.

(ii) Let $\alpha = \beta$, and let $M = \{v_i \mid i \in I\}$ be a basis of V. Then, we have

$$Av_i \in \langle Av_i \rangle = A(\langle v_i \rangle) = \alpha(\langle v_i \rangle) = \beta(\langle v_i \rangle) = B(\langle v_i \rangle) = \langle Bv_i \rangle \text{ for all } i \text{ of } I.$$

It follows that for all i of I, there exists an element λ_i of $K' \setminus \{0\}$ such that $Av_i = \lambda_i Bv_i$. Let j and k be two elements of I. Similarly, we obtain the existence of an element λ of $K' \setminus \{0\}$ such that $A(v_j + v_k) = \lambda B(v_j + v_k)$. It follows that

---

[4]Note that the application λB is linear if and only if λ belongs to the centre $Z(K')$ of K'.

$$0 = (v_j + v_k) - (v_j + v_k)$$
$$= A((v_j + v_k) - (v_j + v_k))$$
$$= Av_j + Av_k - A(v_j + v_k)$$
$$= \lambda_j \ Bv_j + \lambda_k \ Bv_k - \lambda \ B(v_j + v_k)$$
$$= (\lambda_j - \lambda) \ Bv_j + (\lambda_k - \lambda) \ Bv_k.$$

Since $\{Bv_i \mid i \in I\}$ is a basis of V', it follows that $\lambda = \lambda_j = \lambda_k$, that is, $\lambda = \lambda_i$ for all i of I. Since M is a basis of V, we have $A = \lambda \ B$.

It remains to show that λ is an element of $Z(K')$. For, let μ be an element of K'. Let v_0 and v_1 be two elements of M. As above, we get three elements λ_0, λ_1 and λ of K' such that $A(\mu \ v_0) = \lambda_0 \ B(\mu \ v_0)$, $A(v_1) = \lambda_1 \ B(v_1)$ and $A(\mu \ v_0 + v_1) = \lambda \ B(\mu \ v_0 + v_1)$. Considering the equation

$$0 = (\mu \ v_0 + v_1) - (\mu \ v_0 + v_1) = A(\mu \ v_0 + v_1) - A(\mu \ v_0 + v_1)$$
$$= (\lambda_0 - \lambda)B(\mu \ v_0) + (\lambda_1 - \lambda)B(v_1),$$

we get $\lambda_0 = \lambda_1 = \lambda$. It follows that

$$A(\mu \ v_0) = \mu \ A(v_0) = \mu \ \lambda \ B(v_0) \text{ and}$$
$$A(\mu \ v_0) = \lambda \ B(\mu \ v_0) = \lambda \ \mu \ B(v_0).$$

It follows that $\lambda \ \mu = \mu \ \lambda$. $\square$

4.5 Theorem. *Let $P = PG(d, K)$ be a d-dimensional projective space over a vector space V. The group $PGL(d + 1, K)$ operates transitively on the chambers of P.*

Proof. Let $C = \{C_0, C_1, \ldots, C_{d-1}\}$ and $D = \{D_0, D_1, \ldots, D_{d-1}\}$ be two chambers of P with dim C_i = dim D_i = i. Then, there exist two bases $\{v_0, v_1, \ldots, v_d\}$ and $\{w_0, w_1, \ldots, w_d\}$ of V such that $C_i = \langle v_0, v_1, \ldots, v_i \rangle$ and $D_i = \langle w_0, w_1, \ldots, w_i \rangle$.

Define the linear transformation $A \colon V \to V$ by $Av_i := w_i$ for $i = 0, \ldots, d$. If α denotes the projective collineation induced by A, we get $\alpha(C_i) = D_i$, that is, $\alpha(C) = D$. $\square$

Definition. Let P be a d-dimensional projective space. A **frame** of P is an ordered set of $d + 2$ points such that any $d + 1$ of them form a basis of P.

4.6 Theorem. *Let $P = PG(d, K)$ be a d-dimensional projective space over a vector space V.*

(a) *The group $PGL(d + 1, K)$ operates transitively on the frames of P.*
(b) *If K is commutative, the group $PGL(d + 1, K)$ operates sharply transitively on the frames of P. In this case, every projective collineation of P is uniquely determined by its definition on a frame.*

Proof. Let $X = \{x_0, x_1, \ldots, x_{d+1}\}$ and $Y = \{y_0, y_1, \ldots, y_{d+1}\}$ be two frames of **P**. Then, there exist two bases $\{v_0, v_1, \ldots, v_d\}$ and $\{w_0, w_1, \ldots, w_d\}$ of **V** such that $x_i = \langle v_i \rangle$ and $y_i = \langle w_i \rangle$ for $i = 0, \ldots, d$. Furthermore, let v, w be two elements of **V** with $x_{d+1} = \langle v \rangle$ and $y_{d+1} = \langle w \rangle$. Since X and Y are frames, there exist $\lambda_0, \lambda_1, \ldots, \lambda_d$ and $\mu_0, \mu_1, \ldots, \mu_d$ of $K \setminus \{0\}$ such that

$$v = \sum_{i=0}^{d} \lambda_i \, v_i \quad \text{and} \quad w = \sum_{i=0}^{d} \mu_i \, w_i.$$

(a) Existence of a projective collineation α with $\alpha(X) = Y$:

For $i = 0, 1, \ldots, d$, set $Av_i := \lambda_i^{-1} \, \mu_i \, w_i$, and let α be the projective collineation induced by A. For $i = 0, 1, \ldots, d$, one easily computes that $\alpha(x_i) = y_i$. Furthermore, we have

$$\alpha(x_{d+1}) = \langle Av \rangle = \left\langle \sum_{i=0}^{d} \lambda_i \, Av_i \right\rangle = \left\langle \sum_{i=0}^{d} \mu_i \, w_i \right\rangle = \langle w \rangle = y_{d+1}.$$

(b) Uniqueness of α: From now on, we suppose that K is commutative. Let α and β be two projective collineations with $\alpha(X) = \beta(X) = Y$. Then, $\gamma := \alpha \circ \beta^{-1}$ is a projective collineation with $\gamma(X) = X$. Let C be a linear transformation of **V** inducing γ. Then, there exist $\rho_0, \rho_1, \ldots, \rho_d, \rho$ of $K \setminus \{0\}$ such that $Cv_i = \rho_i \, v_i$ $(i = 0, \ldots, d)$ and $Cv = \rho \, v$. By $v = \sum_{i=0}^{d} \lambda_i \, v_i$, it follows that

$$
\begin{aligned}
0 &= v - (\lambda_0 \, v_0 + \lambda_1 \, v_1 + \ldots + \lambda_d \, v_d) \\
&= Cv - C(\lambda_0 \, v_0 + \lambda_1 \, v_1 + \ldots + \lambda_d \, v_d) \\
&= Cv - (\lambda_0 \, Cv_0 + \lambda_1 \, Cv_1 + \ldots + \lambda_d \, Cv_d) \\
&= \rho \, v - (\lambda_0 \, \rho_0 \, v_0 + \lambda_1 \, \rho_1 \, v_1 + \ldots + \lambda_d \, \rho_d \, v_d) \\
&= \rho \, v - (\rho_0 \, \lambda_0 \, v_0 + \rho_1 \, \lambda_1 \, v_1 + \ldots + \rho_d \, \lambda_d \, v_d) \text{ [since K is commutative]} \\
&= \rho \, \lambda_0 \, v_0 + \rho \, \lambda_1 \, v_1 + \ldots + \rho \, \lambda_d \, v_d - (\rho_0 \, \lambda_0 \, v_0 + \rho_1 \, \lambda_1 \, v_1 + \ldots + \rho_d \, \lambda_d \, v_d) \\
&= (\rho - \rho_0) \, \lambda_0 \, v_0 + (\rho - \rho_1) \, \lambda_1 \, v_1 + \ldots + (\rho - \rho_d) \, \lambda_d \, v_d.
\end{aligned}
$$

Since $\lambda_i \neq 0$ for all i, and since $\{v_0, \ldots, v_d\}$ is a basis of **V**, it follows that $\rho = \rho_i$ for all i. Thus, $C = \rho \, id$. It follows that $id = \gamma = \alpha \circ \beta^{-1}$, hence, $\alpha = \beta$. $\qquad\square$

Next, we shall investigate the central collineations of P(**V**).[5]

---

[5] For the definition of central collineations, cf. Sect. 5 of Chap. II.

4.7 Theorem. *Let $P = P(V)$ be a projective space over a vector space V.*

(a) Every homology of P is induced by a semilinear transformation of V.

(b) Let $\delta : P \to P$ be a homology with centre $z = \langle v_0 \rangle$ and axis $H = P(W)$. Observe that every element v of V is of the form $v = a\,v_0 + w$ for some element a of K and some element w of W.

    There exists an element $0 \neq \lambda$ of K such that the semilinear transformation $D : V \to V$ defined by

$$D(a\,v_0 + w) := \rho_\lambda(a)\,v_0 + \lambda\,w \text{ for all } a \text{ of } K \text{ and all } w \text{ of } W$$

induces the homology δ. The automorphism $\rho_\lambda : K \to K$ is defined by $\rho_\lambda(k) := \lambda\,k\,\lambda^{-1}$.

(c) Conversely, let $0 \neq v_0$ be a vector of V, let W be a maximal proper subspace of V such that $V = \langle v_0, W \rangle$, let $0 \neq \lambda$ be an element of K, and let $\rho_\lambda : K \to K$ be defined by $\rho_\lambda(k) := \lambda\,k\,\lambda^{-1}$. Furthermore, let $D : V \to V$ be the semilinear transformation defined by

$$D(a\,v_0 + w) := \rho_\lambda(a)\,v_0 + \lambda\,w \text{ for all } a \text{ of } K \text{ and all } w \text{ of } W.$$

Then, the collineation δ induced by D is a homology with centre $\langle v_0 \rangle$ and axis $H := P(W)$.

Proof. We will prove the assertions (a) and (b) in common.

Step 1. Let $x := \langle w_0 \rangle$ be an element of $H = P(W)$, and let $y := \langle v_0 + w_0 \rangle$ be a point on the line xz different from x and z. Since $\delta : P \to P$ is a homology with centre $z = \langle v_0 \rangle$ and axis $H = P(W)$, the point $\delta(y)$ is on the line xz, hence, there exists an element $0 \neq \lambda$ of K such that $\delta(y) = \langle v_0 + \lambda\,w_0 \rangle$.

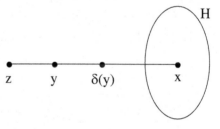

Step 2. The transformation $D : V \to V$ defined above is a semilinear transformation with accompanying automorphism ρ_λ: Obviously, $\rho_\lambda : K \to K$ is an automorphism of K. For all vectors $a\,v_0 + w$, $a_1\,v_0 + w_1$ and $a_2\,v_0 + w_2$ of V and all elements k of K, we have

$$
\begin{aligned}
D(a_1\,v_0 + w_1 + a_2\,v_0 + w_2) &= D((a_1 + a_2)\,v_0 + w_1 + w_2) \\
&= \rho_\lambda(a_1 + a_2)\,v_0 + \lambda\,(w_1 + w_2) \\
&= \rho_\lambda(a_1)\,v_0 + \lambda\,w_1 + \rho_\lambda(a_2)\,v_0 + \lambda\,w_2 \\
&= D(a_1\,v_0 + w_1) + D(a_2\,v_0 + w_2)
\end{aligned}
$$

and

$$D(k(a\ v_0 + w)) = D(ka\ v_0 + k\ w)$$
$$= \rho_\lambda(ka)\ v_0 + \lambda k\ w$$
$$= \rho_\lambda(k)\ \rho_\lambda(a)\ v_0 + \lambda k \lambda^{-1} \lambda\ w$$
$$= \rho_\lambda(k)\ \rho_\lambda(a)\ v_0 + \rho_\lambda(k)\ \lambda\ w$$
$$= \rho_\lambda(k)\ (\rho_\lambda(a)\ v_0 + \lambda\ w)$$
$$= \rho_\lambda(k)\ D(a\ v_0 + w).$$

Step 3. Let α be the collineation induced by D. We have $\alpha = \delta$: By definition, α fixes all points of H. By Theorem 5.3 of Chap. II, α is a central collineation. Since $\alpha(z) = \alpha(\langle v_0 \rangle) = \langle D\ v_0 \rangle = \langle v_0 \rangle = z$ and since z is not contained in H, the point z is the centre of α. Furthermore, we have

$$\alpha(y) = \alpha(\langle v_0 + w_0 \rangle) = \langle D\ (v_0 + w_0) \rangle = \langle v_0 + \lambda\ w_0 \rangle = \delta(y).$$

Thus, α and δ are two central collineations with centre z and axis H and $\alpha(a) = \delta(a)$. By Theorem 5.8 of Chap. II, it follows that $\alpha = \delta$.

(c) The proof is similar to Steps 2 and 3 of the proof of Parts (a) and (b). □

4.8 Theorem. *Let $P = P(V)$ be a projective space over a vector space V.*

(a) Every elation of P is a projective collineation.

(b) Let $\tau : P \to P$ be an elation with axis $H = P(W)$, and let $x = \langle v_0 \rangle$ be a point of P outside of H. Observe that every element v of V is of the form $v = a\ v_0 + w$ for some element a of K and some element w of W.

There exists an element w_0 of W such that the linear transformation $T : V \to V$ defined by

$$T(a\ v_0 + w) := a\ v_0 + a\ w_0 + w \text{ for all } a \text{ of } K \text{ and all } w \text{ of } W$$

induces the elation τ.

(c) Conversely, let $0 \neq v_0$ be a vector of V, let W be a maximal proper subspace of V such that $V = \langle v_0, W \rangle$, and let $0 \neq w_0$ be an element of W. Furthermore, let $T : V \to V$ be the linear transformation defined by

$$T(a\ v_0 + w) := a\ v_0 + a\ w_0 + w \text{ for all } a \text{ of } K \text{ and all } w \text{ of } W.$$

Then, the collineation τ induced by T is an elation with centre $\langle w_0 \rangle$ and axis $H := P(W)$.

Proof. We will prove the assertions (a) and (b) in common.

Step 1. Let $z := \langle w_1 \rangle$ be the centre of τ. Since $\tau : P \to P$ is an elation with centre $z = \langle w_1 \rangle$ and axis $H = P(W)$, the point $\tau(x)$ is on the line xz, hence, there exists an element $0 \neq \lambda$ of K such that $\tau(x) = \langle v_0 + \lambda \, w_1 \rangle$. Set $w_0 := \lambda \, w_1$.

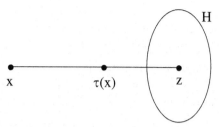

Step 2. The transformation $T: V \to V$ defined above is a linear transformation: For all vectors $a \, v_0 + w$, $a_1 \, v_0 + w_1$ and $a_2 \, v_0 + w_2$ of V and all elements k of K, we have

$$T(a_1 \, v_0 + w_1 + a_2 \, v_0 + w_2) = T((a_1 + a_2) \, v_0 + w_1 + w_2)$$
$$= (a_1 + a_2) \, v_0 + (a_1 + a_2) \, w_0 + w_1 + w_2$$
$$= a_1 \, v_0 + a_1 \, w_0 + w_1 + a_2 \, v_0 + a_2 \, w_0 + w_2$$
$$= T(a_1 \, v_0 + w_1) + T(a_2 \, v_0 + w_2)$$

and

$$T(k(a \, v_0 + w)) = T(ka \, v_0 + k \, w)$$
$$= ka \, v_0 + ka \, w_0 + k \, w$$
$$= k \, (a \, v_0 + a \, w_0 + w)$$
$$= k \, A(a \, v_0 + w).$$

Step 3. Let α be the collineation induced by T. We have $\alpha = \tau$: By definition, α fixes all points of H. By Theorem 5.3 of Chap. II, α is a central collineation. Furthermore, we have

$$\alpha(x) = \alpha(\langle v_0 \rangle) = \langle T(v_0) \rangle = \langle v_0 + w_0 \rangle = \tau(x).$$

Thus, α and τ are two central collineations with centre $z = x \, \tau(x) \cap H$ and axis H and $\alpha(a) = \tau(a)$. By Theorem 5.8 of Chap. II, it follows that $\alpha = \tau$.

(c) The proof is similar to Steps 2 and 3 of the proof of Parts (a) and (b). $\square$

5   The Affine Space $AG(W)$

In the present section, we shall see that any (left) vector space W defines an affine space $A = A(W)$ over the vector space W whose points are the vectors of W (Theorem 5.5). In Theorem 5.6, we shall analyse the subspaces of this affine space $A = A(W)$.

In the remaining part of this section, we shall see that if $A = A(W)$ is an affine space over a vector space W, the projective closure of A is a projective space over some vector space V, and conversely, if $P = P(V)$ is a projective space over some vector space V, the affine space $P \setminus H$ is an affine space over some vector space W for all hyperplanes H of P.

Definition. Let W be a left vector space over a skew field K of dimension at least 2. The geometry $A = A(W) = (X, *, \text{type})$ over the type set {point, line} is defined as follows:

(i) The points of A are the elements of W.
(ii) The lines of A are the cosets of the 1-dimensional subspaces of W.
(iii) A point p of W and a line $g = x + \langle w \rangle$ are called **incident** if p is contained in the coset $x + \langle w \rangle$.

The geometry $A = A(W)$ is called the **affine space over the vector space W**.[6] If W is of finite dimension d, then $A = A(W)$ is denoted by $AG(d, K)$.

Since the points of A are the vectors of W, we shall occasionally identify A and W.

5.1 Theorem. *Let $A = A(W)$ be an affine space over a vector space W. Let x and y be two points of A. Then, $g := x + \langle y - x \rangle$ is the unique line through x and y.*

Proof. Since $x = x + 0 \cdot (y - x)$ and $y = x + 1 \cdot (y - x)$, the points x and y are incident with g, that is, g is a line through x and y.

Conversely, every line h through x and y contains the point x. Thus, the line h is of the form $h = x + \langle w \rangle$. Since y is incident with the line h, there exists an element λ of K such that $y = x + \lambda w$, that is, $\lambda w = y - x$. It follows that $\langle w \rangle = \langle y - x \rangle$, that is, $h = x + \langle y - x \rangle = g$. $\qquad \square$

Definition. Let $A = A(W)$ be an affine space over a vector space W. For two lines $g = x + \langle w \rangle$ and $h = y + \langle v \rangle$, we set $g \parallel h$ if and only if $\langle v \rangle = \langle w \rangle$. Two lines g and h are called **parallel** if $g \parallel h$.

5.2 Theorem. *Let $A = A(W)$ be an affine space over a vector space W. The relation $\parallel$ defines a parallelism on A.*

Proof. Obviously, $\parallel$ is an equivalence relation. If x is a point of A and if $g = y + \langle w \rangle$ is a line of A, $h := x + \langle w \rangle$ is the unique line through x parallel to g. $\qquad \square$

5.3 Theorem. *Let $A = A(W)$ be an affine space over a vector space W.*

(a) *Let $E := x + \langle v, w \rangle$ be a coset of a 2-dimensional subspace $\langle v, w \rangle$ of W. Then, E is a closed subplane of A with respect to $\parallel$.*
(b) *Let E be a closed subplane of A with respect to $\parallel$. Then, there exist elements x, v, w of W such that $E = x + \langle v, w \rangle$.*

Proof. (a) We first shall see that $E = x + \langle v, w \rangle$ is a subspace of A: For, let $a = x + \lambda_1 v + \mu_1 w$ and $b = x + \lambda_2 v + \mu_2 w$ be two points of E, and let g be the line through a and b.

---

[6] We shall see in Theorem 5.5 that $A = A(W)$ is indeed an affine space.

By Theorem 5.1, we have $g = a + \langle b - a \rangle$. Let c be a point on g. Then, there exists an element λ of K with

$$c = a + \lambda\,(b - a)$$
$$= x + \lambda_1 v + \mu_1 w + \lambda\,(x + \lambda_2 v + \mu_2 w - x - \lambda_1 v - \mu_1 w)$$
$$= x + (\lambda_1 + \lambda\,\lambda_2 - \lambda\,\lambda_1)\,v + (\mu_1 + \lambda\,\mu_2 - \lambda\,\mu_1)\,w$$
$$\in x + \langle v, w \rangle.$$

It follows that g is contained in $E = x + \langle v, w \rangle$.

Next, we shall see that E is closed with respect to $||$: For, let $g = x + \langle u \rangle$ be a line of E such that u is contained in $\langle v, w \rangle$, and let $p := x + t$ be a point of E such that t is contained in $\langle v, w \rangle$. By definition of the parallelism $||$, the line $h := x + t + \langle u \rangle$ is the line through p parallel to g. Obviously, the line $h = x + t + \langle u \rangle$ is contained in the plane $E = x + \langle v, w \rangle$.

Since E is generated by the three points x, $x + v$ and $x + w$, the plane E is a closed subplane of A with respect to $||$.

(b) Let x, y and z be three points of A, generating the plane E. Let $a := z - x$ and $b := y - x$.

Step 1. The plane $x + \langle a, b \rangle$ is contained in E: Let g be the line through x and y, and let h be the line through x and z. Then,

$$g = x + \langle y - x \rangle = x + \langle b \rangle \text{ and}$$
$$h = x + \langle z - x \rangle = x + \langle a \rangle.$$

Let $s := x + \lambda a + \mu b$ be an arbitrary point of $x + \langle a, b \rangle$. Since E is a plane through x, y, z, the lines g and h are contained in E. Furthermore, the line $l := x + \lambda a + \langle b \rangle$ is a line through $x + \lambda a$ parallel to g. Since $x + \lambda a$ is incident with h and since E is closed with respect to $||$, it follows that l is contained in E. Thus,

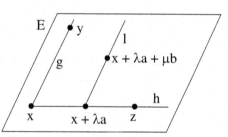

$$s = x + \lambda a + \mu b \in x + \lambda a + \langle b \rangle = l \subseteq E.$$

Step 2. We have $x + \langle a, b \rangle = E$: Since x, y and z are contained in the plane $x + \langle a, b \rangle$ and since E is generated by x, y and z, it follows from (a) that $x + \langle a, b \rangle = E$. □

5.4 Theorem. *Let $A = A(W)$ be an affine space over a vector space W. Furthermore, let $E := x + \langle v, w \rangle$ be a plane of A. Then, E is an affine plane, and two lines g and h of E are parallel if and only if $g \cap h = \oslash$ or $g = h$.*

Proof. Step 1. E is an affine plane: Since any two points of A are incident with exactly one line (cf. Theorem 5.1) and since any line $a + \langle b \rangle$ of A is incident with at least the points a and $a + b$, it remains to verify the parallel axiom.

Let g be a line of E, and let x be a point of E. Since $\|$ is a parallelism of A (cf. Theorem 5.2), there exists exactly one line h of A through x parallel to g. Since, by Theorem 5.3, E is closed with respect to $\|$, the line h is contained in E.

Step 2. Let g and h be two parallel lines of E. Then, by Theorem 5.1 of Chap. I, it follows that $g \cap h = \oslash$ or $g = h$.

Step 3. Any two disjoint lines of E are parallel: Let $g := x + s + \langle u_1 \rangle$ and $h := x + t + \langle u_2 \rangle$ be two non-parallel lines of E, that is, s, t, u_1, u_2 are contained in $\langle v, w \rangle$ and $\langle u_1 \rangle \neq \langle u_2 \rangle$. Then, $\langle v, w \rangle = \langle u_1, u_2 \rangle$, and there exist $\lambda_1, \lambda_2, \mu_1$ and μ_2 of K such that

$$s = \lambda_1 u_1 + \lambda_2 u_2 \text{ and}$$

$$t = \mu_1 u_1 + \mu_2 u_2.$$

It follows that

$$g = x + \lambda_2 u_2 + \langle u_1 \rangle \text{ and}$$

$$h = x + \mu_1 u_1 + \langle u_2 \rangle.$$

Thus, the point $x + \mu_1 u_1 + \lambda_2 u_2$ is incident with g and h. This shows that non-parallel lines of E have a common intersection point. In particular, two disjoint lines of E are parallel.   □

5.5 Theorem. *Let $A = A(W)$ be an affine space over a vector space W. Then, A is an affine space.*

Proof. We shall verify the axioms of an affine space:

Verification of (AS_1): By Theorem 5.1, any two points are incident with exactly one line.

Verification of (AS_2): By Theorem 5.2, the relation $\|$ defines a parallelism on A.

Verification of (AS_3): Let E be an affine subplane of A. By Theorem 5.3, there exists a point x of A and vectors v, w of W such that $E = x + \langle v, w \rangle$. By Theorem 5.4, E is an affine plane, and two lines g and h of E are parallel if and only if $g \cap h = \oslash$ or $g = h$.

Verification of (AS_4): Any line $x + \langle w \rangle$ of A is at least incident with the two points x and $x + w$. Since dim $W \geq 2$, there exist two linearly independent vectors v and w of W. It follows that $\langle v \rangle$ and $\langle w \rangle$ are two distinct lines of A.   □

5.6 Theorem. *Let $A = A(W)$ be an affine space over a vector space W.*

(a) *Let $0 \neq x$ be a vector of W. The 1-dimensional subspace $\langle x \rangle$ of W is the line of A through the points 0 and x.*
(b) *The subspaces of W are exactly the affine subspaces of A through 0.*
(c) *The cosets of the subspaces of W are exactly the affine subspaces of A.*

Proof. (a) By definition, the lines of A are the cosets of the 1-dimensional subspaces of W. The assertion follows.

(b) Step 1. Let X be a subspace of W. Then, X is an affine subspace of A through 0, that is, a subspace of A through 0 which is closed with respect to $||$:

For, let x and y be two points of X. By Theorem 5.1, the line $g = xy$ through x and y is given by $g = x + \langle x - y \rangle$. It follows that g is contained in X.

Let $g = x + \langle y \rangle$ be a line of X, and let z be an arbitrary point of X. We need to show that the line $z + \langle y \rangle$ (which is the line through z parallel to g) is contained in X. Since both x and $x + y$ are points on the line g, they are contained in X. It follows that the point $y = (x + y) - x$ is contained in X. Hence, the line $x + \langle y \rangle$ is contained in X. Thus, X is closed with respect to $||$.

Step 2. Let U be an affine subspace of A through 0. Then, U is a subspace of W:

(i) Let $0 \neq x$ be an element of U. By (a), the subspace $\langle x \rangle$ of W is the line through 0 and x. It follows that $\langle x \rangle$ is contained in U, that is, λx is contained in U for all λ of K.

(ii) Let x and y be two elements of U. If $\langle x \rangle = \langle y \rangle$, $x + y$ is contained in $\langle x \rangle$ and therefore in U. If $\langle x \rangle \neq \langle y \rangle$, the points 0, x and y are three non-collinear points of A generating an affine plane E. Since U is an affine subspace, E is contained in U.

Since 0 is contained in E, it follows from Theorem 5.3 that E is a 2-dimensional subspace of W. It follows that $x + y$ is contained in E and therefore in U.

(c) Step 1. Let X be a subspace of W, and let z be a point of A. Let U be the affine subspace of A through z parallel to X.[7] Then, $U = z + X$:

(i) U is contained in $z + X$: For, let $z \neq u$ be a point of U. Let g be the line through 0 parallel to uz. Since U and X are parallel subspaces, it follows that g is contained in X. Thus, there exists an element $0 \neq x$ of X with $g = \langle x \rangle$. It follows that $uz = z + \langle x \rangle$ is contained in $z + X$. In particular, u is contained in $z + X$.

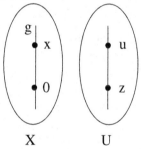

(ii) $z + X$ is contained in U: For, let $y = z + x$ be a point of $z + X$. Then, $z + \langle x \rangle$ is the line through z parallel to $\langle x \rangle$. Since U is the affine subspace through z parallel to X, it follows that $z + \langle x \rangle$ is contained in U. In particular, y is contained in U.

Step 2. Let $x + X$ be a coset of the subspace X of W. Then, $x + X$ is an affine subspace of A: In view of (b), X is an affine subspace of A through the origin. By Step 1, $x + X$ is the affine subspace of A through x parallel to X. □

The rest of this section is devoted to the proof of the following two assertions:

---

[7] By Part (a), X is an affine subspace of A. In view of Theorem 5.8 of Chap. 1, the subspace U exists.

1. Let $P = P(V)$ be a projective space over a vector space V, and let H be a hyperplane of P. Let W be the subspace of V such that $H = P(W)$. Then, $A := P \setminus H$ is an affine space over the vector space W (Corollary 5.9).
2. Let $A = A(W)$ be an affine space over a vector space W, and extend W to a vector space V such that W is a subspace of V of codimension 1. Then, the projective closure of A is a projective space over the vector space V (Theorem 5.10).

At the end of this section, we shall visualize the geometric meaning of the addition of two points (Theorem 5.11).

Definition. Let $P = P(V)$ be a projective space over a vector space V, and let H be a hyperplane of P. Let W be the subspace of V such that $H = P(W)$, and let v_0 be an arbitrary vector of V not contained in W. Every point x of the affine space $A = P \setminus H$ has a unique representation of the form $x = \langle v_0 + w \rangle$ where w is an element of W.

(a) The vector w is called the **vector representation** of x.
(b) The vector space W is called the **vector space associated to** A.
(c) The point $\langle v_0 \rangle$ with the vector representation 0 is called the **origin** of A.

Definition. Via the vector space W associated to $A = P \setminus H$, the affine space A bears the structure of a vector space. More precisely, if x and y are two points of A with vector representations w_x and w_y, then $x + y$ is defined to be the point of A with vector representation w_{x+y}. For an element λ of K the element $\lambda\, x$ is defined to be the point of A with vector representation $w_{\lambda x}$.

It remains to show that A is (isomorphic to) the affine space $A(W)$.

5.7 Theorem. *Let $P = P(V)$ be a projective space over a vector space V. Let H be a hyperplane of P, and let $A := P \setminus H$. Let W be the vector space associated to A.*

(a) *Let $0 = \langle v_0 \rangle$ be the origin of A, and let $x = \langle v_0 + w \rangle$ with the vector representation $0 \neq w \in W$ be a further point of A. Then, the line $0x$ meets the hyperplane H in the point $\langle w \rangle$.*
(b) *Let U be an affine subspace of A through the origin. The set of vectors of W representing the points of U forms a subspace of W.*
(c) *Let X be a subspace of W. The points of A whose vector representations are in X form an affine subspace of A through the origin.*
(d) *The affine subspaces of A through the origin correspond to the subspaces of W.*

Proof. (a) Obviously, the point $\langle w \rangle$ is incident with the line $\langle v_0, v_0 + w \rangle$.
(b) Since U is a subspace through the origin $\langle v_0 \rangle$, there exists a subspace X of W such that $U = \{\langle v_0 + v \rangle \mid v \in X\}$.
(c) The set $U := \{\langle v_0 + v \rangle \mid v \in X\}$ consists of all points of $\langle v_0, X \rangle$ (as a subspace of P) not contained in H.
(d) The assertion follows from (b) and (c).   □

5.8 Theorem. *Let $P = P(V)$ be a projective space over a vector space V. Let H be a hyperplane of P, and let $A := P \setminus H$. Let W be the vector space associated to A, and let $\langle v_0 \rangle$ be the origin of A.*

(a) *Let U be an affine subspace of A. The set of vectors of W representing the points of U forms a coset of a subspace of W.*
(b) *Let $w_0 + X$ be a coset of a subspace X of W. The points of A whose vector representations are in $w_0 + X$ form an affine subspace of A.*
(c) *The affine subspaces of A correspond to the cosets of the subspaces of W.*

Proof. (a) Let x be a point of U not contained in H. Then, there exists a vector w_0 of W such that $x = \langle v_0 + w_0 \rangle$. It follows that there exists a subspace X of W such that $U = \{\langle v_0 + w_0 + w \rangle \mid w \in X\}$. The coset $w_0 + X$ is the set of vectors of W representing the points of U.

(b) Let $x := \langle v_0 + w_0 \rangle$. The points of A with vector representations in $w_0 + X$ are the affine points of the subspace $x + X$.

(c) The assertion follows from (a) and (b).   □

5.9 Corollary. *Let $P = P(V)$ be a projective space over a vector space V. Let H be the hyperplane of P, and let $A := P \setminus H$. Let W be the vector space associated to A. Then, A is isomorphic to the affine space $A(W)$.*

Proof. The assertion follows from Theorems 5.7 and 5.8.   □

5.10 Theorem. *Let $A = A(W)$ be an affine space over a vector space W. Then, there exists a vector space V such that the projective closure P of A is isomorphic to $P(V)$.*

Proof. Extend W to a vector space V such that W is of codimension 1 in V. Let $P^* := P(V)$ be the projective space over V, let H be the hyperplane $\langle W \rangle$ of P^*, and let $A^* := P^* \setminus H$. By Corollary 5.9, the affine spaces A and A^* are isomorphic. By construction, P^* is the projective closure of A^*. Since the projective closure of an affine space is unique, it follows that P and P^* are isomorphic.   □

5.11 Theorem. *Let $A = P \setminus H$ be an affine space defined by a projective space $P = P(V)$. Let $0 := \langle v_0 \rangle$ be the origin of A, and let x and y be two points of A such that 0, x and y are non-collinear.*

Let a and b be the intersection points of the lines 0x and 0y with H. Then, $x+y$ is the intersection point of the lines xb and ya.

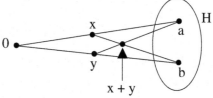

Proof. Let w_x and w_y be the vector representations of x and y. We have

$$x + y = \langle v_0 + w_x + w_y \rangle \in \langle v_0 + w_x, w_y \rangle = xb \text{ and}$$
$$x + y = \langle v_0 + w_x + w_y \rangle \in \langle v_0 + w_y, w_x \rangle = ya.$$

It follows that the point $x + y$ is the intersection point of the lines xb and ya.   □

6   Automorphisms of $A(W)$

The main subjects of the present section are translations and homologies. In particular, we shall see that every translation of the affine space $A = A(W)$ is of the form $\tau(x) = x + y$ for some y of A and that every homology is of the form $\sigma(x) = \lambda x$ for some λ of K (Theorems 6.2 and 6.4).

In addition, we shall see that the translations of A form an abelian group (Theorem 6.3) and that the homologies of A define a skew field isomorphic to K (Theorem 6.5).

Finally, we shall see that for every parallelism-preserving collineation $\alpha : A \to A$, there exists exactly one parallelism-preserving collineation $\beta : A \to A$ with $\beta(0) = 0$ and exactly one translation $\tau : A \to A$ such that $\alpha = \tau\beta$ (Theorem 6.6).

6.1 Theorem. *Let $A = A(W)$ be an affine space over a vector space W, and let $A : W \to W$ be a bijective semilinear transformation. Then, A induces a parallelism-preserving collineation $\alpha : A \to A$ with $\alpha(0) = 0$.*

Proof. For any point x of A, set $\alpha(x) := A(x)$. Since $A : W \to W$ is bijective, α maps the point set of A bijectively onto itself. Let v, w be two vectors of W with $w \neq 0$. Then, for the line $g := v + \langle w \rangle$ of A, we have:

$$\begin{aligned}
\alpha(g) &= \alpha(v + \langle w \rangle) \\
&= A(v + \langle w \rangle) \\
&= A(v) + \langle A(w) \rangle.
\end{aligned}$$

It follows that $\alpha(g)$ is a line. Since A^{-1} is a semilinear transformation, it follows that three points x, y and z of A are collinear if and only if $\alpha(x)$, $\alpha(y)$ and $\alpha(z)$ are collinear.

Finally, let $g_1 := v_1 + \langle w_1 \rangle$ and $g_2 := v_2 + \langle w_2 \rangle$ be two lines of A. Then, we have

$$\begin{aligned}
g_1 \parallel g_2 &\Leftrightarrow v_1 + \langle w_1 \rangle \parallel v_2 + \langle w_2 \rangle \\
&\Leftrightarrow \langle w_1 \rangle = \langle w_2 \rangle \\
&\Leftrightarrow \langle A(w_1) \rangle = \langle A(w_2) \rangle \\
&\Leftrightarrow A(v_1) + \langle A(w_1) \rangle \parallel A(v_2) + \langle A(w_2) \rangle \\
&\Leftrightarrow \alpha(g_1) \parallel \alpha(g_2).
\end{aligned}$$

It follows that α is a parallelism-preserving collineation. $\square$

Definition. Let $A = A(W)$ be an affine space over a vector space W, and let $P = P(V)$ be the projective closure of A. Furthermore, let H be the hyperplane of P such that $A = P \setminus H$.

A collineation $\alpha : A \to A$ is called **affine** if there exists a projective collineation $\beta : P \to P$ such that $\beta(H) = H$ and $\beta|_A = \alpha$.

6.2 Theorem. *Let $A = A(W)$ be an affine space over a vector space W.*

(a) Every translation is an affine collineation.

(b) Let $\tau : A \to A$ be a translation of A. Then, there exists a point y of A such that

$$\tau(x) = x + y \text{ for all } x \text{ of } A.$$

(c) Let y be a point of A, and let $\tau_y : A \to A$ be defined by $\tau_y(x) := x + y$ for all x of A. Then, τ_y is a translation of A.

Proof. Let $P = P(V)$ be the projective closure of A, and let $0 = \langle v_0 \rangle$ be the origin of A.

We will prove the assertions (a) and (b) in common. By Theorem 5.9 of Chap. II, the translation τ can be extended to an elation α of P. By Theorem 4.8, there exists an element y of W such that the linear transformation $T : V \to V$ defined by

$$T(a\, v_0 + x) := a\, v_0 + a\, y + x \text{ for all } a \text{ of } K \text{ and all } x \text{ of } W$$

induces the elation τ. It follows that

$$T(v_0 + x) = v_0 + x + y \text{ for all } x \text{ of } W,$$

hence, $\tau(x) = x + y$ for all x of A.

(c) Let $T : V \to V$ be the linear transformation defined by

$$T(a\, v_0 + x) := a\, v_0 + a\, y + x \text{ for all } a \text{ of } K \text{ and all } x \text{ of } W.$$

By Theorem 4.8, the collineation τ induced by T is an elation with axis $H := P(W)$.

Since $T(\langle v_0 + x \rangle) = \langle v_0 + x + y \rangle$ for all x of W, we have $\alpha|_A = \tau$. By Theorem 5.9 of Chap. II, τ is a translation. $\square$

Definition. Let $A = A(W)$ be an affine space over a vector space W.

(a) For a point x of A, τ_x denotes the transformation $\tau_x : A \to A$ defined by

$$\tau_x(z) := x + z \text{ for all } z \text{ of } A.$$

(b) The set of all translations of A is denoted by T.

6.3 Theorem. *Let $A = A(W)$ be an affine space over a vector space W.*

(a) For any two translations τ_x and τ_y, we have $\tau_x \tau_y = \tau_{x+y}$.

(b) For any two translations τ_x and τ_y, we have $\tau_x \tau_y = \tau_y \tau_x$.

(c) The set T of the translations of A is an abelean group.

(d) For any two elements x and y of A, we have $x + y = \tau_x \tau_y(0)$.

Proof. (a) We have $(\tau_x \tau_y)(z) = \tau_x(\tau_y(z)) = \tau_x(y + z) = x + y + z = \tau_{x+y}(z)$
for all points z of A. It follows that $\tau_x \tau_y = \tau_{x+y}$.

(b) By (a), we have $\tau_x \tau_y = \tau_{x+y} = \tau_{y+x} = \tau_y \tau_x$.

(c) The assertion follows from (a).

(d) We have $(\tau_x \tau_y)(0) = \tau_{x+y}(0) = x + y + 0 = x + y$.

□

6.4 Theorem. *Let $A = A(W)$ be an affine space over a vector space W.*

(a) *Let $\sigma : A \to A$ be a homology with centre 0. There exists an element $0 \neq \lambda$ of K such that*

$$\sigma(x) = \lambda \, x \text{ for all } x \text{ of } A.$$

(b) *Every homology with centre 0 is a parallelism-preserving collineation.*[8]

(c) *For an element $0 \neq \lambda$ of K, let $\sigma_\lambda : A \to A$ be defined by $\sigma_\lambda(x) := \lambda \, x$ for all points x of A. Then, σ_λ is a homology of A with centre 0.*

Proof. Let $P = P(V)$ be the projective closure of A, and let $0 = \langle v_0 \rangle$ be the origin of A.

We will prove the assertions (a) and (b) in common. By Theorem 5.9 of Chap. II, the homology σ can be extended to a homology α of P. By Theorem 4.7, there exists an element λ of K such that the semilinear transformation $D : V \to V$ defined by

$$D(av_0 + x) := \rho_\lambda(a)v_0 + \lambda \, x \text{ for all } a \text{ of } K \text{ and } x \text{ of } W$$

induces the homology σ (note that $\rho_\lambda : K \to K$ is defined by $\rho_\lambda(k) = \lambda k \lambda^{-1}$). It follows that $D(v_0 + x) := v_0 + \lambda x$ for all x of W, hence, $\sigma(x) = \lambda x$ for all x of A.

(c) The assertion follows as in the proof of Theorem 6.2 (c). □

Definition. Let $A = A(W)$ be an affine space over a vector space W.

(a) For an element $0 \neq \lambda$ of K, the symbol σ_λ denotes the transformation $\sigma_\lambda : A \to A$ defined by $\sigma_\lambda(z) := \lambda z$ for all z of A.

(b) The set of all homologies with centre 0 of A is denoted by D_0.

6.5 Theorem. *Let $A = A(W)$ be an affine space over a vector space W.*

(a) *We have $\sigma_\lambda \sigma_\mu = \sigma_{\lambda \mu}$ for all $\lambda, \mu \in K$.*

(b) *We have $\sigma_\lambda + \sigma_\mu = \sigma_{\lambda+\mu}$ for all $\lambda, \mu \in K$.*

(c) *Let $\sigma_0 : A \to A$ be the transformation defined by the relation $\sigma_0(x) := 0$ for all points x of A. Set $\sigma + \sigma_0 := \sigma_0 + \sigma := \sigma$ and $\sigma \sigma_0 := \sigma_0 \sigma := \sigma_0$ for all $\sigma \in D_0$. Then, $D := D_0 \cup \{\sigma_0\}$ is a field isomorphic to K.*

Proof. By definition, we have $\sigma_\lambda(x) = \lambda \, x$ for all λ of K and all x of A. The assertion follows. □

---

[8]One easily sees that a homology with a centre different from 0 is also a parallelism-preserving collineation. For our purposes, homologies with centre 0 are of particular interest.

6.6 Theorem. *Let $A = A(W)$ and $A' = A(W')$ be two affine spaces over two vector spaces W and W', and let $\alpha : A \to A'$ be a parallelism-preserving collineation.*

Then, there exists exactly one parallelism-preserving collineation $\beta : A \to A'$ with $\beta(0) = 0$ and exactly one translation $\tau : A' \to A'$ such that $\alpha = \tau \beta$.

Proof. (i) Existence: Let $x := \alpha(0)$, and let τ be the translation of A' with $\tau(0) = x$. Furthermore, let $\beta := \tau^{-1} \alpha$. Then, $\beta : A \to A'$ is a parallelism-preserving collineation, and we have $\beta(0) = \tau^{-1}(\alpha(0)) = 0$ and $\tau \beta = \alpha$.

(ii) Uniqueness: Let β_1 and β_2 be two parallelism-preserving collineations with $\beta_1(0) = \beta_2(0) = 0$, and let τ_1 and τ_2 be two translations of A' with $\alpha = \tau_1 \beta_1 = \tau_2 \beta_2$. It follows that $\tau_2^{-1} \tau_1 = \beta_2 \beta_1^{-1}$. Thus, $\tau_2^{-1} \tau_1$ is a translation fixing the point 0, that is, $\tau_2^{-1} \tau_1 = id$. It follows that $\tau_1 = \tau_2$ and $\beta_1 = \beta_2$.

$\square$

In Theorem 8.1, we shall see that every parallelism-preserving automorphism $\alpha : A = A(W) \to A' = A(W')$ with $\alpha(0) = 0$ is induced by some semilinear transformation $A : W \to W'$. In view of Theorem 6.6, for any automorphism $\alpha : A = A(W) \to A' = A(W')$, there exists a translation τ of A' and an automorphism $\beta : A \to A'$ induced by some semilinear transformation such that $\alpha = \tau \beta$.

7 The First Fundamental Theorem

The aim of the present section is the proof of the following assertion: If P is a Desarguesian projective space, there exists a vector space V over a (skew-) field K such that $P = P(V)$ (Theorem 7.18).

This result is called the first fundamental theorem of projective geometry. Due to the following reasons, it is one of the most important results about projective spaces:

- Except for the non-Desarguesian projective planes, all projective spaces are completely classified.[9]
- The theorem defines a relation between geometric and algebraic methods. Via the introduction of vector spaces, methods of linear algebra can be applied to projective spaces.

We first give an outline of the different steps of the proof:

Step 1. We will choose a hyperplane H of P and consider the affine space $A := P \setminus H$ with origin 0.

Step 2. Let T the set of the translations of A. We shall show that T is an abelean group (Theorem 7.3), and we shall show that for any element x of A, there is exactly one translation τ_x with $\tau_x(0) = x$ (Theorem 7.1).

---

[9]By Theorem 6.1 of Chap. II, every projective space of dimension $d \geq 3$ is Desarguesian.

Step 3. We shall define on A an addition by $x + y := (\tau_x\,\tau_y)(0)$.

Step 4. Let D_0 be the set of the homologies of A with centre 0. Define on D_0 a multiplication and an addition by $(\sigma_1\,\sigma_2)(x) := \sigma_1(\sigma_2(x))$ and $(\sigma_1 + \sigma_2)(x) := \sigma_1(x) + \sigma_2(x)$.

Step 5. Let $\mathbf{0} : A \to A$ be defined by $\mathbf{0}(x) := 0$ for all x of A. Set $\sigma \cdot \mathbf{0} := \mathbf{0} \cdot \sigma := \mathbf{0}$ and $\sigma + \mathbf{0} := \mathbf{0} + \sigma := \sigma$ for all σ of D. Then, $K := D_0 \cup \{\mathbf{0}\}$ is a skew field (Theorem 7.16).

Step 6. A is a vector space over the field K (Theorem 7.16).

Step 7. Let $A' = A(W)$ be the affine space over the vector space A. By definition, A and A' have the same point sets. We shall see that A and A' are isomorphic (Theorem 7.17). This theorem is called the first fundamental theorem of affine spaces.

Step 8. The projective space P is of the form $P = P(V)$ (Theorem 7.18).

Definition. Let A be a Desarguesian affine space. Let 0 be an arbitrary point of A. In the following, the point 0 will be called the origin of A.

7.1 Theorem. *Let P be a Desarguesian projective space, and let H be a hyperplane of P. Let T be the set of the elations of P with axis H, and let 0 be the origin of $A := P \setminus H$.*

Then, for any point x of A, there exists exactly one elation τ_x of T with $\tau_x(0) = x$.

Proof. First case. Let $x = 0$.

Existence: Obviously, the identity is an elation τ_0 of T with $\tau_0(0) = 0$.

Uniqueness: Let τ_0 be an elation of T with $\tau_0(0) = 0$. Then, τ_0 fixes the point 0 outside of H. By Theorem 5.7 of Chap. II, we have $\tau_0 = id$.

Second case. Let $x \neq 0$.

Existence: Let $z := 0x \cap H$. By Theorem 6.2 of Chap. II, there exists (exactly) one elation τ_x with centre z and axis H such that $\tau_x(0) = x$.

Uniqueness: Let τ be an elation with $\tau(0) = x$. Then, τ has the centre $z := 0x \cap H$ (cf. Theorem 5.7 of Chap. II). The uniqueness of τ follows from Theorem 5.8 of Chap. II.   □

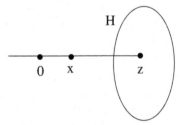

7.2 Theorem. *Let P be a Desarguesian projective space, let H be a hyperplane of P, and let $id \neq \tau_1$, τ_2 be two elations of P with axis H and centres $z_1 \neq z_2$.*

(a) *The product $\tau_1\,\tau_2$ is an elation with axis H.*

(b) *The centre z of $\tau_1\,\tau_2$ is incident with the line $z_1\,z_2$, and we have $z \neq z_1, z_2$.*

(c) *For any point x of $P \setminus H$, we have $(\tau_1\,\tau_2)(x) = \tau_2(x)z_1 \cap \tau_1(x)\,z_2$.*

(d) *We have $\tau_1\,\tau_2 = \tau_2\,\tau_1$.*

Proof.

(a) Obviously, $(\tau_1\,\tau_2)(y) = y$ for all y of H. It follows that $\tau_1\,\tau_2$ is a collineation with axis H. By Theorem 5.3 of Chap. II, $\tau_1\,\tau_2$ is a central collineation.

Let x be a point of $P \setminus H$. Since $\tau_2 \neq id$, we have $\tau_2(x) \neq x$. Since z_2 is the centre of τ_2, the point $\tau_2(x)$ is incident with the line xz_2. Since $\tau_1 \neq id$, the point $\tau_1(\tau_2(x))$ is a point on the line $\tau_2(x)z_1$ distinct from $\tau_2(x)$. In particular, we have $\tau_1(\tau_2(x)) \neq x$ (otherwise, the lines $\tau_2(x)z_1$ and xz_2 would intersect in the points x and $\tau_2(x)$).

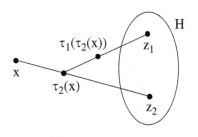

It follows that $\tau_1 \tau_2$ has no fixed point outside of H, that is, $\tau_1 \tau_2$ is an elation.

(b) Let x be a point of $P \setminus H$, and let z be the centre of $\tau_1 \tau_2$. We have $z = x(\tau_1 \tau_2)(x) \cap H$:

Let E be the plane of P generated by x, z_1 and z_2. Then, the point $\tau_2(x)$ is incident with the line xz_2 which is contained in E, and the point $\tau_1(\tau_2(x))$ is incident with the line $\tau_2(x)z_1$ which is also contained in E.

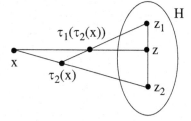

It follows that the line $x\tau_1(\tau_2(x))$ is contained in E.

Thus, $z = x(\tau_1 \tau_2)(x) \cap H$ is contained in $E \cap H = z_1 z_2$.

(c) In the proof of Part (a) we have seen that the point $\tau_1(\tau_2(x))$ is incident with the line $z_1 \tau_2(x)$. Furthermore,

$$z_2 = \tau_1(z_2) \in \tau_1(x\,\tau_2(x)) = \tau_1(x)\tau_1(\tau_2(x)), \text{ that is, } \tau_1(\tau_2(x)) \in z_2\tau_1(x).$$

Altogether, it follows that $\tau_1(\tau_2(x)) = z_1 \tau_2(x) \cap z_2\tau_1(x)$.

(d) By (c), for any point x of $P \setminus H$, we have

$$\tau_1(\tau_2(x)) = z_1 \ \tau_2(x) \cap z_2\tau_1(x) = \tau_2(\tau_1(x)). \qquad \square$$

7.3 Theorem. *Let P be a Desarguesian projective space, and let H be a hyperplane of P. If T is the set of elations of P with axis H, T is an abelean group.*

Proof. T is closed with respect to multiplication: For, let τ_1 and τ_2 be two elations of T with centres z_1 and z_2. If $z_1 \neq z_2$, it follows from Theorem 7.2 that $\tau_1 \tau_2$ is an elation of T. If $z_1 = z_2$, by Theorem 5.2 of Chap. II, $\tau_1 \tau_2$ is an elation of T.

Existence of an inversive element: Let τ be an element of T. If z is the centre of τ, by Theorem 5.2 of Chap. II, the inversive of τ is an elation with axis H and centre z.

Commutativity of T: Let τ_1 and τ_2 be two elations of T with centres z_1 and z_2. If $z_1 \neq z_2$, by Theorem 7.2, $\tau_1 \tau_2 = \tau_2\tau_1$.

If $z_1 = z_2$, choose an elation τ_3 of T with centre $z_3 \neq z_1$. By Theorem 7.2, $\tau_1 \tau_3$ is an elation with centre $z \neq z_1$. Again by Theorem 7.2, the elations τ_1 and τ_2 commute with τ_3 and with $\tau_1 \tau_3$.

It follows that $\tau_1 \tau_2 \tau_3 = \tau_1(\tau_2\tau_3) = \tau_1(\tau_3\tau_2) = (\tau_1\tau_3)\tau_2 = \tau_2(\tau_1\tau_3) = \tau_2\tau_1\tau_3$. Thus, $\tau_1 \tau_2 = \tau_2\tau_1$. $\qquad \square$

Definition. For any two points x and y of A, we denote by $x+y$ the point $\tau_x(\tau_y(0))$ where τ_z is the unique translation τ with $\tau(0) = z$ (cf. Theorem 7.1). The point $x+y$ is called the **sum** of x and y.

7.4 Theorem. *Let A be a Desarguesian affine space, and let P be the projective closure of A. Furthermore, let H be the hyperplane of P such that $A = P \backslash H$.*

Let 0 be the origin of A, and let x and y be two points of A such that the points 0, x and y are non-collinear. Then, we have for $z_x := 0x \cap H$ and $z_y := 0y \cap H$:

$$x + y = xz_y \cap yz_x.$$

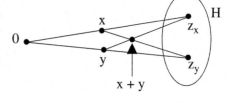

Proof. Let τ_x and τ_y be the translations of A with $\tau_x(0) = x$ and $\tau_y(0) = y$. Then, by Theorem 7.2, we have:

$$x + y = (\tau_x \tau_y)(0)$$
$$= \tau_x(0)z_y \cap \tau_y(0)z_x$$
$$= xz_y \cap yz_x. \qquad \square$$

7.5 Theorem. *Let A be a Desarguesian affine space with origin 0, and let x and y be two points of A such that the points 0, x and y are collinear. Then, the point $x + y$ is incident with the line $0x = 0y$.*

Proof. Let τ_x and τ_y be the translations of A with $\tau_x(0) = x$ and $\tau_y(0) = y$. Then, the translations τ_x and τ_y fix the line $0x = 0y$. Since $x + y = \tau_x(\tau_y(0))$, the point $x + y$ is incident with the line $0x = 0y$. $\qquad \square$

7.6 Theorem. *Let A be a Desarguesian affine space with origin 0. For any point x of A, let τ_x be the translation τ of A with $\tau(0) = x$. Let T be the group of the translations of A.*

(a) We have $\tau_{a+b} = \tau_a \tau_b$ for all a, b of A.
(b) We have $\tau_{-a} = \tau_a^{-1}$ for all a of A.
(c) $(T, +)$ is an abelean group.

Proof. (a) By definition of τ_x, we have $\tau_{a+b}(0) = a + b$. By definition of the addition in A, we have $a + b = \tau_a(\tau_b(0))$.

Altogether, we have $\tau_{a+b}(0) = a + b = \tau_a(\tau_b(0))$. By Theorem 7.1, there exists exactly one translation τ with $\tau(0) = a + b$. It follows that $\tau_{a+b} = \tau_a \tau_b$.

(b) By (a), we have $\tau_a \tau_{-a} = \tau_{a-a} = \tau_0 = id$, Thus, $\tau_{-a} = \tau_a^{-1}$.
(c) The assertion follows from (a) and (b) or from Theorem 7.3. $\qquad \square$

Definition. Let A be a Desarguesian affine space with origin 0. For a point p of A and a line g through 0, we denote by $p + g$ the set $\{p + x \mid x \in g\}$.

7.7 Theorem. *Let A be a Desarguesian affine space with origin 0.*

(a) For any point p of A and every line g through 0, the set $p + g$ is the point set of the line of A through p parallel to g.
(b) For every line g through 0 and any point p on g, we have $p + g = g$.
(c) Two lines $p + g$ and $q + h$ are parallel if and only if $g = h$.

Proof. Let P be the projective closure of A, and let H be the hyperplane of P such that $A = P \backslash H$.

(a) Let $z := g \cap H$ be the point at infinity of g, and let h be the line through p parallel to g, that is, $h = pz$. We shall show that $p + g = h$.

(i) $p+g$ is contained in h: For, let x be a point on $p+g$. Then, there exists a point y on g such that $x = p + y$.

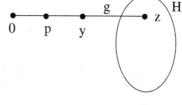

First case. Suppose that p is incident with g. Then, $g = h$. Since $\tau_p(0) = p$ and $\tau_y(0) = y$, the point $z = 0p \cap H = 0y \cap H$ is the centre of τ_p and τ_y. In particular, g is a fixed line of τ_p and τ_y. It follows that $x = p + y = \tau_p(\tau_y(0))$ is incident with $\tau_p(g) = g = h$.

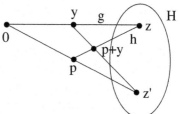

Second case. Suppose that p is not incident with g. Then, the centres $z = 0y \cap H$ of τ_y and $z' := 0p \cap H$ of τ_p are distinct. By Theorem 7.4, it follows that
$$p + y = yz' \cap pz$$
$$= yz' \cap h \in h.$$

(ii) h is contained in $p + g$: For, let x be a point on h.
First case. Suppose that p is incident with g. Then, $g = h$. Since $x - p = \tau_x \left(\tau_p^{-1}(0) \right)$ is incident with $\tau_x(g) = g$, we have

$$x = p + (x - p) \in p + g.$$

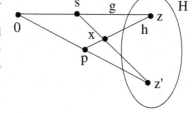

Second case. Suppose that p is not incident with g, and let $z' := p0 \cap H$. Then, the points $0, z, p, x, z'$ are contained in the plane generated by g and h. In particular, the lines g and $z'x$ meet in a point s. By Theorem 7.4, it follows that

$$x = pz \cap sz'$$

$$= p + s \in p + g.$$

(b) follows from (a).
(c) Let $p + g$ and $p + h$ be two lines of
     A. Then, g and h are the lines through 0
     parallel to $p + g$ and $p + h$, respectively.
     It follows that

$$p + g \parallel p + h \Leftrightarrow g \parallel h \Leftrightarrow g = h.$$

$\square$

7.8 Theorem. *Let A be a Desarguesian affine space with origin 0, and let σ be a central collineation of A with centre 0.*
  If p is a point of A and if g is a line through 0, we have $\sigma(p + g) = \sigma(p) + g$.

Proof. Let P be the projective closure of A, and let H be the hyperplane of P such that $A = P \backslash H$. Furthermore, let z be the point at infinity of g, that is, $g = 0z$.

By Theorem 5.9 of Chap. II, the projective closure of σ is a central collineation of P with centre 0 and axis H.[10]

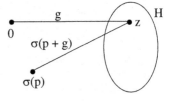

It follows that $\sigma(z) = z$. Since $p + g$ is the line through p parallel to g, we have $p + g = pz$. Hence, $\sigma(p + g) = \sigma(pz) = \sigma(p)z$. Thus, the line $\sigma(p + g)$ is the line through $\sigma(p)$ parallel to g.

By Theorem 7.7, it follows that $\sigma(p + g) = \sigma(p) + g$. $\square$

7.9 Theorem. *Let A be a Desarguesian affine space with origin 0, and let $\mu : A \to A$ be defined as follows: $\mu(x) := -x$ for all points x of A.*

(a) *Let $0 \neq x$ be a point of A. Then, the point $-x$ is incident with the line $0x$.*
(b) *For any point x of A and every line g through 0, we have $\mu(x + g) = -x + g$.*
(c) *The transformation $\mu : A \to A$ is a parallelism-preserving collineation.*
(d) *μ is a homology of A with centre 0.*

Proof. (a) Let $g := 0x$ be the line through 0 and x. Since $\tau_x(0) = x$, we have $\tau_x(g) = g$. Thus, $-x = \tau_x^{-1}(0)$ is incident with $\tau_x^{-1}(g) = g$.
(b) If a is incident with $\mu(x + g)$, there exists a point y on g with $a = \mu(x + y)$. It follows that

$$a = \mu(x + y) = -x - y = -x + (-y) \in -x + g.$$

If a is incident with $-x + g$, there exists a point y on g with $a = -x + y$. It follows that

$$a = -x + y = -(x - y) = \mu(x - y) \in \mu(x + g).$$

---

[10]We denote the projective closure of σ again by σ.

(c) We have $\mu^2(x) = \mu(\mu(x)) = \mu(-x) = x$. Hence, μ is bijective with $\mu^{-1} = \mu$. μ is a parallelism-preserving collineation: Since μ operates bijectively on the point set of A and since $\mu = \mu^{-1}$ maps lines on lines, μ is a collineation.
   For two points x and y of A and two lines g and h through 0, we have:

$$x + g \parallel y + h \Leftrightarrow g = h \Leftrightarrow -x + g \parallel -y + g \Leftrightarrow \mu(x + g) \parallel \mu(y + h).$$

(d) For every line g of A through 0, we have: $\mu(g) = \mu(0 + g) = -0 + g = g$. Thus, μ is a homology of A with centre 0.   □

Our next aim is to show the following: For every parallelism-preserving collineation α with $\alpha(0) = 0$ and for any two points x and y of A, we have $\alpha(x + y) = \alpha(x) + \alpha(y)$ (Theorem 7.12). The next two lemmata prepare the proof of this assertion.

7.10 Lemma. *Let A be a Desarguesian affine space with origin 0. Furthermore, let r and s be two points of A such that 0, r and s are non-collinear.*

(a) The point $r + s$ is neither incident with the line $0r$ nor with the line $0s$.
(b) The points 0, r and $-s$ are non-collinear.
(c) The points 0, $r + s$ and $-s$ are non-collinear.

Proof. (a) Let P be the projective closure of A, and let H be the hyperplane of P such that $A = P \backslash H$. Then, by Theorem 7.4, for $z_r := 0r \cap H$ and $z_s := 0s \cap H$, we have

$$r + s = r z_s \cap s z_r.$$

It follows that the point $r + s$ is neither incident with the line $0r$ nor with the line $0s$.

(b) By assumption, the point r is not incident with the line $0s$. By Theorem 7.9, the point $-s$ is incident with the line $0s$. It follows that the points 0, r and $-s$ are non-collinear.

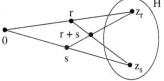

(c) By (a), the point $r + s$ is not incident with the line $0s$. By Theorem 7.9, the point $-s$ is incident with the line $0s$. It follows that the points 0, $r + s$ and $-s$ are non-collinear.   □

7.11 Lemma. *Let A be a Desarguesian affine space with origin 0. Furthermore, let r and s be two points of A such that 0, r and s are collinear. Finally, let t be a point not incident with the line $0r = 0s$.*

(a) If the points 0, $r + t$ and $s - t$ are collinear, we have $s = -r$.
(b) If $s \neq -r$, the points 0, $r + t$ and $s - t$ are non-collinear.

Proof. (a) By Theorem 7.9, the point $-t$ is incident with the line $0t$. Let g be the line through the points 0 and $r + t$. By assumption, the point $s - t$ is incident with g. Let $\mu : A \rightarrow A$ be the homology defined by $\mu(x) := -x$ (cf. Theorem 7.9).

If P is the projective closure of A, and if H is the hyperplane of P such that $A = P\backslash H$, H is the axis of μ.[11] Let $z_r := 0r \cap H$ and $z_t := 0t \cap H$. Then,

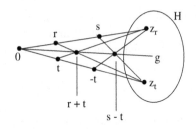

$$\mu(r + t) = \mu(g \cap tz_r)$$
$$= \mu(g) \cap \mu(tz_r)$$
$$= g \cap \mu(t)\mu(z_r)$$
$$= g \cap -t \, z_r$$
$$= s - t.$$

On the other side, we have $\mu(r + t) = -r - t$, that is, $-r - t = s - t$, thus, $r = -s$.

(b) follows from (a).   □

7.12 Theorem. *Let A and A' be two Desarguesian affine spaces. Then, for every parallelism-preserving transformation $\alpha : A \to A'$ with $\alpha(0) = 0$, we have:*

(a) $\alpha(x + y) = \alpha(x) + \alpha(y)$ *for all points x, y of A.*
(b) $\alpha(-x) = -\alpha(x)$ *for all points x of A.*

Proof. We shall prove the assertions (a) and (b) in common: Let P and P' be the projective closures of A and A', and let H and H' be the hyperplanes of P and P' such that $A = P\backslash H$ $A' = P'\backslash H'$. Since α is a parallelism-preserving transformation, α can be extended in a unique way on P (Theorem 4.6 of Chap. II).

Step 1. Let x and y be two points of A such that 0, x and y are not collinear. Then, we have $\alpha(x + y) = \alpha(x) + \alpha(y)$:

Let $z_x := 0x \cap H$ and $z_y := 0y \cap H$. Since $\alpha(0) = 0$, the point $\alpha(x)$ is incident with the line $0\alpha(z_x)$, and the point $\alpha(y)$ is incident with the line $0\alpha(z_y)$.

By Theorem 7.4, we have:

$$\alpha(x + y) = \alpha(xz_y \cap yz_x)$$
$$= \alpha(xz_y) \cap \alpha(yz_x)$$
$$= \alpha(x)\alpha(z_y) \cap \alpha(y)\alpha(z_x)$$
$$= \alpha(x) + \alpha(y).$$

Step 2. We have $\alpha(-x) = -\alpha(x)$ for all x of A:

---

[11] More precisely, the axis of the projective closure of $\mu : P \to P$.

Since $\alpha(0) = 0$, we can assume that $x \neq 0$. Let a be a point which is not incident with the line $0x$. By Theorem 7.9, the point $-x$ is incident with the line $0x$. By Lemma 7.10, the point $x + a$ is not incident with the line $0x$, it follows that the points $0, -x, x + a$ are non-collinear. By Step 1, we have

$$\alpha(a) = \alpha(a + x - x)$$
$$= \alpha(a + x) + \alpha(-x)$$
$$= \alpha(a) + \alpha(x) + \alpha(-x).$$

It follows that $\alpha(-x) = -\alpha(x)$.

Step 3. Let 0, x and y be three collinear points. Then, we have $\alpha(x + y) = \alpha(x) + \alpha(y)$:

For $x = -y$, the assertion follows from Step 2. Let $x \neq -y$. Let a be a point outside the line $0x = 0y$. Then, we have:

The points $0, x + a, y - a$ are non-collinear (application of Lemma 7.11 (b) with $r = x$, $s = y$ and $t = a$). Similarly, the points $0, y, -a$ are non-collinear (application of Lemma 7.10 (b) with $y = r$ and $s = a$). It follows from Step 1 that

$$\alpha(x + y) = \alpha(x + a + y - a)$$
$$= \alpha(x + a) + \alpha(y - a)$$
$$= \alpha(x) + \alpha(a) + \alpha(y) + \alpha(-a)$$
$$= \alpha(x) + \alpha(y).$$

$\square$

7.13 Theorem. *Let A be a Desarguesian affine space with origin 0, and let σ be a homology with centre 0. Then, $-\sigma$ is a homology with centre 0 as well.*

Proof. By Theorem 7.9, the transformation $\mu : A \to A$ with $\mu(x) = -x$ is a homology with centre 0. Since $-\sigma = \mu \, o \, \sigma$, the application $-\sigma$ is also a homology with centre 0. $\square$

7.14 Theorem. *Let A be a Desarguesian affine space with origin 0, and let σ_1 and σ_2 be two homologies with centre 0. Define $\sigma_1 + \sigma_2 : A \to A$ by $(\sigma_1 + \sigma_2)(x) := \sigma_1(x) + \sigma_2(x)$. Then, either $(\sigma_1 + \sigma_2)(x) = 0$ for all points x of A, or $\sigma_1 + \sigma_2$ is a central collineation with centre 0.*

Proof. Step 1. If there is a point $0 \neq x$ of A with $(\sigma_1 + \sigma_2)(x) = 0$, we have $\sigma_1 = -\sigma_2$. In particular, we have $(\sigma_1 + \sigma_2)(y) = 0$ for all points y of A:

Let $0 \neq x$ be a point of A with $(\sigma_1 + \sigma_2)(x) = 0$. Then, $\sigma_1(x) = -\sigma_2(x)$. By Theorem 7.13, σ_1 and $-\sigma_2$ are two homologies with centre 0. By Theorem 5.8 of Chap. II, it follows that $\sigma_1 = -\sigma_2$.

Step 2. For every line g of A through 0 and any point x of A, we have:

$$(\sigma_1 + \sigma_2)(x + g) \subseteq (\sigma_1 + \sigma_2)(x) + g.$$

Let $y = x + a$ be a point on the line $x + g$ such that a is incident with g. Since g is a line through 0 and since 0 is the centre of σ_1 and σ_2, the points $\sigma_1(a)$ and $\sigma_2(a)$ are incident with g. By Theorem 7.5, the point $\sigma_1(a) + \sigma_2(a)$ is incident with g. It follows from Theorem 7.12 that

$$
\begin{aligned}
(\sigma_1 + \sigma_2)(y) &= \sigma_1(y) + \sigma_2(y) \\
&= \sigma_1(x + a) + \sigma_2(x + a) \\
&= \sigma_1(x) + \sigma_1(a) + \sigma_2(x) + \sigma_2(a) \\
&= (\sigma_1 + \sigma_2)(x) + \sigma_1(a) + \sigma_2(a) \in (\sigma_1 + \sigma_2)(x) + g.
\end{aligned}
$$

Step 3. If $(\sigma_1 + \sigma_2)(x) \neq 0$ for all points $0 \neq x$ of A, $\sigma_1 + \sigma_2$ is a homology with centre 0:

Let $0 \neq x$ be an arbitrary point of A, let $\sigma := \sigma_1 + \sigma_2$, and let $y := \sigma(x) = (\sigma_1 + \sigma_2)(x) = \sigma_1(x) + \sigma_2(x)$. Since σ_1 and σ_2 are homologies with centre 0, the points $\sigma_1(x)$ and $\sigma_2(x)$ are on the line $0x$. By Theorem 7.5, the point $y = \sigma_1(x) + \sigma_2(x)$ is on the line $0x$.

By Theorem 6.2 of Chap. II, there exists a homology $\alpha : A \to A$ with centre 0 such that $\alpha(x) = y$.

Let a be a point outside of the line $0x$, and let g be the line through 0 and the point $a - x$. By Theorem 7.7, $h = x + g$ is the line through x parallel to g. (If P is the projective closure of A, and if H is the hyperplane of P with $A = P \backslash H$, the lines g and h meet in a point z of H. If $z_x := 0x \cap H$ and $z_a := 0a \cap H$, $a - x = az_x \cap -xz_a$.)

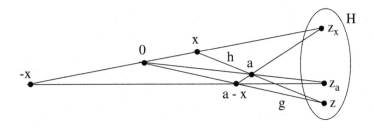

Since the points 0 and $a - x$ are on the line g, the points $x + 0 = x$ and $x + a - x = a$ are on the line $h = x + g$, that is, $h = ax$. In particular, we have $a = 0a \cap h = 0a \cap (x + g)$.

By Step 2, we have

$$
\begin{aligned}
\sigma(a) &= \sigma(0a \cap (x + g)) \\
&= \sigma(0a) \cap \sigma(x + g) \\
&\subseteq 0a \cap \sigma(x) + g
\end{aligned}
$$

$$= 0a \cap (y + g), \text{ that is,}$$

$$\sigma(a) = 0a \cap (y + g).$$

On the other hand, by Theorem 7.8, we have:

$$\alpha(a) = \alpha(0a \cap (x + g))$$

$$= \alpha(0a) \cap \alpha(x + g)$$

$$= 0a \cap (\alpha(x) + g)$$

$$= 0a \cap (y + g), \text{ that is,}$$

$$\alpha(a) = \sigma(a).$$

Using the relation $\sigma(a) = \alpha(a)$ for a point a non-incident with the line $0x$, one can deduce the relation $\sigma(r) = \alpha(r)$ for all points r non-incident with $0a$ in a similar way. It follows that $\sigma = \alpha$, that is, σ is a homology with centre 0. $\quad\square$

Definition. Let A be a Desarguesian affine space with origin 0.

(a) We denote by D_0 the set of the homologies of A with centre 0.
(b) Let $\mathbf{0} : A \to A$ be the transformation defined by $\mathbf{0}(x) := 0$ for all x of A. Set $D := D_0 \cup \{\mathbf{0}\}$.

7.15 Theorem. *Let A be a Desarguesian affine space with origin 0. For two elements σ_1 and σ_2 of D and for a point x of A, set*

$$(\sigma_1 + \sigma_2)(x) := \sigma_1(x) + \sigma_2(x)$$

$$(\sigma_1 \cdot \sigma_2)(x) = \begin{cases} (\sigma_1 \, o \, \sigma_2)(x) = \sigma_1(\sigma_2(x)) \text{ if } \sigma_1, \, \sigma_2 \in D_0 \\ \mathbf{0} \text{ if } \sigma_1 = \mathbf{0} \text{ or } \sigma_2 = \mathbf{0}. \end{cases}$$

Then, $(D, +, \cdot)$ is a skew field.

Proof. Step 1. $(D, +)$ is an abelian group:

Closed by addition: By Theorem 7.14, $\sigma_1 + \sigma_2$ is an element of D for all σ_1, σ_2 of D_0. The assertion follows from $\sigma + \mathbf{0} = \sigma = \mathbf{0} + \sigma = \sigma$ for all σ of D.

Associativity and commutativity: By Theorem 7.6, $(A, +)$ is an abelian group. For σ_1, σ_2, σ_3 of D and x of A, it follows that

$$((\sigma_1 + \sigma_2) + \sigma_3)(x) = \sigma_1(x) + \sigma_2(x) + \sigma_3(x) = (\sigma_1 + (\sigma_2 + \sigma_3))(x) \text{ and}$$

$$(\sigma_1 + \sigma_2)(x) = \sigma_1(x) + \sigma_2(x) = \sigma_2(x) + \sigma_1(x) = (\sigma_2 + \sigma_1)(x).$$

Existence of the neutral elements: By definition of D, $\mathbf{0}$ is an element of D.

Existence of the inverse element: By Theorem 7.13, for each element σ of D, the element $-\sigma$ is an element of D.

Step 2. $(D_0, \cdot)$ is a multiplicative group:

$(D_0, \cdot)$ is the group of the central collineations of A with centre 0.

Step 3. The distributive laws hold:

Let σ_1, σ_2, σ_3 be three elements of D, and let x be a point of A. Then, we have:

$$\begin{aligned}
(\sigma_1 \cdot (\sigma_2 + \sigma_3))(x) &= \sigma_1((\sigma_2 + \sigma_3)(x)) \\
&= \sigma_1(\sigma_2(x) + \sigma_3(x)) \\
&= \sigma_1(\sigma_2(x)) + \sigma_1(\sigma_3(x)) \\
&= (\sigma_1\sigma_2)(x) + (\sigma_1\sigma_3)(x) \\
&= (\sigma_1\sigma_2 + \sigma_1\sigma_3)(x).
\end{aligned}$$

Analogously, it follows that $(\sigma_1 + \sigma_2)\,\sigma_3 = \sigma_1\sigma_3 + \sigma_2\sigma_3$. □

7.16 Theorem. *Let A be a Desarguesian affine space. Then, $(A, +)$ is a vector space over the field D defined in Theorem 7.15.*

Proof. By Theorem 7.6, $(A, +)$ is an abelian group. For x of A and σ of D, we define $\sigma \cdot x := \sigma(x)$. Then, for σ, σ_1, σ_2 of D and x, y of A, we get:

$$(\sigma_1 + \sigma_2) \cdot x = (\sigma_1 + \sigma_2)(x) = \sigma_1(x) + \sigma_2(x) = \sigma_1 \cdot x + \sigma_2 \cdot x$$
$$(\sigma_1 \cdot \sigma_2) \cdot x = (\sigma_1 \cdot \sigma_2)(x) = \sigma_1(\sigma_2(x)) = \sigma_1 \cdot (\sigma_2 \cdot x)$$
$$\sigma \cdot (x + y) = \sigma(x + y) = \sigma(x) + \sigma(y) = \sigma \cdot x + \sigma \cdot y.$$

Finally, we have $id \cdot x = id(x) = x$. □

7.17 Theorem (First Fundamental Theorem for Affine Spaces). *Let A be a Desarguesian affine space. Then, A is an affine space over a vector space.*

Proof. As usual, we denote by 0 the origin of A. Let D be the skew field defined in Theorem 7.15. By Theorem 7.6, $(A, +)$ is a vector space over D. Let A' be the affine space over the vector space A.

Step 1. By definition, A and A' have the same sets of points.

Step 2. The lines of A through 0 are exactly the lines of A' through 0:

For, consider the lines of A and the lines of A' as sets of points. Let g be a line of A through 0, and let x be a point on g distinct from 0. The lines of A' through the point 0 are the 1-dimensional subspaces of the vector space $(A, +)$. Hence

$$h := \langle x \rangle = \{\sigma \cdot x \mid \sigma \in D\}$$

is the line of A' through the points 0 and x. We have $g = h$:

Obviously, the points 0 and x are incident with g and with h.

Let y be a point on g distinct from 0 and x. Since g is a line through 0, by Theorem 6.2 of Chap. II, there exists a homology σ with centre 0 such that $\sigma(x) = y$. It follows that $y = \sigma \cdot x$ is incident with h.

Conversely, let $y = \sigma \cdot x = \sigma(x)$ be a point on h distinct from 0. Since σ fixes the line g, it follows that $y = \sigma(x)$ is incident with g.

Step 3. The lines of A are exactly the lines of A':

Again, consider the lines of A and the lines of A' as sets of points. The lines of A' are exactly the cosets of the 1-dimensional subspaces of the vector space $(A, +)$. Hence, the assertion follows from Theorem 7.7 using Step 2.

Step 4. By Step 1 and Step 3, A and A' are affine spaces with identical point and line sets. It follows that $A = A'$. □

7.18 Theorem (First Fundamental Theorem for Projective Spaces). *Let P be a Desarguesian projective space. Then, P is a projective space over a vector space.*

Proof. Let H be a hyperplane of P, and let $A := P \backslash H$ be the affine space defined by P and H. Then, A is Desarguesian. By the first fundamental theorem for affine spaces (Theorem 7.17), A is an affine space over a vector space.

By Theorem 5.10, the projective closure P of A is a projective space over a vector space. □

8 The Second Fundamental Theorem

The second fundamental theorem for affine spaces states that every parallelism-preserving collineation of a Desarguesian affine space is the product of a collineation induced by a semi-linear transformation and a translation (Theorem 8.2).

For projective spaces, the second fundamental theorem states that every collineation of a Desarguesian projective space is induced by a semi-linear transformation (Theorem 8.3).

8.1 Theorem. *Let $A = A(W)$ and $A' = A(W')$ be two affine spaces over two vector spaces W and W', and let $\alpha : A \to A'$ be a parallelism-preserving transformation with $\alpha(0) = 0$.*

There exists a semilinear transformation $A : W \to W'$ with $A(x) = \alpha(x)$ for all x of W.

Proof. By Theorem 7.12, we have $\alpha(x + y) = \alpha(x) + \alpha(y)$ for all x, y of W. Suppose that W and W' are vector spaces over the skew fields K and K'. It remains to show that there exists an automorphism $\gamma : K \to K'$ such that $\alpha(kx) = \gamma(k)\alpha(x)$ for all x of W and all k of K.

Step 1. For any point $0 \neq x$ of W and any element k of K, there exists an element $\gamma_x(k)$ of K' such that $\alpha(kx) = \gamma_x(k)\alpha(x)$:

The subspace $\langle x \rangle_W$ of W is a line of A. It follows that the points 0, x and kx are collinear. Hence, the points $0 = \alpha(0)$, $\alpha(x)$ and $\alpha(kx)$ are collinear, that is, $\alpha(kx)$ is contained in $\langle \alpha(x) \rangle_{W'}$. Therefore, there exists an element $\gamma_x(k)$ of K' such that $\alpha(kx) = \gamma_x(k)\alpha(x)$.

Step 2. Let x, y be two elements of W with $\langle x \rangle_W \neq \langle y \rangle_W$. Then, $\gamma_x(k) = \gamma_y(k)$ for all k of K:

For k of K, we have

$$\alpha(k(x + y)) = \gamma_{x+y}(k)\, \alpha(x + y)$$
$$= \gamma_{x+y}(k)\, \alpha(x) + \gamma_{x+y}(k)\, \alpha(y).$$

On the other side hand, we have

$$\alpha(k(x + y)) = \alpha(kx) + \alpha(ky)$$
$$= \gamma_x(k)\alpha(x) + \gamma_y(k)\alpha(y).$$

By assumption, 0, x and y are non-collinear. Therefore, $0 = \alpha(0)$, $\alpha(x)$ and $\alpha(y)$ are non-collinear, that is, $\alpha(x)$, $\alpha(y)$ and $\alpha(x) + \alpha(y)$ are linearly independent vectors of W'. Comparison of the coefficients yields $\gamma_x(x) = \gamma_{x+y}(k) = \gamma_y(k)$.

Step 3. Let x and y be two elements of W with $\langle x \rangle_W = \langle y \rangle_W$. Then, $\gamma_x(k) = \gamma_y(k)$ for all k of K:

For, let z be an element of W such that z is not contained in $\langle x \rangle_W = \langle y \rangle_W$. Then, by Step 2, for all k of K, we have:

$$\gamma_x(x) = \gamma_z(k) = \gamma_y(k).$$

Step 4. For an element $0 \neq x$ of W, set $\gamma(k) := \gamma_x(k)$ for all k of K. Then, $\gamma(0) = 0$:

For, let $0 \neq x$ be an element of W. We have $0 = \alpha(0) = \alpha(0x) = \gamma(0)\alpha(x)$. Since $\alpha(x) \neq 0$, we have $\gamma(0) = 0$.

Step 5. $\gamma : K \to K'$ is a field automorphism:

(i) Let k_1 and k_2 be two elements of K, and let $0 \neq x$ be an element of W. Then,

$$\alpha((k_1 + k_2)x) = \alpha(k_1\, x) + \alpha(k_2\, x)$$
$$= \gamma(k_1)\, \alpha(x) + \gamma(k_2)\alpha(x)$$
$$= (\gamma(k_1) + \gamma(k_2))\alpha(x).$$

Furthermore, we have $\alpha((k_1 + k_2)x) = \gamma(k_1 + k_2)\alpha(x)$. It follows that $\gamma(k_1 + k_2) = \gamma(k_1) + \gamma(k_2)$.

(ii) Let k_1, k_2 be two elements of K, and let $0 \neq x$ be an element of W. We have

$$\alpha((k_1 \cdot k_2)x) = \gamma(k_1)\alpha(k_2 x)$$
$$= \gamma(k_1) \cdot \gamma(k_2)\alpha(x).$$

Furthermore, we have $\alpha((k_1 \cdot k_2)x) = \gamma(k_1 \cdot k_2)\, \alpha(x)$. It follows that $\gamma(k_1 \cdot k_2) = \gamma(k_1) \cdot \gamma(k_2)$.

(iii) $\gamma : K \to K'$ is injective:

For, let $\gamma(k_1) = \gamma(k_2)$ for two elements k_1 and k_2 of K. Then, for any element $0 \neq x$ of W, we have

$$\alpha(k_1 x) = \gamma(k_1)\alpha(x) = \gamma(k_2)\alpha(x) = \alpha(k_2 x).$$

It follows from the injectivity of α that $k_1 x = k_2 x$, thus, $k_1 = k_2$.

(iv) $\gamma : K \to K'$ is surjective:

Let k' be an element of K' and let $0 \neq z$ be an element of W'. Since α is surjective, there exist two points x and y of A with $\alpha(x) = z$ and $\alpha(y) = k'z$. Since 0, z and $k'z$ are collinear, the points 0, $x = \alpha^{-1}(z)$ and $y = \alpha^{-1}(k'z)$ are collinear. It follows that $y = kx$ for some element k of K.

We have $k'z = \alpha(y) = \alpha(kx) = \gamma(k)\,\alpha(x) = \gamma(k)\,z$. It follows that $\gamma(k) = k'$.   □

8.2 Theorem (Second Fundamental Theorem for Affine Spaces). *Let A and A' be two Desarguesian affine spaces, and let $\alpha : A \to A'$ be a parallelism-preserving collineation. Then, there exists exactly one translation $\tau : A' \to A'$ and exactly one parallelism-preserving collineation $\beta : A \to A'$ induced by a semi-linear transformation with $\beta(0) = 0$ such that $\alpha = \tau\beta$.*

Proof. By Theorem 7.17, A and A' are two affine space over two vector spaces W and W'. By Theorem 6.6, there exists exactly one parallelism-preserving collineation $\beta : A \to A'$ with $\beta(0) = 0$ and exactly one translation $\tau : A' \to A'$ such that $\alpha = \tau\beta$.

Finally, by Theorem 8.1, the collineation β is induced by a semi-linear transformation.   □

8.3 Theorem (Second Fundamental Theorem for Projective Spaces). *Let P and P' be two Desarguesian projective spaces, and let $\alpha : P \to P'$ be a collineation. Then, α is induced by a semi-linear transformation.*

Proof. By Theorem 7.18, P and P' are two projective spaces over two vector spaces V and V'. Let H and H' be a hyperplane of P and a hyperplane of P', and let $A := P \backslash H$ and $A' := P' \backslash H'$ be the affine spaces defined by P and H and by P' and H'. By Corollary 5.9, A and A' are affine spaces over two vector spaces W and W'. Let $0 = \langle v_0 \rangle$ be the origin of A.

Let $\{v_0\} \cup \{v_i \mid i \in I\}$ be a basis of V such that $\{v_i \mid i \in I\}$ is a basis of W. Let u_0 and $u_i (i \in I)$ be vectors of V' such that $\langle u_0 \rangle = \langle \alpha(v_0) \rangle$ and $\langle u_i \rangle = \langle \alpha(v_i) \rangle$ for all i of I. Since $\alpha : P \to P'$ is a collineation, $\{u_0\} \cup \{u_i \mid i \in I\}$ is a basis of V (Theorem 4.2 of Chap. II).

Let $D : V \to V'$ be the linear transformation defined by $Dv_0 := u_0$ and $Dv_i := u_i$ for all i of I. Then, D induces a projective collineation $\delta : P \to P'$.

For $\beta := \delta^{-1}\alpha$ and for $i = 0, 1, \ldots, d$, we have $\beta(\langle v_0 \rangle) = \langle v_0 \rangle$ and $\beta(\langle v_i \rangle) = \langle v_i \rangle$ for all i of I.

It follows that β is a collineation of P with $\beta(0) = 0$ and $\beta(H) = H$. Thus, β induces a parallelism-preserving collineation $\beta_A : A \to A$ with $\beta_A(0) = 0$.

By Theorem 8.1, there exists a semilinear transformation $B : W \to W$ with accompanying automorphism γ, inducing β_A.

We define the transformation $B' : V \to V$ as follows: Let v be an element of V. Then, there exist an element k of K and an element w of W such that $v = kv_0 + w$. Set

$$B'(v) = B'(kv_0 + w) := \gamma(k)v_0 + B(w).$$

Then, B' is a bijective semi-linear transformation with accompanying automorphism γ inducing a collineation $\beta' : P \to P$. By definition, we have $\beta' \mid_A = \beta_A$, that is, β' is the projective closure of β_A. It follows that $\beta' = \beta$.

Hence, $\alpha := \delta \beta$ is the product of a projective collineation and a collineation induced by a semi-linear transformation. In total, α is induced by a semilinear transformation. □

Chapter 4
Polar Spaces and Polarities

1 Introduction

The present chapter is devoted to the study of the so-called polar spaces. They form an important part of modern incidence geometry.

In Sect. 2, polar spaces are introduced. Due to the famous Theorem of Buekenhout and Shult [21], polar spaces can be endowed with a structure of subspaces. The subspaces of polar spaces are projective spaces, although the polar spaces themselves are not projective spaces.

The diagram of a polar space is the subject of Sect. 3.

If π is a polarity of a projective space P, a polar space S_π is defined by π. The connection between polarities and polar spaces is the subject of Sect. 4.

The algebraic description of polarities is the topic of Sect. 5. It will be shown that to any polarity there exists a reflexive sesquilinear form and vice versa. The main result of Sect. 5 is the Theorem of Birkhoff and von Neumann [7] stating that every polarity is induced by a symmetric, anti-symmetric, hermitian or anti-hermitian sesquilinear form.

Closely related to the sesquilinear forms are the pseudo-quadratic forms. Every pseudo-quadratic form of a vector space V defines a pseudo-quadric in the corresponding projective space $P = P(V)$. The points and the lines of a pseudo-quadric define a polar space. Pseudo-quadratic forms and pseudo-quadrics are investigated in Sect. 6.

The third and last family of polar spaces is constructed in Sect. 7 by using the lines of a 3-dimensional projective space. A polar space of this kind is called a Kleinian polar space.

The three families of polar spaces constructed in the present chapter all have the property that the (projective) subspaces are Desarguesian. A substantial result of Veldkamp [56] (for Char $K \neq 2$) and Tits [50] says that the converse is also true, that is, every polar space with Desarguesian projective subspaces stems from a polarity, a pseudo-quadric or a Kleinian polar space.

J. Ueberberg, *Foundations of Incidence Geometry*, Springer Monographs in Mathematics, 123
DOI 10.1007/978-3-642-20972-7_4, © Springer-Verlag Berlin Heidelberg 2011

Finally, in Sect. 8, we will present the Theorem of Buekenhout [17] and Parmentier [37] characterizing polar spaces as linear spaces with polarities.

2   The Theorem of Buekenhout–Shult

Definition. Let Γ be a geometry of rank 2 over the type set {point, line}.

(a) Two points of Γ are called **collinear** if they are on a common line.[1]
(b) Γ is called a **partial linear space** if Γ fulfils the following two conditions:

(PL_1) Any two points are incident with at most one line.
(PL_2) Any line is incident with at least two points.

Definition. A **generalized quadrangle** is a set geometry Q of rank 2 over the type set {point, line} satisfying the following conditions:

(V_1) Any two points are incident with at most one line.
(V_2) Let g be a line, and let x be a point which is not on g. Then, there exists exactly one point y on g such that x and y are on a common line.
(V_3) Any line of Q is incident with at least two points. Any point of Q is incident with at least two lines.

Definition. (a) A **polar space** is a set geometry S of rank 2 over the type set {point, line} satisfying the following conditions.[2]

(P_1) Let g be a line, and let x be a point not on g. Then, there exists either exactly one point on g collinear with x, or all points of g are collinear with x.
(P_2) Every line is incident with at least three points.

(b) If the geometry S satisfies condition (P_1) and the weaker condition

(P'_2) Every line is incident with at least two points,

the geometry S is called a **generalized polar space**.
(c) A (generalized) polar space S is called **nondegenerate** if S fulfils the following additional condition:

(P_3) For every point x of S, there exists a point y such that x and y are not collinear.

(d) Let S be a (generalized) polar space. For a point x of S, we denote by $x^{\perp}$ (say "x perp") the set of points which are collinear with x.

---

[1] In particular, a point x is collinear with itself.
[2] The letter S reminds us that polar spaces as defined above originally were defined by Buekenhout and Shult [21] as **Shult-spaces**.

(e) Let S be a (generalized) polar space, and let U be a set of points of S. U is called a **subspace** of S if any two points of U are collinear and if every line of S, containing at least two points of U is completely contained in U.

(f) A (generalized) polar space S is called **of finite rank** n if there exists a natural number n such that for every chain $\varnothing \neq U_1 \subset U_2 \subset \ldots \subset U_r$ of subspaces $U_1, \ldots, U_r$, the relation $r \leq n$ holds and if there is at least one chain of length n. Otherwise, the polar space is called **of infinite rank**.

(g) A subspace U of S is called **of finite rank** n if there exists a natural number n such that for every chain $\varnothing \neq U_1 \subset U_2 \subset \ldots \subset U_r = U$ of proper subspaces $U_1, \ldots, U_r$, the relation $r \leq n$ holds and if there is at least one chain of length n. Otherwise, the subspace U is called **of infinite rank**. The empty set is by definition a subspace of rank 0.

Note that in the literature subspaces are often called **singular subspaces**.

There is the following relation between partial linear spaces, generalized quadrangles and polar spaces: Every nondegenerate polar space S is a partial linear space, that is, any two points of S are incident with at most one line (cf. Theorem 2.9).

Furthermore, the generalized quadrangles are exactly the generalized nondegenerate polar spaces of rank 2 (cf. Theorem 2.20).

2.1 Theorem. *Let S be a polar space.*

(a) The intersection of an arbitrary family of subspaces of S is a subspace.

(b) Let M be a maximal set of pairwise collinear points. Then, M is a subspace.

Proof. (a) Let $(U_i)_{i \in I}$ be a family of subspaces of S, and let x and y be two points contained in U_i for all i of I. Since U_i is a subspace for all i of I, the line xy is contained in U_i for all i of I.

(b) Let M be a maximal set of pairwise collinear points, and let g be a line, containing two points x and y of M. Let z be a point on g distinct from x and y. We shall show that z is collinear with all points of M:

For, let m be a point of M. If m lies on g, the points m and z are collinear. If m does not lie on g, m is collinear with the points x and y. Hence, m is collinear with two points on g and therefore, by Axiom (P_1), collinear with all points on g. It follows that m and z are collinear.

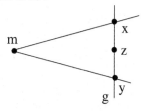

If z were not contained in M, $M \cup \{z\}$ would be a set of pairwise collinear points, in contradiction to the maximality of M. It follows that M is a subspace. □

2.2 Theorem. *Let S be a polar space. Every subspace of S is contained in a maximal subspace of S.*

Proof. Let U be a subspace of S, and let $M := \{W \text{ subspace of } S \mid U \subseteq W\}$ be the set of subspaces of S containing U. Let $(W_i)_{i \in I}$ be a chain in M. The set $\bigcup_{i \in I} W_i$ is a subspace of S and thus an upper bound in M. By the Lemma of Zorn (Lemma 4.6 of Chap. 1), there is a maximal element in M.   □

Definition. Let S be a polar space. Let M be a set of pairwise collinear points of S, and let

$$\langle M \rangle := \cap \, U \mid U \text{ is a subspace of } S \text{ containing } M.$$

$\langle M \rangle$ is the smallest subspace of S containing M. It is called **the subspace generated by** M.

2.3 Theorem. *Let S be a polar space. Any set of pairwise collinear points generates a subspace.*

Proof. Let X be a set of pairwise collinear points. By the Lemma of Zorn (Lemma 4.6 of Chap. 1), X is contained in a maximal set of pairwise collinear points. By Theorem 2.1 (b), X is contained in a subspace. Let U be the intersection over all subspaces, containing X. Then, U is generated by X.   □

For the investigation of polar spaces, the notion of a projective hyperplane is very useful. The definition of a projective hyperplane is motivated by the following property of a hyperplane H of a projective space P: Every line of P is either contained in H or it has exactly one point in common with H.

Definition. Let S be a polar space, and let U be a subspace of S. A proper subspace H of U is called a **projective hyperplane** of U if every line of U has at least one point in common with H.

2.4 Theorem. *Let S be a polar space, and let U be a subspace of S. If H is a projective hyperplane of U, H is a maximal subspace of U.*

Proof. The assertion follows as in Theorem 3.4 of Chap. 1. (Note that Theorem 3.4 of Chap. 1 deals with linear spaces, whereas U might contain pairs of points which are contained in more than one line. This case is only excluded in Theorem 2.9.)   □

2.5 Theorem. *Let S be a polar space, and let U be a subspace of S. Furthermore, let p be a point of S that is not collinear with all points of U.*

(a) *The set $U_p := p^\perp \cap U$ is a projective hyperplane of U.*

(b) *If U is a maximal subspace of S, the subspace $\langle p, \, U_p \rangle$ is also a maximal subspace of S.*

(c) *If U is a maximal subspace of S, the subspace $\langle p, \, U_p \rangle$ consists of the points on the lines through p that have at least one point in common with U_p.[3]*

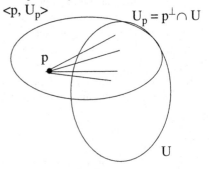

---

[3]In this case, the line through p has exactly one point in common with U_p.

Proof. (a) We first shall show that $U_p = p^\perp \cap U$ is a subspace: For, let x and y be two points of U_p. Since U is a subspace, the points x and y are collinear, and there is a line g through x and y. Since p is not contained in U (otherwise, p would be collinear with all points of U), p is not incident with g. Since the points x and y are contained in U_p, they are collinear with p. By Axiom (P_1), all points of g are collinear with p, it follows that g is contained in $p^\perp$. In summary, $U_p = p^\perp \cap U$ is a subspace.

Let g be a line of U. Then, there exists a point x on g collinear with p. It follows that x is contained in $p^\perp \cap U$. Hence, $U_p = p^\perp \cap U$ is a projective hyperplane of U.

(b) By assumption, there exists a point z of U which is not collinear with p.

Step 1. We have $U = \langle z, U_p \rangle$: By (a), U_p is a projective hyperplane of U. By Theorem 2.4, U_p is a maximal subspace of U. Since z is not contained in U_p, it follows that $U = \langle z, U_p \rangle$.

Step 2. The subspace $\langle p, U_p \rangle$ is a maximal subspace of S: Assume that $\langle p, U_p \rangle$ is not maximal. Let W be a subspace of S, containing $\langle p, U_p \rangle$ properly. Then, there exists in W a line g through p disjoint from U. Let x be a point on g collinear with the point z.

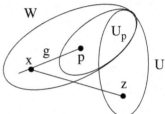

Since x is contained in W, it follows that x is collinear with every point of U_p. By construction, x and z are collinear. It follows that the set $\{x, z\} \cup U_p$ consists of pairwise collinear points. Hence, $\langle x, z, U_p \rangle$ is a subspace of S containing the subspace $U = \langle z, U_p \rangle$ properly, in contradiction to the maximality of U.

(c) By assumption, there exists a point z of U not collinear with p.

Step 1. Let $W := \langle p, U_p \rangle$, and let $W_z := z^\perp \cap W$. Then, $W_z = U_p$: Let x be a point of U_p. Since x is contained in U, it follows that x and z are collinear. Hence, x is contained in $z^\perp \cap W = W_z$. It follows that U_p is contained in W_z.

We shall see that W_z is contained in U_p: Assume that there is a point x of $W_z \setminus U_p$. Since U_p is a projective hyperplane of U and hence a maximal subspace of U, it follows that $U = \langle z, U_p \rangle$.

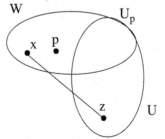

Since z and x are collinear, $\langle z, x, U_p \rangle$ is a subspace through $\langle z, U_p \rangle = U$, in contradiction to the maximality of U.

Step 2. $\langle p, U_p \rangle$ consists of the points on the lines through p which have at least one point in common with U_p: By (a), W_z is a projective hyperplane of W. In view of Step 1, we have $U_p = W_z$. Hence, U_p is a projective hyperplane of W. It follows that every line of W through p intersects the subspace U_p in a point. □

The following lemma prepares the proof of Theorem 2.7.

2.6 Lemma. *Let S be a polar space, and let p and q be two non-collinear points of S. Let x be a point of $p^\perp \cap q^\perp$ that is collinear with all points of $p^\perp \cap q^\perp$. Then, x is collinear with all points of $p^\perp$ and with all points of $q^\perp$.*[4]

Proof. Step 1. The point x is collinear with all points of $p^\perp$: Let a be a point of $p^\perp$. We need to show that x and a are collinear. Since x is contained in $p^\perp \cap q^\perp$, there exists a line g through p and x. Furthermore, since a is contained in $p^\perp$, there exists a line h through a and p. If the point x is on the line h, a and x are incident with h, hence, they are collinear.

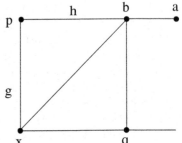

Assume that x is not incident with h. Since S is a polar space, there exists a point b on h, collinear with q.[5] It follows that b is collinear with p and q. By assumption, b is collinear with x as well.

Since x is collinear with the two points p and b of h, x is collinear with all points on h, in particular with a.

Step 2. The point x is collinear with all points of $q^\perp$: Due to the symmetry of p and q, the proof of the assertion is identical to the proof of Step 1.   □

2.7 Theorem. *Let S be a nondegenerate polar space, and let p and q be two non-collinear points of S. If $p^\perp \cap q^\perp$ contains at least two lines, $p^\perp \cap q^\perp$ is a nondegenerate polar space.*

Proof. We shall verify Axioms (P_1) to (P_3):

Verification of (P_1): Let g be a line of $p^\perp \cap q^\perp$, and let x be a point of $p^\perp \cap q^\perp$ which is not on g.

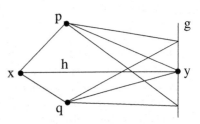

Let h be a line through x intersecting g in a point y. We shall show that h is contained in $p^\perp \cap q^\perp$:

Since g is contained in $p^\perp \cap q^\perp$, the points of g are collinear with p and q. In particular, the point y is collinear with p and q.

Furthermore, the point x of $p^\perp \cap q^\perp$ is collinear with the points p and q. Therefore, all points of h are collinear with p and q. It follows that h is contained in $p^\perp \cap q^\perp$.

Verification of (P_2): Since any line of S is incident with at least three points, any line of $p^\perp \cap q^\perp$ is incident with at least three points.

---

[4]In Theorem 2.7, we shall see that in nondegenerate polar spaces, there is no point of $p^\perp \cap q^\perp$ which is collinear with all points of $p^\perp \cap q^\perp$.

[5]The point q is not on h, since otherwise p and q would be collinear, in contradiction to the assumption.

Verification of (P_3): Assume that there is a point x of $p^\perp \cap q^\perp$ which is collinear with all points of $p^\perp \cap q^\perp$.

We shall show that the point x is collinear with all points of S, in contradiction to (P_3). For, let a be an arbitrary point of S. If a is contained in $p^\perp$ or if a is contained in $q^\perp$, the points x and a are collinear by Lemma 2.6. Hence, we can assume that a is neither collinear with p nor with q.

Since x is contained in $p^\perp \cap q^\perp$, there exists a line g through x and p and a line h through x and q. Since a and p are non-collinear, the point a is not on g. It follows that there exists a point b on g such that a and b are joined by a line l. If $b = x$, the points x and a are collinear, and the assertion is shown. Thus, we can assume that $x \neq b$.

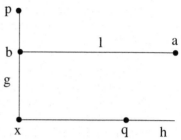

The points b and q are non-collinear since, otherwise, q would be collinear with the points x and b on g. This would imply that q is collinear with all points of g, in particular with p, in contradiction to the assumption.

Step 1. The point x is collinear with all points of $q^\perp \cap b^\perp$: For, let w be a point of $q^\perp \cap b^\perp$. In particular, the point w is contained in $q^\perp$. By Lemma 2.6, the point x is collinear with all points of $q^\perp$. In particular, x and w are collinear.

Step 2. The points x and a are collinear: By Step 1, x is collinear with all points of $q^\perp \cap b^\perp$. By Lemma 2.6, the point x is collinear with all points of $b^\perp$. Since a and b are collinear, the points x and a are collinear.

Since a is an arbitrary point of S, it follows that x is collinear with all points of S, in contradiction to Axiom (P_3).

It remains to show that $p^\perp \cap q^\perp$ is a geometry: For, let g be a line contained in $p^\perp \cap q^\perp$. Let x be a point of $p^\perp \cap q^\perp$. If x is incident with g, then $\{x, g\}$ is a chamber through x. If x is not incident with g, there exists a line h through x intersecting g in a point. Above (Verification of (P_1)), we have seen that h is contained in $p^\perp \cap q^\perp$. Hence, x is contained in the chamber $\{x, h\}$. Obviously, every line of $p^\perp \cap q^\perp$ is contained in a chamber.   □

2.8 Theorem. *Let S be a nondegenerate polar space, let g be a line of S and let a be a point on g. Then, there exists a point b outside of g such that b is collinear with a, but not collinear with any further point on g.*

Proof. Assume that there exists a point a on g such that for any point b outside of g, we have: If a and b are collinear, b is collinear with at least one further point on g, that is, b is collinear with all points of g.

For a point x of $g \setminus \{a\}$, let

$$\Pi_x := \{u \in S \setminus g \mid u \text{ is collinear with } x, \text{ but with no further point on } g\}.$$

Step 1. Let x and y be two points of $g \setminus \{a\}$. For any two elements u of Π_x and v of Π_y, the points u and v are non-collinear:

Otherwise, there are two points u of Π_x and v of Π_y which are joined by a line h.

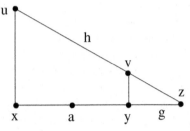

(i) We have $h \cap g = \varnothing$: Assume that there exists a point z contained in $h \cap g$. If $z \neq x$, z is (besides x) a second point on g collinear with u, in contradiction to $u \in \Pi_x$. If $z = x$, $z \neq y$ is a second point on g, collinear with v, in contradiction to the fact that v is contained in Π_y.

(ii) No point on h is collinear with all points on g: Assume that there exists a point z on h collinear with all points on g. Then, z is collinear with the point y. It follows that y is collinear with the points z and v and therefore with all points of h. Hence, u and y are collinear, in contradiction to the fact that u is contained in Π_x.

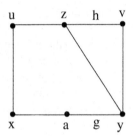

(iii) We shall see that the assumption that there exists a line h through u and v, yields a contradiction: Since the lines g and h are disjoint, the point a is not on h. It follows that there exists a point s on h which is collinear with a. By (ii), the point s is collinear with a, but with no further point on g. This contradicts our assumption with respect to a.

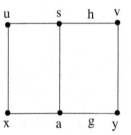

Step 2. We shall see that our assumption at the beginning of our proof yields a contradiction: For, let x be a point on g distinct from a. By Axiom (P_3), there exists a point v of S not collinear with x. It follows that there exists exactly one point y on g collinear with v. We have $y \neq a$, since, otherwise, the point v would be collinear with a, but with no further point on g, in contradiction to our assumption that no such point exists.

Let u be a point of S that is not collinear with y. As above there is a point $x' \in g \setminus \{a\}$ such that u is collinear with x', but with no further point on g.

It follows that u is contained in $\Pi_{x'}$ and that v is contained in Π_y. Let l be a line through x' and u. Since v and u are non-collinear (Step 1), there exists exactly one point z on l ($z \neq u$) collinear with v. If z is contained in $\Pi_{x'}$, by Step 1, z and v would be non-collinear, a contradiction. It follows that z is collinear with all points of g, in particular with the point a.

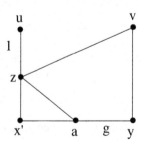

Hence, a is collinear with the points x' and z and therefore with all points on l. It follows that a and u are collinear, a contradiction. □

2.9 Theorem. *Let S be a nondegenerate polar space. Then, any two points of S are incident with at most one line of S.*

Proof. Assume that there exist two points a and b incident with two lines g and h.

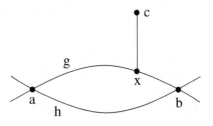

Step 1. There is a point c which is not collinear with any point of $g \cap h$:

Since $g \neq h$, there exists a point x on g or h not contained in $g \cap h$.[6] W.l.o.g., let x be a point incident with g. By Theorem 2.8, there exists a point c of S which is not on g and which is collinear with x, but with no further point on g. In particular, c is not collinear with any point of $g \cap h$.

Step 2. On the line h, there exists exactly one point y collinear with c. The point y is not incident with any line through c and x.

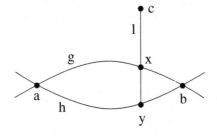

Since c is not on h, there exists a point y on h, collinear with c. If there were a further point y' on h collinear with c, c would be collinear with two and hence with all points on h. This is a contradiction to the fact that c is neither collinear with a nor with b.

Assume that the point y is on a line l through c and x. Then, the points a and b would be collinear with x and y and hence with all points on l, in particular with c, in contradiction to the construction of c.

Step 3. The points x and y are collinear: Since y is collinear with the points a and b on g, the point y is collinear with all points on g, in particular with x.

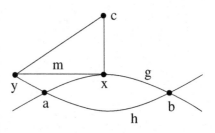

Step 4. Let m be a line through x and y. Then, the line m is contained in $a^{\perp} \cap c^{\perp}$: The points x and y are collinear with a and c, it follows that all points of m are collinear with a and c, that is, the line m is contained in $a^{\perp} \cap c^{\perp}$.

Step 5. Let u be a point of $a^{\perp}$. If u is collinear with x or y, u is collinear with all points on m: W.l.o.g., let u and x be collinear. Then, u is collinear with a and x and hence with all points on g. In particular, u and b are collinear. Thus, u is collinear

---

[6]Observe that the existence of the point x is due to the fact that polar spaces are by definition set geometries.

with a and b and hence with all points on h. In particular, u and y are collinear. Since u is collinear with x and y, u is collinear with all points on m.

Step 6. We shall see that the assumption that there are two lines g and h through the points a and b, yields a contradiction:

Since the points a and c are non-collinear, $S' := a^\perp \cap c^\perp$ is, by Theorem 2.7, a nondegenerate polar space. By Step 4, the line m is contained in S'. By Theorem 2.8, there exists a point u of S' which is not on m such that u is collinear with x, but with no further point of m.

By Step 5, u is collinear with all points of m, a contradiction.

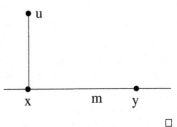

$\square$

2.10 Theorem. *Let S be a polar space, and let M be a maximal subspace of S. Let H and H' be two projective hyperplanes of M. For every point x of H not contained in $H \cap H'$, we have $H = \langle x, H \cap H' \rangle$.*

Proof. Obviously, $\langle x, H \cap H' \rangle$ is contained in H. Conversely, let z be a point of H. Since H' is a projective hyperplane, the line zx meets H' in a point y. Since z and x are contained in H, the point y is also contained in H, hence y is contained in $H \cap H'$. It follows that

$$z \in xy \subseteq \langle x, H \cap H' \rangle.$$

$\square$

2.11 Theorem. *Let S be a nondegenerate polar space, and let M be a maximal subspace of S. Let x_1 and x_2 be two points of S outside of M such that $H_1 := x_1^\perp \cap M$ and $H_2 := x_2^\perp \cap M$ are two (distinct) projective hyperplanes of M. Let x be a point of M outside of $H_1 \cap H_2$. Then, there exists a projective hyperplane of M through $H_1 \cap H_2$ and x.*

Proof.

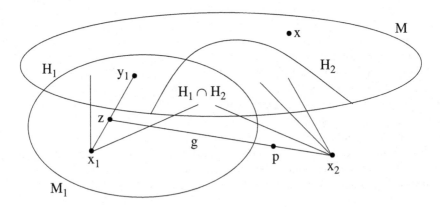

Step 1. By Theorem 2.5, the subspace $M_1 := \langle x_1, H_1 \rangle$ is a maximal subspace of S. Since H_1 and H_2 are two different hyperplanes of M, there exists a point y_1 of H_1 not contained in $H_1 \cap H_2$. The line $x_1 y_1$ is disjoint to the subspace $H_1 \cap H_2$.

By Axiom (P_1), there exists a point z on the line $x_1 y_1$ which is collinear with the point x_2. The points x_2 and z are different since, otherwise, we would have $H_2 = x_2^\perp \cap M = z^\perp \cap M = H_1$, a contradiction.

Step 2. Denote by g the line through z and x_2. There is a point p on g which is collinear with the point x. The point p is not contained in M: Assume that p is contained in M. Then, the point p is contained in

$$\left(x_2^\perp \cap M\right) \cap \left(z^\perp \cap M\right) = \left(x_2^\perp \cap M\right) \cap \left(x_1^\perp \cap M\right) = H_1 \cap H_2.$$

This implies that the point x_2 is contained in the subspace M_1, since the line zp is contained in M_1; a contradiction.

Step 3. The subspace $H_1 \cap H_2$ is contained in $p^\perp \cap M$, hence $p^\perp \cap M$ is a projective hyperplane of M through $H_1 \cap H_2$ and x: For, let r be a point of $H_1 \cap H_2$. Since r is collinear with the points z and x_2, the point r is collinear with all points on the line $g = x_2 z$. In particular, r is collinear with p. $\qquad \square$

2.12 Theorem. *Let S be a nondegenerate polar space, and let M be a maximal subspace of S. Let x be a point of M, and let g be a line of M such that x and g are not incident. There exists a hyperplane H of M such that g is contained in H, but x is not contained in H.*

Proof. Let $E := \langle x, g \rangle$ be the subspace of M generated by x and g. Let a and b be two arbitrary points on g, and let h be the line of E through the points a and x. By Theorem 2.8, there exists a point x' of S which is collinear with the point a but not with the point x.

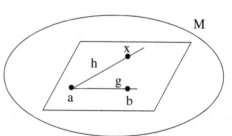

Let $H_1 := x'^\perp \cap M$. In view of Theorem 2.5, H_1 is a projective hyperplane of M containing the point a but not the point x.

If the line g is contained in H_1, the theorem is proven. So we may assume that the line g meets the projective hyperplane H_1 in the point a. The line xb meets H_1 in a point y. Since $H_1 \cap g = a$, we have $y \neq b$. Since the point x is not contained in H_1, we also have $y \neq x$. By Theorem 2.8, there is a point y' of S which is collinear with a but not with y (consider the line through a and y). Let $H_2 := y'^\perp \cap M$. Since the point y is contained in H_1 but not in H_2, the hyperplanes H_1 and H_2 are different. By Theorem 2.11, there exists a projective hyperplane H through $H_1 \cap H_2$ and the point b. Since the point a is contained in $H_1 \cap H_2$, the line $g = ab$ is contained in H.

The point x is not contained in H: Assume on the contrary that x is contained in H. Then, the line xb is contained in H, hence the point y is contained in H. Since the point y is contained in H_1 but not in $H_1 \cap H_2$, it follows from Theorem 2.10 that

$$H_1 = \langle H_1 \cap H_2, y \rangle \subseteq H.$$

Since H and H_1 are projective hyperplanes of M, it follows that $H_1 = H$ in contradiction to the fact that the point x is contained in H but not in H_1. $\qquad \square$

2.13 Theorem. *Let S be a nondegenerate polar space, and let M be a maximal subspace of S. Let x be a point of M, and let g be a line of M such that x and g are not incident. Let H be a projective hyperplane of S such that the line g is contained in H, but the point x is not contained in H. If $E := \langle x, g \rangle$ is the subspace of M generated by x and g, it follows that $H \cap E = g$.*

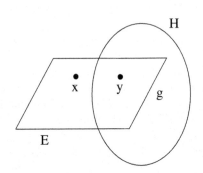

Proof. Assume that there exists a point y of $H \cap E$ such that y is not incident with g. By Theorem 2.12, there exists a projective hyperplane H_1 of M through g not containing y.

The point x is not contained in H_1: Assume that x is contained in H_1. Then, the plane $E = \langle x, g \rangle$ is contained in H_1, implying that the point y is contained in H_1 which yields a contradiction.

Since the point y is contained in H but not in H_1, we have $H \neq H_1$. By Theorem 2.11, there exists a projective hyperplane H_2 through $H \cap H_1$ and x. Since the point x is contained in H_2 but neither in H or H_1, it follows that $H_2 \neq H$ and $H_2 \neq H_1$. Since x and g are contained in H_2, the whole plane E is contained in H_2, hence the point y is contained in H_2. It follows from Theorem 2.10 that

$$H = \langle H \cap H_1, y \rangle = H_2,$$

a contradiction.   □

2.14 Theorem (**Buekenhout**). *Let S be a nondegenerate polar space. Every subspace of S containing at least two lines is a projective space.*

Proof. Let U be a subspace of S containing at least two lines. By Theorem 2.2, the subspace U is contained in a maximal subspace M of S. We shall show that M is a projective space.

Verification of (PS_1): By Theorem 2.9, every two points of M are incident with exactly one line of M.

Verification of (PS_2): Let p, a, b, x, y be five points of M such that the lines ab and xy intersect in the point p. We shall show that the lines ax and by intersect in a point: For, let E be the plane of M generated by the points p, a and x.

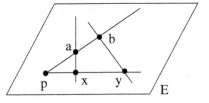

By Theorem 2.12, there exists a projective hyperplane H containing the line ax but not the point p. By Theorem 2.13, we have $H \cap E = ax$. The line by is not contained in H since, otherwise, H would contain the point p. Since H is a projective hyperplane of M, the line by meets H in a point z. Obviously, the point z is contained in the plane E. It follows that

$$ax \cap by = (H \cap E) \cap by = E \cap (H \cap by) = z.$$

Verification of (PS_3): By assumption, M contains at least two lines and every line of M contains at least three points.   □

Remark. The Theorem of Buekenhout–Shult (Theorem 2.18) for polar spaces of finite rank was proven in 1973 [21], whereas Theorem 2.14 was proven only in 1990 by Buekenhout [16] using the ideas of [21] and Teirlinck [48]. Independently, Theorem 2.14 was also proven by Johnson [32] and Percsy [40] also using [21] and [48].

From now on, we will concentrate on polar spaces of finite rank.

2.15 Theorem. *Let S be a nondegenerate polar space of finite rank. If U is a subspace of S, there exists a maximal subspace of S disjoint to U.*

Proof. By Theorem 2.2, there exists a maximal subspace M. If $M \cap U = \varnothing$, the assertion is shown. Therefore, we may assume that $M \cap U \neq \varnothing$. In what follows we shall construct a maximal subspace M' such that $M' \cap U$ is a proper subspace of $M \cap U$. Since S is of finite rank, it follows by successive application of this construction that there is a maximal subspace disjoint to U.

Step 1. By Axiom (P_3), there is a point z of $M \cap U$ and a point p of S which are non-collinear.

Step 2. Construction of the maximal subspace M': By Theorem 2.5, the subspace $p^\perp \cap (M \cap U)$ is a projective hyperplane of $M \cap U$. Again by Theorem 2.5, the subspace $M' := \langle p, \ p^\perp \cap M \rangle$ is a maximal subspace of S.

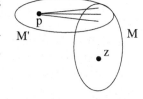

Step 3. We have $M = \langle p^\perp \cap M, z \rangle$: Since $p^\perp \cap M$ is a projective hyperplane of M, by Theorem 2.4, $p^\perp \cap M$ is a maximal subspace of M. Since p and z are non-collinear, it follows that z is contained in $M \setminus (p^\perp \cap M)$. Hence, $M = \langle p^\perp \cap M, z \rangle$.

Step 4. $M' \cap U$ is contained in $M \cap U$:

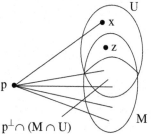

Assume that there is a point x of $M' \cap U$ which is not contained in $M \cap U$. Then, the points of $M \cap U$, the points of $p^\perp \cap M$ and the point x are pairwise collinear: For, if a is contained in $M \cap U$ and b is contained in $p^\perp \cap M$, a and b are contained in M, hence, they are collinear.

If a is contained in $M \cap U$, x and a are contained in U and therefore collinear. If, finally, b is contained in $p^\perp \cap M$, x and b both are contained in M' and are collinear.

Let $W := \langle M \cap U, \ p^\perp \cap M, \ x \rangle$ be the subspace generated by $M \cap U$, $p^\perp \cap M$ and x. By construction, $p^\perp \cap M$ and the points x and z are contained in W. By Step 3, it follows that $M = \langle p^\perp \cap M, z \rangle$ is contained in W. Furthermore, x is a point in W which is not in M. Hence, M is a proper subspace of W, in contradiction to the maximality of M.

Step 5. The subspace $M' \cap U$ is a proper subspace of $M \cap U$: By Step 4, $M' \cap U$ is contained in $M \cap U$. Furthermore, by Step 1, the points z contained in $M \cap U$ and p are non-collinear. It follows that z is not contained in M'. In summary, $M' \cap U$ is a proper subspace of $M \cap U$. $\square$

2.16 Theorem. *Let S be a nondegenerate polar space of finite rank, and let U and W be two subspaces of S such that U is contained in W. Then, there exists a maximal subspace M of S such that $M \cap W = U$.*

Proof. We shall prove the assertion by induction on $r := \operatorname{rank} U$. If $U = \emptyset$, the assertion follows from Theorem 2.15.

$r - 1 \to r$: Let $U \neq \emptyset$. By Theorem 2.5, U contains a projective hyperplane H. Since rank $H \leq \operatorname{rank} U - 1 = r - 1$, by induction, there exists a maximal subspace M' such that $M' \cap W = H$.

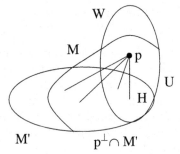

Let p be a point of $U \backslash H$. Then, by Theorem 2.5, $M := \langle p, p^\perp \cap M' \rangle$ is a maximal subspace of S containing the subspace $U = \langle p, H \rangle$. It follows that $M \cap W$ contains U.

It remains to show that $M \cap W$ is contained in U: For, let x be a point of $M \cap W$. Since p and x are contained in W, there exists the line $g := px$ through p and x. By Theorem 2.5, $M = \langle p, p^\perp \cap M' \rangle$ consists of the points on the lines through p which have at least one point in common with $p^\perp \cap M'$. Since x is contained in M, there is a point a of $p^\perp \cap M'$ on $g = px$. Since $g = px$ is contained in W, the point a is contained in W.

Altogether, the point a is contained in $(p^\perp \cap M') \cap W \subseteq M' \cap W = H$. In particular, a is contained in U. Since p is contained in U, it follows that $g = pa$ is contained in U. Since x is a point on g, the point x is contained in U. $\square$

2.17 Theorem. *Let S be a nondegenerate polar space of finite rank, and let U be a subspace of S. There exist two maximal subspaces M_1 and M_2 such that $U = M_1 \cap M_2$.*

Proof. By Theorem 2.2, there exists a maximal subspace M_1 of S through U. By Theorem 2.16,[7] there exists a maximal subspace M_2 with $M_1 \cap M_2 = U$. $\square$

Definition. Let S be a polar space. For a subspace U of S, we denote by dim U the **dimension** of U as a projective space.

As for projective geometries, for a subspace U of a polar space of finite rank, we have rank $U = \dim U + 1$.

2.18 Theorem (Buekenhout, Shult). *Let S be a nondegenerate polar space of finite rank n.*

---

[7]Choose $U = U$ and $M_1 = W$.

(a) *Every maximal subspace of S containing at least two lines is an $(n-1)$-dimensional projective space. In particular, every subspace of S is a projective space of dimension d with $2 \leq d \leq n-1$.*

(b) *The intersection of any two subspaces is a subspace.*

(c) *Let U be a maximal subspace of S, and let p be a point of S outside of U. Then, there exists a uniquely determined subspace W through p with $\dim(W \cap U) = n - 2$. Furthermore, the points of U collinear with p are exactly the points of $W \cap U$.*

(d) *There are two disjoint maximal subspaces of S.*

Proof. (a) Step 1. Every maximal subspace of S is a projective space: This assertion follows from Theorem 2.14.

Step 2. There is a maximal subspace U of S with $\dim U = n - 1$. For any further (maximal) subspace W of S, we have: $\dim W \leq n - 1$: The assertion follows from the assumption that rank $S = n$.

Step 3. For any maximal subspace W of S, we have $\dim W = n - 1$: By Step 2, there exists a maximal subspace U of S with $\dim U = n - 1$. Assume that there is a maximal subspace W with $\dim W \neq n-1$. It follows from Step 2 that $\dim W < n - 1$.

(i) There is a point p contained in $W \setminus U$: Otherwise, W is contained in U. Since W is maximal, it follows that $W = U$, in contradiction to $n - 1 = \dim U = \dim W < n - 1$.

(ii) We have $\dim(W \cap U) < n-2$: Otherwise, we have $\dim W \geq \dim\langle p, W \cap U \rangle = n - 2 + 1 = n - 1$.

(iii) There is a maximal subspace M such that $\dim M = n - 1$ and such that $\dim (W \cap M) > \dim (W \cap U)$:

Let $U_p := p^{\perp} \cap U$. Then, by Theorem 2.5, U_p is a projective hyperplane of U. In particular, we have $\dim U_p = n - 2$. Furthermore, $M := \langle p, U_p \rangle$ is a maximal subspace with $\dim M = n-1$. Finally, $M \cap W$ contains $U \cap W$: For, let x be a point contained in $U \cap W$. Then, x is contained in W.

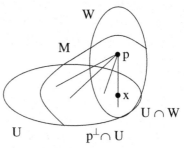

In particular, x and p are collinear. Since x is contained in U, the point x is contained in $p^{\perp} \cap U = U_p \subseteq \langle p, U_p \rangle = M$.

Since p is contained in $M \cap W$ and since p is not contained in $U \cap W$, the subspace $U \cap W$ is properly contained in $M \cap W$.

(iv) We shall see that the assumption $\dim W < n-1$ yields a contradiction: By successive application of Step (iii), we obtain a sequence $M_1, M_2, \ldots$ of maximal subspaces with $\dim M_i = n - 1$ and $\dim (W \cap M_i) > \dim(W \cap M_{i-1})$.

It follows that there exists an index r such that dim $M_r = n-1$ and dim $(W \cap M_r) = n-2$. It follows from dim $W < n-1$ that W is contained in M_r, in contradiction to the maximality of W.

(b) The assertion follows from Theorem 2.1.

(c) Let U be a maximal subspace of S, and let p be a point of S outside of U.

   (i) Existence of a subspace W through p with dim $(W \cap U) = n-2$: By (a), we have dim $U = n-1$. By Theorem 2.5, $p^{\perp} \cap U$ is a projective hyperplane of U, that is, dim $(p^{\perp} \cap U) = n-2$. Finally, by Theorem 2.5, the subspace $W := \langle p, p^{\perp} \cap U \rangle$ is maximal. It follows that dim $W = n-1$.

   (ii) Uniqueness of a subspace W through p with dim $(W \cap U) = n-2$: Let $W := \langle p, p^{\perp} \cap U \rangle$, and let X be a further subspace through p with dim $(X \cap U) = n-2$. Since the points of X are collinear with p, it follows that $X \cap U$ is contained in $p^{\perp} \cap U$. Since dim $(X \cap U) = n-2 = $ dim$(p^{\perp} \cap U)$, it follows that $X \cap U = p^{\perp} \cap U$. Furthermore, the point p is contained in X implying that the subspace $W = \langle p, p^{\perp} \cap U \rangle$ is contained in X. Due to the maximality of W, it follows that $W = X$.

   (iii) By construction, we have $U \cap W = p^{\perp} \cap U$. It follows that $U \cap W$ consists exactly of the points of U collinear with p.

(d) The assertion follows from Theorem 2.17 with $U := \emptyset$. $\square$

Tits [50] introduced polar spaces as geometries, fulfilling the properties (a) to (d) of Theorem 2.18. Only by the Theorem of Buekenhout–Shult has the definition of polar spaces assumed its current form. The Theorem of Buekenhout–Shult states that a polar space in the sense of Buekenhout–Shult is a polar space in the sense of Tits. The converse direction is the content of the following theorem.

2.19 Theorem. *Let $\Gamma = (X, *, type)$ be a set geometry[8] of rank n over the type set {point, line, ..., hyperplane[9]} fulfilling the following properties:*

 (i) *Let H be a hyperplane of Γ. Then, the residue $\Gamma|_H$ is an $(n-1)$-dimensional projective space.*

 (ii) *For any two subspaces U and W of Γ, the set $U \cap W$ is a subspace.*

 (iii) *Let H be a hyperplane, and let p be a point of Γ outside of H. Then, there exists exactly one hyperplane L of Γ through p with dim $(H \cap L) = n-2$. L consists of all points of H which are collinear with p.*

 (iv) *There are two disjoint hyperplanes of Γ.*

Then, the points and lines of Γ form a nondegenerate polar space.

Proof. Verification of (P_1): Let g be a line, and let x be a point which is not on g.

---

[8]Remember that a set geometry is a geometry where all subspaces can be seen as subsets of the set of points.

[9]Note that we have two sorts of hyperplanes: A hyperplane of Γ is a subspace of type $n-1$ of the geometry Γ. As indicated in Part (i), this hyperplane is a projective space. A *projective* hyperplane is a hyperplane of this projective space, hence a subspace of type $n-2$ of Γ.

First case. The point x and the line g are contained in a common hyperplane H: Then, x and g are a point and a line in a projective space, it follows that x is collinear with all points on g.

Second case. The point x and the line g are not contained in any common hyperplane: Since Γ is a geometry, g is contained in a chamber. In particular, g is contained in a hyperplane H.

By assumption, x and g are not contained in a common hyperplane, hence, x is not contained in H. By assumption (iii), there exists a hyperplane L through x with dim $(H \cap L) = n - 2$. If g were contained in L, x and g would be contained in the common hyperplane L, a contradiction. Since $H \cap L$ is a hyperplane of H, the line g meets $H \cap L$ in a point y. Since the points x and y are contained in L, they are collinear.

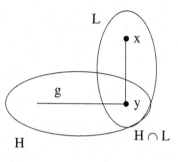

Since, by assumption (iii), $H \cap L$ consists of all points of H collinear with x, the point y is the only point on g collinear with x.

Verification of (P_2): Let g be a line. Since Γ is a geometry, g is contained in a chamber and therefore in particular in a hyperplane H. Thus, g is a line of a projective space. Therefore, g contains at least three points.

Verification of (P_3): Let x be a point of Γ. Since there are two disjoint hyperplanes, there exists a hyperplane H not containing x. Let L be the hyperplane of Γ through x with dim $(H \cap L) = n - 2$. Since, by assumption (iii), L contains all points of H that are collinear with x, for any point y of $L \backslash (H \cap L)$, it holds that x and y are non-collinear. $\qquad\square$

We conclude this section by proving the following relation between polar spaces of rank 2 and generalized quadrangles.

2.20 Theorem. *The generalized quadrangles are exactly the generalized nondegenerate polar spaces of rank 2.*

Proof. Step 1. Let Q be a generalized quadrangle. We shall show that Q is a generalized nondegenerate polar space of rank 2:

Axiom (P_1) follows from Axiom (V_2).
Axiom (P_2') follows from Axiom (V_3).

Verification of (P_3): Let x be a point of Q. By Axiom (V_3), there is a line g through x. Since there are at least two points on g, there exists a point y on g different from x. Since there are at least two lines through y, there is a line h through y different from g.

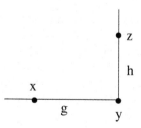

By Axiom (V_1), the point x is not on h. Let z be a point on h distinct from y. Since y is the only point on h collinear with x, the points x and z are non-collinear.

Altogether, Q is a generalized nondegenerate polar space. Since there are no non-incident point-line pairs (x, g) such that p is collinear with all points on g, it follows that Q is of rank 2.

Step 2. Conversely, let S be a generalized nondegenerate polar space of rank 2. We shall show that S is a generalized quadrangle:

Axiom (V_1) follows from Theorem 2.9.

Verification of (V_2): Let g be a line, and let x be a point outside of g. By Axiom (P_1), x is collinear with exactly one or with all points of g. Assume that x is collinear with all points of g. Then, by Theorem 2.1, x and g would generate a plane, in contradiction to rank $S = 2$.

Verification of (V_3): By Axiom (P'_2), on every line, there are at least two points. Let x be a point of S. Since S is of rank 2, there is at least one line g in S.

First case. The point x is not on g: Then, there exists a line h through x intersecting g in exactly one point y. By (P_3), there exists a point z that is non-collinear with y. It follows that z is not on h, and there is a line l through z intersecting the line g in a point $a \neq y$. If x were on l, the point x would be collinear with two points of g, namely a and y, a contradiction.

Since the point x is not incident with the line l, there exists a line m through x intersecting the line l in a point b. Assuming $m = h$ the point b would be incident with the two lines h and l which intersect the line g in the points y and a. Hence, b would be collinear with two points of g, a contradiction.

It follows that x is incident with the two lines m and h.

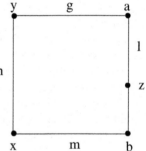

Second case. The point x is on g: By Axiom (P_3), there exists a point y which is non-collinear with x. Let h be a line through y intersecting the line g in a point z. If $z = x$, x is incident with the two lines g and h. Otherwise, h is a line which is not incident with x, and we can apply Case 1.   □

3   The Diagram of a Polar Space

The idea of the diagram of a geometry is to summarize classes of rank-2-geometries in a pictogram. The diagram of a geometry of rank $n \geq 3$ is determined by the diagrams of its rank-2-residues (Sect. 7 of Chap. 1).

In Sect. 7 of Chap. 1, we have already introduced some diagrams of rank-2-geometries. In the present section, we shall introduce the following diagrams for generalized quadrangles and for the 4-gon (geometry of the ordinary quadrangle):

$$\circ\!\!=\!\!=\!\!=\!\!\circ \text{ or } \circ\!\!-\!\!\overset{4}{-}\!\!-\!\!\circ$$

Altogether we obtain the following list of diagrams of rank-2-geometries:

Rank-2-geometry	Diagram
Arbitrary rank-2-geometry	
Generalized digon	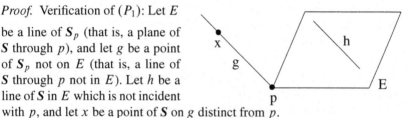
Linear space	
Affine plane	
Projective plane or 3-gon	
Generalized quadrangle or 4-gon	

3.1 Theorem. *Let S be a nondegenerate polar space of rank $n \geq 3$. Then, for any point p of S, the residue S_p is a nondegenerate polar space of rank $n-1$.*

Proof. Verification of (P_1): Let E
be a line of S_p (that is, a plane of
S through p), and let g be a point
of S_p not on E (that is, a line of
S through p not in E). Let h be a
line of S in E which is not incident
with p, and let x be a point of S on g distinct from p.

First case. The point x is collinear with all points on h: Then, the set $\{z \in S \mid z \in h\} \cup \{x\} \cup \{p\}$ consists of pairwise collinear points. By Theorem 2.1, this set is contained in a common subspace. Therefore, in S_p every point of E is collinear with g.

Second case. The point x is collinear with exactly one point on h: Let z be this point. Then, the line pz of S is the only point of S_p on E collinear with g.

Verification of (P_2): Let E be a line of S_p. In S, E is a plane through p. Since E is a projective plane, there are at least three lines of S through p in E. These three lines are three points on E in S_p.

Verification of (P_3): Let g be a point of S_p,
that is, a line of S through p. By Theorem 2.8,
there exists a point x of S such that x is
collinear with p, but non-collinear with any
further point of g. Let $h := px$ be the line
through x and p.

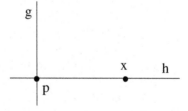

Since the lines g and h are not contained in a common plane, the points g and h of S_p are not collinear in S_p. □

We recall that, for a geometry Γ with a linear diagram, we always denote the types of Γ by point, line, plane, ..., hyperplane.

3.2 Theorem. *Let S be a nondegenerate polar space of rank $n \geq 2$, and let Γ be the geometry defined by the subspaces of S.*

(a) Γ *is a set geometry.*
(b) Γ *is residually connected.*
(c) Γ *has the diagram*

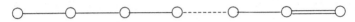

Proof. (a) is obvious.

(b) We shall prove the assertion by induction on n.

Step 1. Let $n = 2$. We need to show that generalized quadrangles are connected. For, let a and b be two elements of a generalized quadrangle. We exemplarily consider the case that a and b are two points. By Axiom (V_3), the point b is incident with a line g.

If the point a is on g, a-g-b is a path from a to b.

Otherwise, by Axiom (V_2), there exists a line h through a intersecting g in a point c. It follows that a-h-c-g-b is a path from a to b.

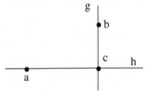

Step 2. Let $n \geq 3$. Let F be a flag such that the residue Γ_F is of rank at least 2. Let U and V be two elements of Γ_F.

First case. The flag F contains a point x. Then, the elements U and V are contained in the residue Γ_x. By Theorem 3.1, the residue Γ_x is a polar space of rank n−1. By induction, Γ_x is residually connected, hence, in Γ_F, there is a path from U to V.

Second case. The flag F contains a hyperplane H. Since H is a projective space, by Theorem 7.9 of Chap. 1, Γ_H is residually connected, hence there is a path from U to V in Γ_F.

Third case. The flag F neither contains a point nor a hyperplane. Then, U and V are either two points or two hyperplanes or a point and a hyperplane. Let W be an arbitrary element of F.

If U and V are two points, let H be a hyperplane of Γ contained in Γ_F. It follows that the subspace W is incident with U, V and H. Hence, U and V are contained in H, it follows that U-H-V is a path from U to V in Γ_F.

If U and V are two hyperplanes, let x be a point of Γ contained in Γ_F. It follows that the subspace W is incident with U, V and x. Hence, x is contained in U and V, it follows that U-x-V is a path from U to V in Γ_F.

If U and V are a point x and a hyperplane H, it follows that the subspace W is incident with x and H. Thus, x and H are incident, that is, x-H is a path from U to V in Γ_F.

(c) Again, we shall prove the assertion by induction on n.

Step 1. Let $n = 2$. By Theorem 2.20, Γ is a generalized quadrangle. By definition, it has the diagram

Step 2. Let $n \geq 3$. We shall apply Theorem 7.5 of Chap. 1: Let x be a point of Γ. Then, by Theorem 3.1, the residue Γ_x is a polar space. By induction, Γ_x has the diagram

Let H be a hyperplane of Γ. Then, the residue Γ_H is an $(n-1)$-dimensional projective space and has, by Theorem 7.6 of Chap. 1, the diagram

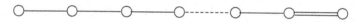

If F is a flag of type $\{1, 2, \ldots, n-2\}$, the residue Γ_F consists of points and hyperplanes such that every point is incident with every hyperplane, it follows that Γ_F has the diagram

It follows from Theorem 7.5 of Chap. 1 that Γ has the diagram

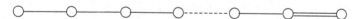

□

By Theorem 7.15 of Chap. 1, every residually connected geometry Γ with the diagram

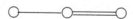

and the property that on each line of Γ, there are at least three points, is a projective geometry.

For polar spaces, one could guess that every residually connected geometry with the diagram

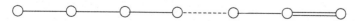

and the property that on each line of Γ, there are at least three points, is a polar space. In fact, the situation for polar spaces is more complicated: There exists a geometry Γ with diagram

such that on each line of Γ, there are at least three points, whose points and lines do not form a polar space (Theorem 3.11).

In order to obtain a polar space from a geometry Γ with diagram two further conditions are necessary. This fact motivates the following definition of a polar geometry.

Definition. Let $\Gamma = (X, *, \text{type})$ be a residually connected geometry of rank $n \geq 2$ with the diagram

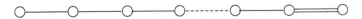

Then, Γ is called a **polar geometry of rank** n if Γ fulfils the following conditions:

(PoG$_1$) Any two points of Γ are incident with at most one line.
(PoG$_2$) For any two elements U and V of Γ with $U \neq V$, we have:

$$\{x \text{ point of } \Gamma \mid x * U\} \neq \{x \text{ point of } \Gamma \mid x * V\}.$$

(PoG$_3$) Any line of Γ is incident with at least three points.

3.3 Theorem. *Let S be a nondegenerate polar space of rank n. Then, the subspaces of S together with the incidence relation induced by S form a polar geometry Γ of rank n.*

Proof. By Theorem 3.2, Γ defines a residually connected geometry with the diagram

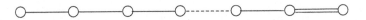

The properties (PoG$_1$) and (PoG$_3$) follow of Theorem 2.9 and from Axiom (P_2), respectively.

Property (PoG$_2$) follows from the fact that Γ is a set geometry.   □

3.4 Theorem. *Let $\Gamma = (X, *, \text{type})$ be a polar geometry of rank $n \geq 3$, and let p be a point of Γ. Then, the residue Γ_p is a polar geometry of rank $n-1$.*

Proof. Since Γ is residually connected, Γ_p is residually connected as well. By Theorem 7.5 of Chap. 1, the geometry Γ_p has the diagram

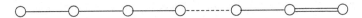

Verification of (PoG$_1$): Let g and h be two points of Γ_p, that is, two lines of Γ through p. Assume that there are two planes E and F through g and h. Since Γ has the diagram

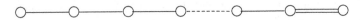

E and F are projective planes.

By (PoG$_2$), there exists a point x in E, not contained in F.[10] Let l be a line of E through x which is not incident with p. Since E is a projective plane, the two lines l and g meet in a point a, and the two lines l and h meet in a point b. In the (projective) plane F there is a line m through the two points a and b. Since the point x is on the line l, but not in the plane F, we have $l \neq m$.

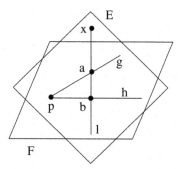

Hence, the points a and b are incident with two distinct lines l and m, in contradiction to (PoG$_1$).

Verification of (PoG$_2$): Let U and V be two subspaces of Γ_p, that is, two subspaces of Γ through p. By assumption (PoG$_2$), there exists a point x in U not contained in V. Since U is a projective space, there exists a line g in U through p and x. Since x is not contained in V, the line g is not contained in V. Thus, g is a point of Γ_p, contained in U, but not in V.

Verification of (PoG$_3$): The assertion follows from the fact that in a projective plane, any point is incident with at least three lines.   □

3.5 Theorem. *Let $\Gamma = (X, *, \text{type})$ be a polar geometry of rank $n \geq 2$.*

(a) The points and lines of Γ form a nondegenerate polar space S.
(b) The subspaces of S correspond to the elements of Γ.

Proof. We shall prove the assertions (a) and (b) by induction on n.

For $n = 2$, the assertion follows from the definition of the diagram ○═══○ .
Let $n \geq 3$.

Step 1. Let H be a hyperplane of Γ. Then, the residue Γ_H is an $(n-1)$-dimensional projective space: By Theorem 7.5 of Chap. 1, Γ_H is a residually connected geometry with the diagram

By Theorem 7.15 of Chap. 1, Γ_H is an $(n-1)$-dimensional projective space.

Step 2. Let x be a point of Γ, and let H be a hyperplane of Γ, not containing x. Then, there exists a hyperplane L and an $(n-2)$-dimensional subspace U of Γ such that x is contained in L and such that U is contained in L and in H:

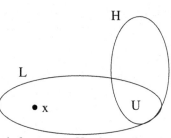

By Theorem 7.10 of Chap. 1, there exists a path from x to H where all nodes except the first node is of type $n-2$ or $n-1$. Let $W : x = X_0, X_1, X_2, \ldots, X_r = H$ be such a path of minimal length with dim $X_1 = n - 1$. We shall show that $r = 3$:

---

[10]Or a point which is in F but not in E.

Assume that $r > 3$. Since W is a path of minimal length, the subpath $x = X_0$, X_1, X_2, X_3, X_4, X_5 is a path of minimal length from x to X_5. Set $H_1 := X_1$, $H_2 := X_3$, $H_3 := X_5$, $U_1 := X_2$ and $U_2 := X_4$. Then, H_1, H_2 and H_3 are hyperplanes, and U_1 and U_2 are subspaces of dimension $n-2$.

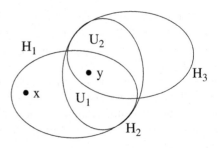

Since U_1 and U_2 both are incident with H_2 and since H_2 is an $(n-1)$-dimensional projective space (in view of Step 1), it follows that

$$\dim (U_1 \cap U_2) = \dim U_1 + \dim U_2 - \dim\langle U_1, U_2 \rangle$$
$$= n - 2 + n - 2 - (n - 1) = n - 3 \geq 0.$$

It follows that there exists a point y in $U_1 \cap U_2$. Since the points x and y both are contained in the hyperplane H_1, there exists a line g through x and y. By Theorem 3.4, the residue Γ_y is a polar geometry of rank $n-1$. It follows that the points and lines of Γ_y define a polar space S_y whose subspaces are by induction (Part (b)) exactly the subspaces of Γ_y.

In S_y, g is a point and H_3 is a hyperplane. By Theorem 2.5, there exists in S_y a hyperplane L through g such that $L \cap H_3$ is a hyperplane of H_3.[11] In Γ, L is a hyperplane through $g = xy$ such that $\dim (L \cap H_3) = n - 2$. It follows that x-L-$L \cap H_3$-H_3 is a path of length 3 from x to H_3, in contradiction to the minimality of the path $x = X_0$-X_1-X_2-X_3-X_4-$X_5 = H_3$.

Step 3. Let x be a point of Γ, and let g be a line of Γ not incident with x. There exists at least one point on g collinear with x: If x and g are in a common hyperplane, x is collinear with all points on g. Therefore, we may assume that x and g are not contained in a common hyperplane.

Let H be a hyperplane of Γ through g. By Step 2, there exists a hyperplane L and an $(n-2)$-dimensional subspace U of Γ such that x is contained in L and such that U is contained in L and in H.

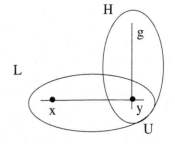

Since U is a hyperplane of H, the line g meets U in a point y.[12] In L, there exist a line through x and y.

---

[11]Choose $L := \langle g, g^{\perp} \cap H_3 \rangle$.

[12]The line g is not contained in U since otherwise x and g would be contained in the hyperplane L.

Step 4. Let x be a point of Γ, and let g be a line of Γ not incident with x. Let a and b be two points on g collinear with x. Then, the point x and the line g are contained in a common plane of Γ:

Let E be a plane through g. W.l.o.g., we may suppose that x is a point outside of E. By assumption, there exist a line h through x and a and a line l through x and b. In Γ_a, h is a point and E is a line not containing h.

Since the points and lines of the residue Γ_a form by induction a polar space, there exists in Γ_a a line F_a through h intersecting the line E in a point h'. In Γ, the line F_a is a plane through h which has a line h' in common with the plane E. Similarly, there exists a plane F_b through the line l intersecting the plane E in a line l'.

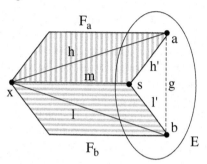

Since the lines h' and l' both are contained in the (projective) plane E, they meet in a point s. Since the points x and s both are in the planes F_a and F_b, there exists a line m_a in F_a and a line m_b in F_b through x and s. By assumption (PoG1), through any two points of Γ there is at most one line, it follows that $m := m_a = m_b$, that is, the two planes F_a and F_b have the line m in common.

In the residue Γ_s, the point m is joined with the point h' by the line F_a, and the point m is joined with the point l' by the line F_b. Since Γ_s is by induction a polar space, there exists a subspace V in Γ_s through the points m, h' and l'.[13] In Γ, V is either a 2-dimensional or a 3-dimensional subspace through the lines m, h' and l'. In particular, the point x and the line g are contained in V.[14] It follows that the point x and the line g are contained in a common plane.

Step 5. Verification of (P_1): Let x be a point of Γ, and let g be a line of Γ not containing x. By Step 3, there exists at least one point on g collinear with x. By Step 4, x is either collinear with exactly one or with all points on g.

Step 6. Verification of (P_2): Axiom (P_2) follows from Axiom (PoG3).

Step 7. Verification of (P_3): Let x be a point of Γ. Let g be a line through x, and let p be a point on g different from x. By induction, the residue Γ_p is a nondegenerate polar space. It follows that there exists in Γ_p a point h such that g and h are two non-collinear points. In Γ, g and h are two lines through p not contained in any common plane.

Let y be a point on h distinct from p. Assume that x and y are collinear.

---

[13]More precisely, the points and lines of Γ_s define a polar space S. In S, there exists a subspace V through the points m, h' and l'. By induction (applied on Part (b)), V is a subspace of Γ_s.

[14]The point x is on m, hence it is contained in V. By (PoG_1), there exists at most one line through the points a and b. The line through a and b in V is therefore the line g.

Then, the point x is collinear with the two points p and y on h, and by Step 4, x and h are contained in a common plane E.

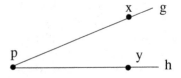

By (PoG$_1$), the line $g = px$ is contained in the (projective) plane E. This contradicts the fact that the lines g and h are not contained in any common plane. It follows that x and y are non-collinear.

Step 8. Verification of (b): By Steps 5 to 7, the points and lines of Γ form a polar space S of rank n. For a subspace V of Γ, we denote by V' the set of points incident with V (if V is a point, set $V' := V$). We shall show that to every subspace U of S there is an element V of Γ such that $U = V'$, and conversely that to every subspace V of Γ, the set V' is a subspace of S. Note that it follows from (PoG$_2$) that for any two subspaces V, W of Γ with $V \neq W$, we have $V' \neq W'$.

(i) Let E be a plane of S. There exists a subspace F of Γ such that $E = F'$: For, let x be a point of E and let g be a line of E which is not incident with x. By Step 4, there exists a plane F of Γ through x and g. Since E and F' are two projective planes through x and g, it follows from (PoG$_1$) that $E = F'$.

(ii) Let F be a plane of Γ. Then, F' is a plane of S: Since F is a projective plane consisting of points and lines of S, it follows that F' is a plane of S.

(iii) Let U be a subspace of S. There exists a subspace V of Γ such that $U = V'$: Let x be a point of U. It follows from (i) and (ii) that the points and lines of Γ_x (lines and planes of Γ through x) are the points and lines of the polar space S_x. By induction, there exists a subspace V of Γ_x such that $U = V'$ (in S_x). It follows that $U = V'$ in S.

(iv) Let V be a subspace of Γ. Then, V' is a subspace of S: In Γ_x, V' is by induction a subspace of S_x. It follows that V' is a subspace of S.

□

In order to construct a geometry Γ (the so-called Neumaier-geometry) with the diagram

$$\circ\!\!-\!\!\!-\!\!\!-\!\!\!-\!\!\circ\!\!=\!\!=\!\!=\!\!\circ$$

which is not a polar space, we first shall introduce the construction of the smallest generalized quadrangle. Afterwards, we shall prove some results about the projective plane of order 2. The construction of the Neumaier-geometry will be the subject of Theorem 3.11.

3.6 Theorem. *Let $M := \{1, 2, 3, 4, 5, 6\}$, and let Γ be the geometry of rank 2 defined as follows:*

(i) The points of Γ are the subsets of M with exactly two elements.

(ii) The lines of Γ are the partitions of M consisting of three subsets of M each with two elements.

(iii) A point $\{x, y\}$ of Γ is on a line $\{\{a,b\}, \{c,d\}, \{e,f\}\}$ of Γ if $\{x,y\}$ is contained in the set $\{\{a,b\}, \{c,d\}, \{e,f\}\}$.

Γ *is a generalized quadrangle. On any line of Γ, there are exactly three points. Through any point of Γ, there are exactly three lines. Γ has 15 points and 15 lines.*

Proof. Obviously, there are exactly three points on any line of Γ and through any point of Γ there are exactly three lines. Furthermore, Γ has exactly 15 points and 15 lines.

Let $\{a, b\}$ be a point of Γ, and let g be a line of Γ which is not incident with x. Then, g is of the form $\{\{a,c\}, \{b,d\}, \{e,f\}\}$ with $\{a, b, c, d, e, f\}=\{1, 2, 3, 4, 5, 6\}$. The line $\{\{a, b\}, \{c,d\}, \{e,f\}\}$ is the only line through $x = \{a, b\}$ intersecting g in a point, namely in the point $\{e, f\}$.   □

3.7 Lemma. *Every projective plane of order 2 is Desarguesian.*

Proof. Let K be the field of two elements, and let V be the 3-dimensional vector space over K. Since every 2-dimensional subspace of V has exactly three 1-dimensional subspaces, the projective plane $PG(2, K)$ is a finite projective plane of order 2. Since, by Theorem 8.7 of Chap. 1, there is exactly one projective plane of order 2 (up to isomorphism), any projective plane of order 2 is Desarguesian.   □

3.8 Lemma. *Let P be the projective plane of order 2, and let G be the full automorphism group of P.*

(a) The group G consists of 168 elements.
(b) The group G is generated by the elations[15] of P.

Proof. In view of Theorem 8.7 of Chap. 1, every projective plane of order 2 is of the form

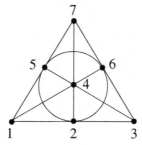

(a) Step 1. Let α be a collineation of P. The collineation α is uniquely determined by its definition on three non-collinear points: For, let x_1, x_2 and x_3 be three non-collinear points, and let α and β be two collineations of P with $\alpha(x_i) = \beta(x_i)$ for $i = 1, 2, 3$. W.l.o.g., let $x_1 = 1$, $x_2 = 3$, and $x_3 = 7$.[16]

---

[15]Remember that an elation is a central collineation where the centre and the axis are incident.
[16]Since the automorphism group of P operates transitively on the frames of P (Theorem 4.6 of Chap. 3), it is no restriction to choose $x_1 = 1$, $x_2 = 3$ and $x_3 = 7$.

If $\gamma := \beta^{-1}\alpha$, it follows that $\gamma(1) = 1$, $\gamma(3) = 3$, and $\gamma(7) = 7$. Hence $\gamma(2) = 2$, $\gamma(5) = 5$, and $\gamma(6) = 6$. Finally, it follows that $\gamma(4) = 4$. It follows from $\gamma = id$ that $\alpha = \beta$.

Step 2. The group G consists of 168 elements: In view of Step 1 and Theorem 4.6 of Chap. 3, the number of elements of G equals the number of ordered triangles of P. It follows that $|G| = 7 \times 6 \times 4 = 168$.

(b) We shall show that every collineation α of P is the product of at most three elations: Let x_1, x_2 and x_3 be three non-collinear points of P, and let $y_i := \alpha(x_i)$ for $i = 1, 2, 3$. In view of Step 1 of the proof of Part (a), it suffices to construct three elations σ_1, σ_2 and σ_3 such that $\sigma_3(\sigma_2(\sigma_1(x_i))) = y_i$ for $i = 1$, 2, 3.

Step 1. Construction of the first elation σ_1:

If $x_1 = y_1$, let $\sigma_1 := id$.

If $x_1 \neq y_1$, let z_1 be the third point on the line $x_1 y_1$. Let l be a line through z_1 different from the line $x_1 y_1$, and let σ_1 be the elation with centre z_1 and axis l such that $\sigma_1(x_1) = y_1$.[17] Let $x'_2 := \sigma_1(x_2)$ and $x'_3 := \sigma_1(x_3)$.

Step 2. Construction of the second elation σ_2:

If $x'_2 = y_2$, let $\sigma_2 := id$.

If $x'_2 \neq y_2$, let z_2 be the third point on the line $x'_2 y_2$. Since the points x_1, x_2 and x_3 are non-collinear, the points $y_1 = \sigma_1(x_1)$, $x'_2 = \sigma_1(x_2)$ and $x'_3 = \sigma_1(x_3)$ are also non-collinear. In particular, we have $x'_2 \neq y_1$.

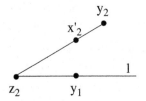

In addition, we have $y_1 \neq y_2$. Hence, there is a line l through z_2 and y_1 different from the line $x'_2 y_2$ (such a line also exists if $y_1 = z_2$). Let σ_2 be the elation with centre z_2 and axis l such that $\sigma_2(x'_2) = y_2$. We have $\sigma_2(\sigma_1(x_i)) = y_i$ for $i = 1, 2$. Let $x''_3 := \sigma_2(x'_3)$.

Step 3. Construction of the third elation σ_3:

If $x''_3 = y_3$, let $\sigma_3 := id$.

If $x''_3 \neq y_3$, let z_3 be the third point on the line $x''_3 y_3$. Since $y_1 = \sigma_2(\sigma_1(x_1))$, $y_2 = \sigma_2(\sigma_1(x_2))$, $x''_3 = \sigma_2(\sigma_1(x_3))$ and y_1, y_2, y_3 are two sets of non-collinear points, the lines $x''_3 y_3$ and $y_1 y_2$ meet in a point $z_3 \neq x''_3$, y_3.

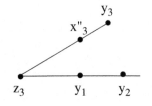

Let σ_3 be the elation with centre z_3 and axis $y_1 y_2$ such that $\sigma_3(x''_3) = y_3$. We have $\sigma_3(\sigma_2(\sigma_1(x_i))) = y_i$ for $i = 1, 2, 3$.   □

In the following, we shall denote by S_n the symmetric group on n elements and by A_n the alternating group on n elements.

---

[17]This central collineation exists in view of Theorem 6.2 of Chap. 2 and Lemma 3.7.

3.9 Lemma. *Let **P** be the projective plane of order 2, and let G be the full automorphism group of **P**. The group G is a subgroup of the group A_7.*

Proof. Let σ be an arbitrary elation of **P** with centre z and axis l.

With a suitable numbering, we have $z = 1$ and $l = \{1, 2, 3\}$. Hence $\sigma = (4\ 6)(5\ 7)$. Since σ is the product of two transpositions, σ is an element of A_7. Since G is generated by elations (Lemma 3.8), G is a subgroup of A_7.   □

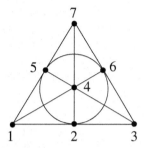

3.10 Lemma. *Let $X := \{1, 2, 3, 4, 5, 6, 7\}$.*

(a) There are 30 distinct ways to endow the set X with lines (as subsets of $\{1, 2, 3, 4, 5, 6, 7\}$) such that a projective plane of order 2 is constructed.

(b) Under the operation of the alternating group A_7 the 30 planes constructed in (a) split into two orbits each containing 15 planes.

Proof. (a) Let **P** be a projective plane with the point set $X = \{1, 2, 3, 4, 5, 6, 7\}$ where the lines of **P** are subsets of X. Altogether, there exist $|S_7| = 7!$ permutations on the set X. By Lemma 3.8, the full automorphism group of **P** consists of 168 elements. It follows that there exist $7!/168 = 30$ projective planes of order 2 on the point set X (any two of them are isomorphic).

(b) Since the automorphism group of **P** is a subgroup of A_7 (Lemma 3.9) and since $|S_7 : A_7| = 2$, the 30 planes constructed in (a) split up into two orbits each containing 15 planes under the operation of the group A_7.   □

Definition. Let $X := \{1, 2, 3, 4, 5, 6, 7\}$, and let Γ be the geometry defined as follows:

(i)   The points of Γ are the elements of X.

(ii) The lines of Γ are the subsets of X containing three elements.

(iii) The planes of Γ are the 15 projective planes of an orbit under the operation of the group A_7 of Lemma 3.10.

(iv) A point x of X and a line g are incident if x is contained in g.

(v)   Every point of Γ and every plane of Γ are incident.

(vi) A line g of Γ and a plane E of Γ are incident if g is a line of E.

Γ is called the **Neumaier-geometry** [35].

3.11 Theorem (Neumaier). *Let Γ be the Neumaier-geometry.*

(a) Γ is a residually connected geometry with the diagram

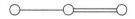

(b) The geometry Γ cannot be constructed as the set of the points, lines and planes of a polar space of rank 3.

Proof. (a) Obviously, Γ is a connected geometry.

Step 1. If E is a plane of Γ, the residue Γ_E is by construction a projective plane of order 2. In particular, Γ_E is connected.

Step 2. If g is a line of Γ, the residue Γ_g is a generalized digon since already in Γ, every point is incident with every plane. In particular, Γ_g is connected.

Step 3. Let x be a point of Γ. The residue Γ_x is a generalized quadrangle:

  (i) For, let $M := X \setminus \{x\}$. Every point of Γ_x (that is, every line of Γ through x) is a subset of X of cardinality 3 containing x. Conversely, every subset of X of cardinality 3 containing x is a point of Γ_x. The points of Γ_x can be understood as the subsets of $M = X \setminus \{x\}$ of cardinality 2.

  (ii) Let E be a line of Γ_x (that is, a plane of Γ through x). Then, the three lines of E through x define a partition of M.

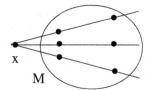

  (iii) Conversely, let $\{\{a,b\}, \{c,d\}, \{e,f\}\}$ be a partition of $M = X \setminus \{x\}$. Then, there are exactly two ways in which the lines $\{x,a,b\}$, $\{x,c,d\}$ and $\{x,e,f\}$ can be continued to a projective plane of order 2:

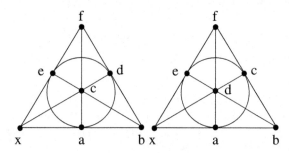

  Since the first plane can be transformed into the second plane by the permutation (c,d), the two planes are in different orbits under the group A_7. It follows that only one the two planes is a plane of Γ. Therefore, the partitions of $M = X \setminus \{x\}$ each consisting of three subsets of cardinality 2 of M can be identified with the set of planes of Γ through x, that is, with the set of lines of Γ_x.

  (iv) By Theorem 3.6, Γ_x is a generalized quadrangle. In particular, Γ_x is connected.

  (b) Any two points of Γ are incident with five lines. In view of Theorem 2.9, Γ cannot stem from a polar space.   □

4 Polarities

In the present section, we shall investigate the polarities of a projective space and show that every polarity defines a polar space (Theorem 4.7).

For that purpose, we need the notion of dual projective spaces.

Definition. Let P be a projective space, and let $P^* = (X^*, *, \text{type}^*)$ be the geometry over the type set {point, line} defined as follows:

(i) The points of P^* are the subspaces of P of codimension 1. The lines of P^* are the subspaces of P of codimension 2.

(ii) Two elements U and W of P^* are incident in P^* if they are incident in P.

P^* is called the **dual space of P**.

4.1 Theorem. *Let P be a projective space, and let P^* be the dual space of P.*

(a) The dual space P^* of P is a projective space.

(b) We have $\dim(P^*) = \dim(P)$.

(c) Let P be of finite dimension d. For every subspace U of P, we have

$$\dim_P *(U) = d - 1 - \dim_P(U).$$

Proof. (a) We shall verify Axioms (PS_1) to (PS_3).

Verification of (PS_1): **Linearity**: Since any two hyperplanes of P meet in exactly one subspace of codimension 2, any two points of P^* are incident with exactly one line.

Verification of (PS_2): **Veblen–Young**: Let H_x, H_y, H_a, H_b and H_p be five distinct hyperplanes of P such that $H_x \cap H_y \subseteq H_p$ and $H_a \cap H_b \subseteq H_p$ and $H_x \cap H_y \neq H_a \cap H_b$.[18]

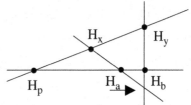

Since $H_x \cap H_y$ and $H_a \cap H_b$ are both hyperplanes of H_p, they meet in a subspace U of H_p of codimension 2, that is, in a subspace U of P of codimension 3. It follows that there are two points r of $H_x \cap H_a$ and s of $H_y \cap H_b$ such that

$$H_x \cap H_a = \langle U, r \rangle \text{ and } H_y \cap H_b = \langle U, s \rangle.$$

Hence, $\langle (H_x \cap H_a), (H_y \cap H_b) \rangle = \langle U, r, s \rangle$ is a subspace of codimension 1. It follows that the lines $H_x \cap H_a$ and $H_y \cap H_b$ meet in P^* in the point $\langle (H_x \cap H_a), (H_y \cap H_b) \rangle$.

Verification of (PS_3): Let U be a subspace of P of codimension 2, and let g be line of P disjoint to U. Let x, y and z be three different points on g. The subspaces

---

[18] The condition $H_x \cap H_y \neq H_a \cap H_b$ says that in P^*, the lines $H_x \cap H_y$ and $H_a \cap H_b$ are distinct.

$\langle U, x \rangle$, $\langle U, y \rangle$ and $\langle U, z \rangle$ are three hyperplanes of P through U, that is, on every line of P^*, there are at least three points.

(b) and (c) are easy to verify.   □

We shall turn to the study of dualities and polarities. Remember that a duality δ of a projective geometry P has been defined as an autocorrelation of P such that δ induces on the type set of P a permutation of order 2. A polarity of P is a duality of P of order 2. We shall extend the definition of a duality and of a polarity to projective spaces of arbitrary dimension:

Definition. Let P be a projective space, and let P^* be the dual space of P.

(a) A collineation $\alpha : P \to P^*$ from P onto P^* is called a **duality** of P.
(b) A **polarity** of P is a duality of P of order 2.

4.2 Theorem. *Let P be a projective space, and let δ be a bijective transformation of the set of points of P onto the set of hyperplanes of P. The following conditions are equivalent:*

 (i) The transformation δ is a duality of P.
(ii) Three points x, y and z are collinear if and only if the hyperplanes $\delta(x)$, $\delta(y)$ and $\delta(z)$ meet in a common subspace of codimension 2, that is, if

$$\delta(x) \cap \delta(y) = \delta(y) \cap \delta(z) = \delta(x) \cap \delta(z) = \delta(x) \cap \delta(y) \cap \delta(z).$$

Proof. The proof follows from the definition of a duality.   □

4.3 Theorem. *Let P be a d-dimensional projective space, and let δ be a duality of P. Then, for any subspace U of P, we have the relation $\dim \delta(U) = d - 1 - \dim U$.*

Proof. By Theorem 7.5 of Chap. 1, P has the diagram

By definition, a duality δ induces a permutation δ_I of order 2 on the type set $I :=$ $\{0, 1, \ldots, d - 1\}$. Since, by Theorem 2.3 of Chap. 2, δ_I fixes the diagram of P, it follows that $\delta_I(j) = d - 1 - j$, for $j = 0, \ldots, d - 1$.   □

4.4 Theorem. *Let P be a projective space, and let δ be a duality of P. For any subspace U of P, we have*

$$\delta(U) = \bigcap_{x \in U} \delta(x).$$

Proof. Let U be a subspace of P. Since $\delta : P \to P^*$ is a collineation, the set $M := \{\delta(x) \mid x \in U\}$ is a subspace of P^*. It follows that

$$\delta(U) = M = \langle \delta(x) \mid x \in U \rangle_{P*} = \bigcap_{x \in U} \delta(x).$$

                                                                          □

4.5 Theorem. *Let **P** be a projective space, and let $id \neq \pi$ be a bijective transformation of the set of points into the set of hyperplanes of **P**. The following two statements are equivalent:*

(i) The transformation π is a polarity.
*(ii) For any two points x and y of **P**, the relation "$x \in \pi(y)$" implies the relation "$y \in \pi(x)$".*

Proof. (i) $\Rightarrow$ (ii): Let π be a polarity, and let x and y be two points of **P** such that x is contained in $\pi(y)$. Let $*$ be the incidence relation of the geometry defined by **P**. Since π is a polarity, it follows from $x^*\pi(y)$ that $\pi(x)^*\pi(\pi(y))$. It follows from $\pi(\pi(y)) = \pi^2(y) = y$ that $\pi(x)^*$ y, that is, y is contained in $\pi(x)$.

(ii) $\Rightarrow$ (i): Step 1. π is a duality: We shall apply Theorem 4.2:

(α) Let a, b and c be three collinear points of **P**, and let x be a point of $\pi(a) \cap \pi(b)$. By assumption (ii), it follows that a and b are contained in $\pi(x)$. Since $\pi(x)$ is a hyperplane of **P**, the line ab is contained in $\pi(x)$, in particular, the point c is contained in $\pi(x)$. Again by assumption (ii), it follows that x is contained in $\pi(c)$.

Altogether, $\pi(a) \cap \pi(b)$ is contained in $\pi(c)$. Due to the symmetry of the points a, b and c, it follows that

$$\pi(a) \cap \pi(b) = \pi(b) \cap \pi(c) = \pi(a) \cap \pi(c).$$

Thus, $\pi(a) \cap \pi(b) \cap \pi(c)$ is a subspace of codimension 2 of **P**.

(β) Conversely, let a, b and c be three points of **P** such that the subspace $\pi(a) \cap \pi(b) \cap \pi(c)$ is of codimension 2.

Let x be a point of $\pi(c) \setminus (\pi(a) \cap \pi(b) \cap \pi(c))$, and let $c' := \pi(x) \cap ab$ be the intersection point of the hyperplane $\pi(x)$ with the line ab.

From (α) follows that $\pi(a) \cap \pi(b)$ is contained in $\pi(c')$. Since c' is contained in $\pi(x)$, by assumption (ii), the point x is contained in $\pi(c')$. It follows that $\pi(c') = \langle \pi(a) \cap \pi(b), x \rangle = \pi(c)$.

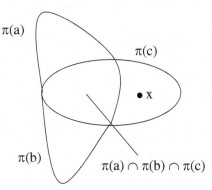

Since π is bijective, it follows that $c = c'$. In particular, the points a, b and c are collinear.

Step 2. π is of order 2: Let x be a point of **P**, and let $y := \pi^2(x)$. Let z be a point of $\pi(x)$. Since π is a duality (Step 1), $\pi(z)$ contains $\pi(\pi(x)) = y$. It follows that z is contained in $\pi(y)$. Since z is an arbitrary point of $\pi(x)$, it follows that $\pi(x)$ is contained in $\pi(y)$ and therefore $\pi(x) = \pi(y)$. Since π is bijective, we have $x = y$, hence, $\pi^2(x) = x$. Since, by assumption, $id \neq \pi$, it follows that π is of the order 2. $\square$

Definition. Let P be a projective space, and let π be a polarity of P.

(a) A point x of P is called **absolute with respect to** π if the point x is contained in the hyperplane $\pi(x)$.
(b) A subspace U of P is called **absolute with respect to** π if dim $U \leq$ dim $\pi(U)$ and if U is contained in $\pi(U)$.[19]

4.6 Theorem. *Let P be a projective space, and let π be a polarity of P.*

(a) If U is an absolute subspace of P, every point of U is absolute.
(b) Let x and y be two absolute points of P. Then, the line xy through x and y is absolute if and only if y is contained in $\pi(x)$.[20]

Proof. (a) For any point z of U, we have $z \in U \subseteq \pi(U) \subseteq \pi(z)$.
(b) (i) Let $g := xy$ be an absolute line. By Theorem 4.4, we have $y \in g \subseteq \pi(g) \subseteq \pi(x)$.
    (ii) Let x and y be two absolute points of P such that y is contained in $\pi(x)$. Since π is a polarity, the point x is contained in $\pi(y)$. Since x and y are absolute, it follows that x is contained in $\pi(x)$ and that y is contained in $\pi(y)$. It follows that $g = xy$ is contained in $\pi(x) \cap \pi(y)$.

    Let z be a point on g different from x and y. Since π is a polarity, it follows from Theorem 4.2 that $\pi(x) \cap \pi(y) \subseteq \pi(z)$. It follows that $\pi(g) = \bigcap_{z \in g} \pi(z) \supseteq \pi(x) \cap \pi(y) \supseteq g$. Hence, g is absolute. □

Remark. In general, the converse of Theorem 4.6 (a) is not true. There are polarities π such that all points of a line g are absolute, without g being absolute with respect to π.

4.7 Theorem. *Let P be a projective space, and let π be a polarity of P such that there exists at least one absolute line with respect to π. The absolute points and the absolute lines with respect to π define a polar space.*

Proof. We shall verify Axioms (P_1) and (P_2):

Verification of (P_1): Let g be an absolute line of P, and let x be an absolute point of P not on g. Let y be an arbitrary point on g. By Theorem 4.6, the line xy is absolute if and only if y is contained in $\pi(x)$.

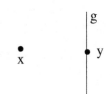

Since $\pi(x)$ is a hyperplane of P, either the line g and the hyperplane $\pi(x)$ meet in a point, or g is contained in $\pi(x)$.

---

[19]For an absolute subspace U of a d-dimensional projective space P, we always have dim $U \leq 1/2$ $(d-1)$.
[20]Since π is a polarity, it follows that x is also contained in $\pi(y)$.

It follows that either there exists exactly one absolute line through x intersecting the line g, or all lines through x intersecting the line g are absolute.

Verification of (P_2): The fact that on every line, there are at least three points follows from the fact that on every line of P, there are at least three points. □

Definition. Let P be a projective space, and let π be a polarity of P such that there exists at least one absolute line with respect to π. Furthermore, let S_π be the geometry over the type set {point, line} defined as follows:

(i) The points and lines of S_π are exactly the absolute points and absolute lines with respect to π.

(ii) A point and a line of S_π are incident in S_π if they are incident in P.

S_π is called the **polar space defined by** π.

4.8 Theorem. *Let P be a projective space, let π be a polarity of P, and let S_π be the polar space defined by π. We have $x^\perp = \pi(x) \cap S_\pi$ for all points x of S_π.*[21]

Proof. By definition, we have

$$x^\perp = \{z \in S_\pi \mid x \text{ and } z \text{ are collinear}\}$$
$$= \{z \in S_\pi \mid x \text{ and } z \text{ are joined by an absolute line}\} \cup \{x\}$$
$$= \{z \in S_\pi \mid z \in \pi(x)\} \cup \{x\} \quad \text{(by Theorem 4.6)}$$
$$= \pi(x) \cap S_\pi.$$

□

Remark. Let P be a projective space, and let π be a polarity of P. Often, the notation $x^\perp := \pi(x)$ is used for all points x of P. In view of Theorem 4.8, this definition is in accordance with the definition of $x^\perp$ for a point x of the polar space S_π.

5 Sesquilinear Forms

Dualities can be described algebraically by so-called sesquilinear forms (cf. Theorem 5.6). Polarities can be described algebraically by so-called reflexive sesquilinear forms (cf. Theorem 5.7). In the present section, we shall investigate the relation between dualities, polarities and sesquilinear forms.

In Sect. 6, we shall introduce pseudo-quadrics and pseudo-quadratic forms which are strongly related to sesquilinear forms. We will follow the outline of Tits [50]. Since Tits considers vector spaces in [50] as right vector spaces and since we want to make it easier for the reader to follow the exposition of Tits, all vector spaces in the present chapter are right vector spaces.

---

[21]Note that $\pi(x)$ is defined for all points of P, whereas $x^\perp$ is only defined for the points x of S_π (absolute points of P).

Definition. Let K be a skew field. A transformation $\sigma : K \to K$ is called an **anti-automorphism** if σ fulfils the following two conditions:

(i) σ is bijective.
(ii) We have $\sigma(\lambda + \mu) = \sigma(\lambda) + \sigma(\mu)$ for all λ, μ of K.
(iii) We have $\sigma(\lambda\mu) = \sigma(\mu)\sigma(\lambda)$ for all λ, μ of K.

5.1 Example. (a) If K is a commutative field, every automorphism is an anti-automorphism.

(b) Let Q be the quaternion skew field, that is,

$$Q = \{a + ib + jc + kd \mid a,\, b,\, c,\, d \in \mathbf{R}\}$$

with $i^2 = j^2 = k^2 = -1$ and $ij = k$, $jk = i$, $ki = j$. Then, the transformation $\sigma : Q \to Q$ with

$$\sigma(a + ib + jc + kd) := a - ib - jc - kd$$

is an anti-automorphism.

Definition. Let V be a right vector space over a skew field K, and let $\sigma : K \to K$ be an anti-automorphism.

(a) A σ-**sesquilinear form** is a transformation $f : V \times V \to K$ fulfilling the following conditions:

    (i) We have $f(v_1 + v_2,\, w_1 + w_2) = f(v_1,\, w_1) + f(v_1,\, w_2) + f(v_2,\, w_1) + f(v_2,\, w_2)$ for all v_1, v_2, w_1, w_2 of V.
    (ii) We have $f(v\,\lambda,\, w\,\mu) = \sigma(\lambda)\, f(v,\, w)\, \mu$ for all λ, μ of K and v, w of V.

(b) A σ-sesquilinear form $f : V \times V \to K$ is called **nondegenerate** if for any vector $0 \neq v$ of V, there exists a vector w such that $f(v, w) \neq 0$ and if for any vector $0 \neq w'$ of V, there exists a vector v' such that $f(v', w') \neq 0$.

(c) A σ-sesquilinear form $f : V \times V \to K$ is called a **bilinear form** if the field K is commutative and $\sigma = id$, that is, $f(v\,\lambda,\, w\,\mu) = \lambda\, f(a,\, b)\, \mu$ for all λ, μ of K and all v, w of V.

(d) A bilinear form $f : V \times V \to K$ is called **symmetric** if $f(v, w) = f(w, v)$ for all v, w of V.

Definition. Let P be a projective space over a right vector space V, and let P^* be the dual space of P. Let $\delta : P \to P^*$ be a duality. Let $f : V \times V \to K$ be a σ-sesquilinear form such that

$$\delta(\langle v \rangle) = \{w \in V \mid f(v,\, w) = 0\} \text{ for all } v \text{ of } V.$$

Then, the duality δ is called the **duality of P induced by** f. It is denoted by δ_f.

5.2 Theorem. *Let P be a d-dimensional projective space over a vector space V, and let δ be a duality of P. Furthermore, suppose that there exists a σ-sesquilinear form $f : V \times V \to K$ such that $\delta = \delta_f$.*

For any two points $x = \langle v \rangle$ *and* $y = \langle w \rangle$ *of* **P**, *where* $0 \neq v$, w *are two elements of* **V**, *the point* y *is contained in* $\delta(x)$ *if and only if* $f(v, w) = 0$.

Proof. The proof follows from the fact that $\delta(x) = \{w \in V \mid f(v, w) = 0\}$.   □

In Theorem 5.6, we shall see that there exists for any duality δ a σ-sesquilinear form f such that $\delta = \delta_f$. For the proof of this assertion, we shall make use of the following information about opposite fields and dual vector spaces:

Definition. Let K be a skew field. The **opposite field** K^* **of** K is defined as follows:

(i) The elements of K^* are the elements of K.
(ii) The addition in K^* is the addition in K.
(iii) For two elements λ, μ of K^*, let $\lambda^* \mu := \mu \lambda$. The operation $*$ is the multiplication in K^*.

5.3 Theorem. *Let* K *be a skew field. The opposite field* K^* *of* K *is also a skew field.*

Proof. The proof is obvious.   □

Definition. Let V be a right vector space over a skew field K. The dual right vector space V^* of V over the opposite field K^* of K is defined as follows:

- The elements of V^* are the linear transformations $\varphi : V \to K$.
- For two elements φ_1, φ_2 of V^*, let $(\varphi_1 + \varphi_2)(v) := \varphi_1(v) + \varphi_2(v)$.
- For an element φ of V^* and an element λ of K, let $(\varphi\lambda)(v) := \lambda\varphi(v)$.[22]

5.4 Theorem. *Let* V *be a right vector space over a skew field* K. *Then, the dual vector space* V^* *of* V *is a right vector space over the opposite field* K^*.

Proof. The proof is obvious. Note that

$$(\varphi(\lambda * \mu))(v) = (\lambda * \mu)\varphi(v) = (\mu\lambda)\varphi(v) = \mu(\lambda\varphi)(v) = \mu(\varphi\lambda)(v) = ((\varphi\lambda)\mu)(v),$$

hence $\varphi(\lambda * \mu) = (\varphi\lambda)\mu$.   □

5.5 Theorem. *Let* **P** *be a projective space over a right vector space* V. *Let* **P*** *be the dual space of* **P**, *and let* V^* *be the dual space of* V. *Then,* **P*** *and* **P**(V^*) *are isomorphic.*

Proof. Let $\alpha : P(V^*) \to P^*$ be defined as follows: For an element $0 \neq \varphi$ of V^*, set

$$\alpha(\varphi) := \text{Kern } \varphi.$$

Step 1. $\alpha : P(V^*) \to P^*$ is well-defined: For, let $0 \neq \varphi$ be an element of V^*, and let $0 \neq \lambda$ be an element of K^*. Set $\psi := \varphi\lambda$. Obviously, $\varphi(v) = 0$ if and only if $\psi(v) = (\varphi\lambda)(v) = \lambda\varphi(v) = 0$. Hence $\alpha(\varphi) = \text{Kern } \varphi = \text{Kern } \psi = \alpha(\psi)$.

Step 2. $\alpha : P(V^*) \to P^*$ is injective: For, let $0 \neq \varphi$ and $0 \neq \psi$ be two elements of V^* with $\alpha(\varphi) = \text{Kern } \varphi = \text{Kern } \psi = \alpha(\psi)$. Let $\{v_0\} \cup \{v_i \mid i \in I\}$ be a

---

[22]Note that $\varphi(v)$ is an element of K. Hence, the expression $\lambda \varphi(v)$ is well-defined.

basis of V such that $\{v_i \mid i \in I\}$ is a basis of Kern φ = Kern ψ. It follows that $\varphi(v_i) = \psi(v_i) = 0$ for all i of I. Since $0 \neq \varphi$ and $0 \neq \psi$, we have $\varphi(v_0) \neq 0$, and $\psi(v_0) \neq 0$. Since $\varphi(v_0)$ and $\psi(v_0)$ are elements of K, there exists an element λ of K such that $\varphi(v_0) = \lambda\,\psi(v_0)$. It follows that $\psi\,\lambda = \varphi$. Hence, $\langle\varphi\rangle = \langle\psi\rangle$.

Step 3. $\alpha : P(V^*) \to P^*$ is surjective: Let H be a hyperplane of P. There exists a hyperplane W of V such that $H = P(W)$. Let $\{v_0\} \cup \{v_i \mid i \in I\}$ be a basis of V such that $\{v_i \mid i \in I\}$ is a basis of W. Define the transformation $\varphi : V \to K$ by $\varphi(v_0) = 1$ and $\varphi(v_i) = 0$ for all i of I. It follows that $\alpha(\varphi) = $ Kern$\varphi = W$.

Step 4. $\alpha : P(V^*) \to P^*$ is a collineation:

(i) Let $\langle\varphi\rangle$, $\langle\psi\rangle$ and $\langle\chi\rangle$ be three collinear elements of $P(V^*)$. W.l.o.g., there exist two elements λ and μ of K such that $\varphi = \psi\lambda + \chi\mu$. Let v be an element of Kern $\psi \cap$ Kern χ. It follows that

$$\varphi(v) = (\psi\lambda + \chi\mu)(v) = (\psi\lambda)(v) + (\chi\mu)(v) = \lambda\psi(v) + \mu\chi(v) = 0.$$

Hence, Kern $\psi \cap$ Kern$\chi \subseteq$ Kern φ. It follows that $\alpha(\varphi)$, $\alpha(\psi)$ and $\alpha(\chi)$ are collinear in P^*.

(ii) Let $\alpha(\varphi)$, $\alpha(\psi)$ and $\alpha(\chi)$ be three collinear elements of P^*. W.l.o.g., let Kern $\psi \cap$ Kern χ be contained in Kern φ. Let $\{v_0\} \cup \{v_1\} \cup \{v_i \mid i \in I\}$ be a basis of V such that $\{v_i \mid i \in I\}$ is a basis of Kern $\psi \cap$ Kern χ, such that $\{v_0\} \cup \{v_i \mid i \in I\}$ is a basis of Kern ψ and such that $\{v_1\} \cup \{v_i \mid i \in I\}$ is a basis of Kern χ.

Let $b := \psi(v_1)$ and $c := \chi(v_0)$. It follows that $b \neq 0$ and $c \neq 0$. Let $a_0 := \varphi(v_0)$, and let $a_1 := \varphi(v_1)$. Finally, set $\lambda := a_1\,b^{-1}$ and $\mu := a_0\,c^{-1}$. It follows that

$$(\psi\lambda + \chi\mu)(v_0) = (\psi\lambda)(v_0) + (\chi\mu)(v_0) = \lambda\psi(v_0) + \mu\chi(v_0) = 0 + \mu\,c = a_0 = \varphi(v_0)$$
$$(\psi\lambda + \chi\mu)(v_1) = (\psi\lambda)(v_1) + (\chi\mu)(v_1) = \lambda\psi(v_1) + \mu\chi(v_1) = \lambda b + 0 = a_1 = \varphi(v_1)$$
$$(\psi\lambda + \chi\mu)(v_i) = (\psi\lambda)(v_i) + (\chi\mu)(v_i) = \lambda\psi(v_i) + \mu\chi(v_i) = 0 \text{ for all } i \text{ of I.}$$

Hence, $\varphi = \psi\lambda + \chi\mu$, that is φ, ψ and χ are collinear. $\qquad\square$

5.6 Theorem. *Let P be a projective space over a right vector space V, and let δ be a duality of P. There exists a σ-sesquilinear form $f : V \times V \to K$ inducing δ, that is, $\delta = \delta_f$.*

Proof. Let P^* be the dual space of P, and let V^* be the dual space of V. Note that, by Theorem 5.4, the vector space V^* is a right vector space over the opposite field K^*.

By Theorem 5.5, P^* and $P(V^*)$ are isomorphic. Since $\delta : P(V) \to P(V^*)$ is a collineation, it follows from Theorem 8.3 of Chap. 3 that there exists a semilinear transformation $A : V \to V^*$ with accompanying isomorphism $\sigma : K \to K^*$ inducing δ. Obviously, σ induces an anti-automorphism $K \to K$ which we also denote by σ.

Define the transformation $f : V \times V \to K$ by $f(v, w) := (A(v))(w)$. Then, f is a σ-sesquilinear form:

Let v, w, v_1, v_2, w_1 and w_2 be some arbitrary elements of V, and let λ and μ be two elements of K. We have

$$f(v_1 + v_2, w_1 + w_2) = (A(v_1 + v_2))(w_1 + w_2)$$
$$= (A(v_1 + v_2))(w_1) + (A(v_1 + v_2))(w_2)$$
$$= (A(v_1))(w_1) + (A(v_2))(w_1) + (A(v_1))(w_2) + (A(v_2))(w_2)$$
$$v = f(v_1, w_1) + f(v_2, w_1) + f(v_1, w_2) + f(v_2, w_2)$$
$$f(v\lambda, w\mu) = (A(v\lambda))(w\mu)$$
$$= ((A(v\,\lambda))(w))\mu$$
$$= ((A(v)\sigma(\lambda))(w))\mu$$
$$= \sigma(\lambda)(A(v)(w))\mu$$
$$= \sigma(\lambda) f(v, w)\mu.$$

Hence, $f : V \times V \to K$ is a σ-sesquilinear form. $\square$

In what follows, we shall investigate the relation between sesquilinear forms and polarities.

Definition. Let V be a right vector space over a skew field K, and let $f : V \times V \to K$ be a nondegenerate σ-sesquilinear form. f is called **reflexive** if for all elements v, w of V, the relation

$$f(v, w) = 0 \Leftrightarrow f(w, v) = 0$$

holds.

5.7 Theorem. *Let P be a projective space over a vector space V, and let δ be a duality of P. Furthermore, let $f : V \times V \to K$ be a σ-sesquilinear form such that $\delta = \delta_f$.*

(a) δ is a polarity if and only if f is reflexive.
(b) If δ is a polarity, the σ-sesquilinear form f is nondegenerate.

Proof. (a) Let δ be a polarity, and let $x = \langle v \rangle$ and $y = \langle w \rangle$ be two points of P. Then, we have

$$\delta \text{ polarity} \Leftrightarrow (x \in \delta(y) \Leftrightarrow y \in \delta(x)) \qquad \text{(Theorem 4.5)}$$
$$\Leftrightarrow (f(w, v) = 0 \Leftrightarrow f(v, w) = 0) \qquad \text{(Theorem 5.2)}$$
$$\Leftrightarrow f \text{ is reflexive.}$$

(b) Since δ is induced by f, we have $\delta(\langle v \rangle) = \{w \in V \mid f(v, w) = 0\}$.

Step 1. Let $0 \neq v_0$ be a vector of V. Since $\delta(\langle v_0 \rangle)$ is a hyperplane, we have $\{w \in V \mid f(v_0, w) = 0\} \neq P$, hence, there exists a vector w_0 of V such that $f(v_0, w_0) \neq 0$.

Step 2. Let $0 \neq w_0$ be a vector of V. In view of Step 1, there exists a vector v_0 of V such that $f(w_0, v_0) \neq 0$. Since δ is a polarity, f is reflexive, and it follows that $f(v_0, w_0) \neq 0$.

From Steps 1 and 2, it follows that f is nondegenerate. $\square$

5.8 Theorem. *Let V be a right vector space over a skew field K with $d \geq 2$, and let $f : V \times V \to K$ be a nondegenerate σ-sesquilinear form. The following two conditions are equivalent:*

(i) f is reflexive.
(ii) There is an element $0 \neq \varepsilon$ of K such that

$$f(w, v) = \sigma(f(v, w))\varepsilon \text{ for all } v, w \text{ of } K.$$

Proof. (ii) $\Rightarrow$ (i): Let $f(v, w) = 0$ for two vectors v, w of V. Then,

$$f(w, v) = \sigma(f(v, w))\varepsilon = \sigma(0)\varepsilon = 0.$$

It follows that f is reflexive.

(i) $\Rightarrow$ (ii): Let f be reflexive. Since f is reflexive and since $\sigma : K \to K$ is an anti-automorphism, we have $\{w \in V \mid f(v, w) = 0\} = \{w \in V \mid \sigma^{-1}(f(w, v)) = 0\}$.

For an element $0 \neq v$ of V, let $f_v : V \to K$ and $g_v : V \to K$ be the transformations $f_v(w) := f(v, w)$ and $g_v(w) := \sigma^{-1}(f(w, v))$. Obviously, f_v is a linear transformation. Since

$$\begin{aligned}
g_v(w\lambda) &= \sigma^{-1}(f(w\lambda, v)) \\
&= \sigma^{-1}(\sigma(\lambda)f(w, v)) \\
&= \sigma^{-1}(f(w, v))\sigma^{-1}(\sigma(\lambda)) \\
&= \sigma^{-1}(f(w, v))\lambda \\
&= g_v(w)\lambda
\end{aligned}$$

g_v is a linear transformation as well. Since f_v and g_v are both linear applications from V into K and since Kern f_v = Kern g_v, it follows that there exist elements γ_v of K, such that $g_v(w) = \gamma_v \, f_v(w)$ for all w of V.

Let $\varepsilon_v := \sigma(\gamma_v)$. Then,

$$\sigma^{-1}(f(w, v)) = g_v(w) = \gamma_v \, f_v(w) = \gamma_v \, f(v, w) \text{ for all } w \text{ of } V.$$

It follows that $f(w, v) = \sigma(\gamma_v \, f(v, w)) = \sigma(f(v, w))\sigma(\gamma_v) = \sigma(f(v, w))\varepsilon_v$ for all w of V.

Let u, v be two elements of V. We need to show that $\varepsilon_u = \varepsilon_v$.

First case. Let u and v be linearly independent. For an arbitrary vector w of V, we have:

$$f(w, u + v) = \sigma(f(u + v, w))\varepsilon_{u+v} = \sigma(f(u, w))\varepsilon_{u+v} + \sigma(f(v, w))\varepsilon_{u+v}.$$

At the same time we have

$$f(w, u + v) = f(w, u) + f(w, v) = \sigma(f(u, w))\varepsilon_u + \sigma(f(v, w))\varepsilon_v.$$

It follows that

$$\sigma(f(u,\ w))(\varepsilon_{u+v} - \varepsilon_u) = \sigma(f(v,\ w))(\varepsilon_v - \varepsilon_{u+v}).$$

Since f is nondegenerate and since u and v are linearly independent, there exists a vector w of V such that $f(u,\ w) \neq 0$ and $f(v,\ w) = 0$. It follows that $\varepsilon_{u+v} = \varepsilon_u$. Analogously, it follows that $\varepsilon_{u+v} = \varepsilon_v$, hence $\varepsilon_v = \varepsilon_u$.

Second case. Let u and v be linearly dependent. Since, by assumption, dim $V \geq 2$, there exists a vector w, linearly independent from u and v. By Case 1, it follows that $\varepsilon_v = \varepsilon_w = \varepsilon_u$.   □

5.9 Corollary. *Let P be a projective space over a right vector space V, and let π be a polarity of P.*

(a) There exists a reflexive σ-sesquilinear form f such that π is induced by f, that is, $\pi = \pi_f$.
(b) There exists an element ε of K such that $f(w,\ v) = \sigma(f(v,\ w))\varepsilon$ for all v, w of V.
(c) If f is a σ-sesquilinear form inducing π, then f is reflexive and nondegenerate.

Proof. (a) follows from the Theorems 5.6 and 5.7.
(b) follows from Theorem 5.8.
(c) follows from Theorem 5.7.   □

Definition. Let V be a right vector space over a skew field K. A nondegenerate σ-sesquilinear form $f : V \times V \to K$ is called $(\sigma,\ \varepsilon)$-**hermitian** if there exists an element ε of K such that

$$f(w,\ v) = \sigma(f(v,\ w))\varepsilon \text{ for all } v,\ w \text{ of } V.$$

By Theorem 5.8, for dim $V \geq 2$, every nondegenerate reflexive σ-sesquilinear form is $(\sigma,\ \varepsilon)$-hermitian and vice versa. In this sense, the notions reflexive and $(\sigma,\ \varepsilon)$-hermitian are equivalent.

5.10 Theorem. *Let V be a right vector space over a skew field K, and let f be a $(\sigma,\ \varepsilon)$-hermitian sesquilinear form. Then, we have*

(a) $\sigma(\varepsilon) = \varepsilon^{-1}$.
(b) $\sigma^2(\varepsilon) = \varepsilon$.
(c) $\sigma^2(\lambda) = \varepsilon \lambda \varepsilon^{-1}$ for all λ of K.

Proof. We shall prove Parts (a) to (c) in common. Let λ be an element of K. Since f is nondegenerate, there exist two elements v, w of V such that $f(v,\ w) = \lambda$. It follows that

$$\lambda = f(v,\ w) = \sigma(f(w,\ v))\varepsilon = \sigma(\sigma(f(v,\ w))\varepsilon)\varepsilon = \sigma(\varepsilon)\sigma^2(\lambda)\varepsilon, \text{ that is,}$$

$$\lambda = \sigma(\varepsilon)\sigma^2(\lambda)\varepsilon. \tag{*}$$

For $\lambda = 1$, it follows that $1 = \sigma(\varepsilon)\sigma^2(1)\varepsilon = \sigma(\varepsilon)\varepsilon$, hence, $\sigma(\varepsilon) = \varepsilon^{-1}$. This proves Part (a). Using $\sigma(\varepsilon) = \varepsilon^{-1}$, we obtain from equation (*):

$$\lambda = \varepsilon^{-1}\sigma^2(\lambda)\varepsilon, \text{ that is,}$$
$$\sigma^2(\lambda) = \varepsilon\lambda\varepsilon^{-1}.$$

This proves (b) and (c).   □

Definition. Let V be a right vector space over a skew field K, and let f be a (σ, ε)-hermitian sesquilinear form.

(a) f is called **alternating** if the following conditions are fulfilled:

  (i) K is a field.
  (ii) We have $f(w, v) = -f(v, w)$ for all v, w of V, that is, $\sigma = id$ and $\varepsilon = -1$.
  (iii) We have $f(v, v) = 0$ for all v of V.

(b) f is called **symmetric** if the following conditions are fulfilled:

  (i) K is a field.
  (ii) We have $f(w, v) = f(v, w)$ for all v, w of V, that is, $\sigma = id$ and $\varepsilon = 1$.

(c) f is called **hermitian** if the following conditions are fulfilled:

  (i) We have $\sigma^2 = id$ and $\sigma \neq id$.
  (ii) We have $f(w, v) = \sigma(f(v, w))$ for all v, w of V, that is, $\varepsilon = 1$.

(d) f is called **anti-hermitian** if the following conditions are fulfilled:

  (i) We have $\sigma^2 = id$ and $\sigma \neq id$.
  (ii) We have $f(w, v) = -\sigma(f(v, w))$ for all v, w of V, that is, $\varepsilon = -1$.

Given a polarity π of a projective space P, in general, there exists more than one reflexive σ-sesquilinear form $f : V \times V \to K$ inducing π. This is due to the fact that the polarity π is defined on the 1-dimensional subspaces of V, whereas f is defined on the elements of V (more precisely on the pairs (v, w) where v, w are two elements of V). "Proportional" sesquilinear forms define the same polarity.

  The following Theorem of Birkhoff and von Neumann says that every polarity is induced by an alternating, a symmetric, a hermitian or an anti-hermitian sesquilinear form.

5.11 Theorem (Birkhoff, von Neumann). *Let P be a projective space over a right vector space V, and let π be a polarity of P. There exists an alternating, a symmetric, a hermitian or an anti-hermitian sesquilinear form inducing the polarity π.*

Proof. By Corollary 5.9, there exists a reflexive nondegenerate σ-sesquilinear form f such that π is induced by f, that is, $\pi = \pi_f$.

  First case. Let $\sigma = id$ and let $\varepsilon = -1$. Since $\sigma = id$, K is a field. It follows from $\varepsilon = -1$ that $f(w, v) = -f(v, w)$ for all v, w of V.

  If Char $K \neq 2$, it follows from $f(v, v) = -f(v, v)$ for all v of V that $f(v, v) = 0$ for all v of V. It follows that f is alternating.

If Char $K = 2$, $f(w, v) = f(v, w)$ for all v, w of V, and f is symmetric.

Second case. Let $\varepsilon = -1$ and let $\sigma \neq id$. Since $\sigma^2(\lambda) = \varepsilon \lambda \varepsilon^{-1}$ for all λ of K (Theorem 5.10), it follows that $\sigma^2 = id$. Hence, f is anti-hermitian.

Third case. Let $\varepsilon \neq -1$. Then, $\varepsilon^{-1} \neq -1$, and it follows that $\mu := 1 + \varepsilon^{-1} \neq 0$.

In the following Steps 1–5, we will define a sesquilinear form g and show that g is symmetric or hermitian. In the final Step 6, we will show that the polarity π is induced by g.

Step 1. Let $\rho : K \to K$ be defined by $\rho(\lambda) = \mu \, \sigma(\lambda) \, \mu^{-1}$ for all λ of K. Then, ρ is an anti-automorphism of K: We have

$$
\begin{aligned}
\rho(\lambda v) &= \mu \, \sigma(\lambda v) \mu^{-1} \\
&= \mu \, \sigma(v) \sigma(\lambda) \mu^{-1} \\
&= \mu \, \sigma(v) \mu^{-1} \mu \sigma(\lambda) \mu^{-1} \\
&= \rho(v) \rho(\lambda) \text{ for all } \lambda, \, v \text{ of } K.
\end{aligned}
$$

Step 2. We define the transformation $g : V \times V \to K$ by $g(v, w) := \mu \, f(v, w)$. Then, g is a ρ-sesquilinear form: We have

$$
\begin{aligned}
g(v\lambda, w) &= \mu \, f(v \, \lambda, w) \\
&= \mu \, \sigma(\lambda) f(v, w) \\
&= \mu \, \sigma(\lambda) \, \mu^{-1} \, \mu f(v, w) \\
&= \rho(\lambda) \, g(v, w).
\end{aligned}
$$

Step 3. g is nondegenerate: Let $0 \neq v$ be an element of V. Since f is nondegenerate, there exists a vector w of V such that $f(v, w) \neq 0$. It follows from $\mu \neq 0$ that $g(v, w) = \mu \, f(v, w) \neq 0$.

Step 4. g is (ρ, δ)-hermitian with $\delta := \mu \, \sigma(\mu)^{-1} \, \varepsilon$: We have

$$
\begin{aligned}
g(w, v) &= \mu \, f(w, v) \\
&= \mu \, \sigma(f(v, w)) \varepsilon \\
&= \mu \, \mu^{-1} \mu \sigma(f(v, w)) \mu^{-1} \mu \varepsilon \\
&= \mu \, \mu^{-1} \rho(f(v, w)) \mu \varepsilon \\
&= \rho(\mu^{-1} \, g(v, w)) \, \mu \varepsilon \\
&= \rho(g(v, w)) \rho(\mu^{-1}) \mu \varepsilon \\
&= \rho(g(v, w)) \mu \sigma(\mu^{-1}) \, \mu^{-1} \mu \varepsilon \\
&= \rho(g(v, w)) \mu \sigma(\mu^{-1}) \, \varepsilon \\
&= \rho(g(v, w)) \delta \text{ for all } v, \, w \text{ of } V.
\end{aligned}
$$

Step 5. g is symmetric or hermitian: Since g is a (ρ, δ)-hermitian sesquilinear form, we need to show that $\rho^2 = id$ and $\delta = 1$: We have

$$\begin{aligned}
\sigma(\mu) &= \sigma(1 + \varepsilon^{-1}) \\
&= \sigma(1) + \sigma(\varepsilon)^{-1} \\
&= 1 + \varepsilon \qquad\qquad\qquad \text{(Theorem 5.10)} \\
&= \varepsilon\varepsilon^{-1} + \varepsilon \\
&= \varepsilon(\varepsilon^{-1} + 1) \\
&= \varepsilon\mu.
\end{aligned}$$

It follows that

$$\begin{aligned}
\delta &= \mu\,\sigma(\mu^{-1})\,\varepsilon \\
&= \mu\,\sigma(\mu)^{-1}\,\varepsilon \\
&= \mu\,(\varepsilon\,\mu)^{-1}\,\varepsilon \\
&= \mu\,\mu^{-1}\,\varepsilon^{-1}\,\varepsilon = 1.
\end{aligned}$$

By Theorem 5.10, we have $\rho^2(\lambda) = \delta\,\lambda\,\delta^{-1} = \lambda$ for all λ of K. It follows that $\rho^2 = id$. For $\rho = id$, g is symmetric. For $\rho \neq id$, ρ is hermitian.

Step 6. The polarity π is induced by g, that is, $\pi = \pi_g$: Indeed, for any point $x = \langle v \rangle$ of P, where $0 \neq v$ is an element of V, we have

$$\begin{aligned}
\pi_g(x) &= \{w \in V \mid g(v, w) = 0\} \\
&= \{w \in V \mid \mu\,f(v, w) = 0\} \\
&= \{w \in V \mid f(v, w) = 0\} \qquad \text{(since } \mu \neq 0) \\
&= \pi(x).
\end{aligned}$$

It follows that $\pi = \pi_g$.   □

6   Pseudo-Quadrics

Closely related to the sesquilinear forms are the pseudo-quadratic forms introduced by Tits [50]. They are used to construct the so-called pseudo-quadrics. The pseudo-quadrics provide a further class of polar spaces (cf. Theorem 6.5).

The difference between a polar space defined by a pseudo-quadric and a polar space defined by a polarity is not obvious. We shall discuss these aspects at the end of the present section.

Definition. Let K be a skew field, let $\sigma : K \to K$ be an anti-automorphism, and let $0 \neq \varepsilon$ be an element of K such that $\sigma(\varepsilon) = \varepsilon^{-1}$, and $\sigma^2(\lambda) = \varepsilon\lambda\varepsilon^{-1}$ for all λ of K. The set

$$K_{\sigma,\varepsilon} := \{\lambda - \sigma(\lambda)\varepsilon \mid \lambda \in K\}$$

is called the $(\sigma,\,\varepsilon)$-**subgroup** of K.

6.1 Theorem. *Let K be a skew field, and let $K_{\sigma,\varepsilon}$ be a (σ, ε)-subgroup of K. Then, $K_{\sigma,\varepsilon}$ is an additive subgroup of K.*

Proof. Let $x := \lambda - \sigma(\lambda)\,\varepsilon$ and $y = \mu - \sigma(\mu)\,\varepsilon$ be two elements of $K_{\sigma,\varepsilon}$. Then, we have

$$x - y = (\lambda - \sigma(\lambda)\,\varepsilon) - (\mu - \sigma(\mu)\,\varepsilon)$$
$$= \lambda - \mu - (\sigma(\lambda) - \sigma(\mu))\,\varepsilon$$
$$= \lambda - \mu - \sigma(\lambda - \mu)\,\varepsilon \in K_{\sigma,\varepsilon}.$$

$\square$

6.2 Theorem. *Let K be a skew field, and let $K_{\sigma,\varepsilon}$ be a (σ, ε)-subgroup of K.*

(a) For $\sigma = id$ and $\varepsilon = 1$, we have $K_{\sigma,\varepsilon} = \{0\}$.
(b) The following statements are equivalent:

(i) We have $K_{\sigma,\varepsilon} = K$.
(ii) We have $\sigma = id$, $\varepsilon = -1$ and Char $K \neq 2$.

Proof. (a) For $\sigma = id$ and $\varepsilon = 1$, we have $K_{\sigma,\varepsilon} = \{\lambda - \sigma(\lambda)\,\varepsilon \mid \lambda \in K\} = \{0\}$.
(b) (i) $\Rightarrow$ (ii): Let $x := \lambda - \sigma(\lambda)\,\varepsilon$ be an element of $K_{\sigma,\varepsilon} = K$. Then,

$$\sigma(x) = \sigma(\lambda - \sigma(\lambda)\,\varepsilon)$$
$$= \sigma(\lambda) - \sigma(\sigma(\lambda)\,\varepsilon)$$
$$= \sigma(\lambda) - \sigma(\varepsilon)\,\sigma^2(\lambda)$$
$$= \sigma(\lambda) - \varepsilon^{-1}\,\varepsilon\,\lambda\,\varepsilon^{-1} \qquad \text{(definition of } K_{\sigma,\varepsilon})$$
$$= \sigma(\lambda) - \lambda\,\varepsilon^{-1}$$
$$= \sigma(\lambda)\,\varepsilon\,\varepsilon^{-1} - \lambda\,\varepsilon^{-1}$$
$$= (\sigma(\lambda)\,\varepsilon - \lambda)\,\varepsilon^{-1}$$
$$= -(\lambda - \sigma(\lambda)\,\varepsilon)\,\varepsilon^{-1}$$
$$= -x\,\varepsilon^{-1}.$$

From $K_{\sigma,\varepsilon} = K$, it follows that ε is an element of $K_{\sigma,\varepsilon}$. It follows that $\sigma(\varepsilon) = -\varepsilon\,\varepsilon^{-1} = -1$, that is, $\varepsilon = \sigma^{-1}(-1) = -1$. In particular, we have $\sigma(x) = -x\,\varepsilon^{-1} = x$ for all x of K. Hence, we have $\sigma = id$.

Assume that Char $K = 2$. Then, $K_{\sigma,\varepsilon} = \{\lambda + \lambda \mid \lambda \in K\} = \{0\}$, in contradiction to $K_{\sigma,\varepsilon} = K$.
(ii) $\Rightarrow$ (i): From $\sigma = id$ and $\varepsilon = -1$, it follows that

$$K_{\sigma,\varepsilon} = \{\lambda - \sigma(\lambda)\,\varepsilon \mid \lambda \in K\}$$
$$= \{\lambda + \lambda \mid \lambda \in K\}$$
$$= \{2\lambda \mid \lambda \in K\}$$
$$= K \text{ if Char } K \neq 2.$$

$\square$

Definition. Let V be a right vector space over a skew field K, and let $K_{\sigma,\varepsilon}$ be a (σ, ε)-subgroup of K such that $K_{\sigma,\varepsilon} \neq K$. A transformation $q : V \to K/K_{\sigma,\varepsilon}$ is called a **pseudo-quadratic form with respect to** σ and ε if q fulfils the following two conditions:

(i) We have $q(v\,\lambda) = \sigma(\lambda)\,q(v)\,\lambda$ for all v of V and all λ of K.
(ii) There exists a (σ, ε)-sesquilinear form $f : V \times V \to K$ such that

$$q(v + w) = q(v) + q(w) + (f(v,\ w) + K_{\sigma,\varepsilon}).$$

We say that f **belongs to** q.

If $K = K_{\sigma,\varepsilon}$, then $K/K_{\sigma,\varepsilon} = 0$. This is the reason why this case is excluded in the definition of a pseudo-quadratic form.

Pseudo-quadratic forms are generalizations of quadratic forms:

Definition. Let V be a vector space over a commutative field K, and let $f : V \times V \to K$ be a symmetric bilinear form. The set

$$\mathrm{Rad}(f) := \{v \in V \mid f(v,\ w) = 0 \text{ for all } w \in V\}$$

is called the **radical** of f.

Definition. Let V be a vector space over a commutative field K.

(a) A transformation $q : V \to K$ is called a **quadratic form** if q fulfils the following two conditions:

  (i) We have $q(v\,\lambda) = \lambda^2\,q(v)$ for all v of V and all λ of K.
  (ii) The transformation $f : V \times V \to K$ defined by

$$f(v,\ w) := q(v + w) - q(v) - q(w)$$

  is a symmetric bilinear form. We say that f **belongs to** q.
(b) The quadratic form $q : V \to K$ is called **nondegenerate** or non-singular if

$$\mathrm{Rad}(f) \cap q^{-1}(0) = \{0\},$$

that is, if for every element $0 \neq v$ of V with $q(v) = 0$, there exists an element w of V with $f(v,\ w) \neq 0$.

The following theorem says that pseudo-quadratic forms are a generalization of quadratic forms.

6.3 Theorem. *Let V be a right vector space over a skew field K, and let $K_{\sigma,\varepsilon}$ be a (σ, ε)-subgroup of K. Furthermore, let $q : V \to K/K_{\sigma,\varepsilon}$ be a pseudo-quadratic form with respect to σ and ε. If $\sigma = id$ and $\varepsilon = 1$, q is a quadratic form.*

Proof. Let $\sigma = id$ and $\varepsilon = 1$. By Theorem 6.2, we have $K_{\sigma,\ \varepsilon} = \{0\}$. It follows that $K/K_{\sigma,\varepsilon} = K$. Let $f : V \times V \to K$ be the (σ, ε)-hermitian sesquilinear form belonging to q. Since $\sigma = id$ and since $\varepsilon = 1$, we obtain the equations

$$f(v\,\lambda,\ w\,\mu) = \lambda\,f(v,\ w)\,\mu \text{ for all } v,\ w \text{ of } V,\ \lambda,\ \mu \text{ of } K \text{ and}$$
$$f(w,\ v) = f(v,\ w) \text{ for all } v,\ w \text{ of } V.$$

Since σ is an anti-automorphism and since $\sigma = id$, it follows that

$$\lambda\,\mu = \sigma(\lambda\,\mu) = \sigma(\mu)\,\sigma(\lambda) = \mu\,\lambda$$

for all elements λ and μ of K. Hence, K is commutative. Thus, the $(\sigma,\ \varepsilon)$-hermitian sesquilinear form f is a symmetric bilinear form. □

6.4 Theorem. *Let V be a right vector space over a skew field K, let $K_{\sigma,\varepsilon}$ be a $(\sigma,\ \varepsilon)$-subgroup of K such that $K_{\sigma,\varepsilon} \neq K$, and let $q : V \to K/K_{\sigma,\varepsilon}$ be a pseudo-quadratic form with sesquilinear form f.*

(a) Let v be an element of V with $q(v) = \bar{0}$.[23] Then, $f(v,\ v) = 0$.

(b) Let $\langle v,\ w\rangle$ be a 2-dimensional subspace of V. Then, the following two statements are equivalent:

(i) We have $q(z) = \bar{0}$ for all z of $\langle v,\ w\rangle$.
(ii) We have $q(v) = \bar{0}$, $q(w) = \bar{0}$ and $f(v,\ w) = 0$.

Proof. (a) Assume that there exists an element $0 \neq v$ of V with $q(v) = \bar{0}$ and $f(v,\ v) \neq 0$. By assumption on a pseudo-quadratic form, we have $K \neq K_{\sigma,\varepsilon}$, that is, there exists an element μ of $K\backslash K_{\sigma,\varepsilon}$. Let $\lambda := f(v,\ v)^{-1}\,\mu$. Then,

$$f(v,\ v\,\lambda) = f(v,\ v)\,\lambda = f(v,\ v)\,f(v,\ v)^{-1}\,\mu = \mu.$$

On the other hand, we have

$$\begin{aligned}
\bar{0} &= q(v) \\
&= \sigma(1 + \lambda)\,q(v)\,(1 + \lambda) &\text{(since } q(v) = \bar{0}) \\
&= q(v(1 + \lambda)) \\
&= q(v + v\,\lambda) \\
&= q(v) + q(v\,\lambda) + (f(v,\ v\,\lambda) + K_{\sigma,\varepsilon}) \\
&= q(v) + \sigma(\lambda)\,q(v)\,\lambda + (f(v,\ v)\,\lambda + K_{\sigma,\varepsilon}) \\
&= \bar{0} + \bar{0} + (f(v,\ v)\,f(v,\ v)^{-1}\,\mu + K_{\sigma,\varepsilon}) \\
&= (\mu + K_{\sigma,\varepsilon}) \neq \bar{0}, &\text{(since } \mu \notin K_{\sigma,\varepsilon})
\end{aligned}$$

a contradiction.

(b) (i) $\Rightarrow$ (ii): Let $q(z) = \bar{0}$ for all z of $\langle v,\ w\rangle$. Obviously, we have $q(v) = \bar{0}$ and $q(w) = \bar{0}$. Assume that $f(v,\ w) \neq 0$. As in the proof of (a), let $\lambda := f(v,\ w)^{-1}\,\mu$ for some μ of $K\backslash K_{\sigma,\varepsilon}$. Then,

---

[23]We denote by $\bar{0}$ the neutral element of the factor group $K/K_{\sigma,\varepsilon}$.

$$\bar{0} = q(v + w\lambda)$$
$$= q(v) + q(w\lambda) + (f(v, w\lambda) + K_{\sigma,\varepsilon})$$
$$= q(v) + \sigma(\lambda)q(w)\lambda + (f(v, w)\lambda + K_{\sigma,\varepsilon})$$
$$= \bar{0} + \bar{0} + (f(v, w)f(v, w)^{-1}\mu + K_{\sigma,\varepsilon})$$
$$= (\mu + K_{\sigma,\varepsilon}),$$

in contradiction to $\mu \notin K_{\sigma,\varepsilon}$.

(ii) $\Rightarrow$ (i): Let $q(v) = q(w) = \bar{0}$, and let $f(v, w) = 0$. Let μ be an arbitrary element of K. We have

$$q(v + w\mu) = q(v) + q(w\mu) + (f(v, w\mu) + K_{\sigma,\varepsilon})$$
$$= q(v) + \sigma(\mu)q(w)\mu + (f(v, w)\mu + K_{\sigma,\varepsilon})$$
$$= \bar{0} + \bar{0} + (0 + K_{\sigma,\varepsilon}) = \bar{0} \text{ for all } v + w\mu \text{ of } \langle v, w\rangle.$$

□

Definition. (a) Let P be a projective space over a right vector space V, and let $q : V \to K/K_{\sigma,\varepsilon}$ be a pseudo-quadratic form.

The set $Q := \{x = \langle v\rangle \in P \mid q(v) = \bar{0}\}$ is called a **pseudo-quadric with respect to** q.

(b) Let P be a projective space over a vector space V over a commutative field K, and let $q : V \to K$ be a quadratic form.

The set $Q := \{x = \langle v\rangle \in P \mid q(v) = 0\}$ is called a **quadric with respect to** q. If P is a projective plane, a quadric is also called a **conic**.

(c) A line g of P is called a line of a quadric or of a pseudo-quadric Q if all points on g are contained in Q.

6.5 Theorem. *Let P be a projective space, and let Q be a pseudo-quadric in P. If Q contains a line, the points and the lines of Q define a polar space.*

Proof. Let $x = \langle v\rangle$ and $g := \langle w_1, w_2\rangle$ be a point and a line of P such that x is not on g. We shall show that either exactly one point of g or all points of g are joined with x by a line of Q.

Let q be the pseudo-quadratic form with respect to Q, and let f be the (σ, ε)-sesquilinear form belonging to q. Furthermore, let $f_v : V \to K$ be the linear transformation defined by $f_v(w) := f(v, w)$. Note that $q(v) = \bar{0}$ and $q(u) = \bar{0}$ for all u of $\langle w_1, w_2\rangle$, since $x = \langle v\rangle$ is contained in Q and since $g = \langle w_1, w_2\rangle$ is contained in Q. For a point $y = \langle u\rangle$ on $g = \langle w_1, w_2\rangle$, we have:

The point $y = \langle u\rangle$ is joined with the point $x = \langle v\rangle$ by a line of Q

$\Leftrightarrow \langle v, u\rangle \subseteq Q$

$\Leftrightarrow q(v) = \bar{0}, q(u) = \bar{0}$ and $f(v, u) = 0$.   (Theorem 6.4)

$\Leftrightarrow f(v, u) = 0$ (since $q(v) = q(u) = \bar{0}$)

$\Leftrightarrow u \in \text{Kern } f_v$.

Since Kern $f_v \cap \langle w_1, w_2 \rangle$ is either of dimension 1 or 2, either exactly one point of g is joined with x by a line of Q or x is joined with all points of g by a line of Q.

Obviously, on every line of Q, there are at least three points.   □

Definition. Let P be a projective space, and let Q be a pseudo-quadric of P defined by a pseudo-quadratic form q such that Q contains at least one line. Furthermore, let S_q be the geometry over the type set {point, line} defined as follows:

 (i) The points and lines of S_q are exactly the points and lines of P contained in Q.
 (ii) A point and a line of S_q are incident in S_q if they are incident in P.

S_q is called the **polar space defined by** q.

In the rest of this section, we shall explain the relation between polar spaces defined by polarities and polar spaces defined by a pseudo-quadratic form.

By Theorem 6.3, pseudo-quadratic forms are generalizations of quadratic forms. The pseudo-quadrics stemming from a quadratic form are called quadrics. Quadrics and quadratic forms are the subject of Chap. 5.

Let q be a nondegenerate quadratic form with (nondegenerate) symmetric bilinear form f. Since $f(v, w) = f(w, v)$, the bilinear form f is reflexive. Suppose that there exists a polarity π such that π is induced by f. The polarity π defines itself a polar space S_π (Theorem 4.7). On the other hand, the quadratic form q defines a polar space S_q by Theorem 6.5.

We shall see in the following theorem that $S_q = S_\pi$ if and only if Char $K \neq 2$. For Char $K = 2$, the bilinear form f and the quadratic form q define different polar spaces.

Moreover, for Char $K = 2$, there exists a symmetric bilinear form f such that there is no quadratic form q with the property that f belongs to q.

6.6 Theorem. *Let V be a vector space over a commutative field K, and let $P = P(V)$ be the projective space defined by V.*

(a) *Let $q : V \to K$ be a nondegenerate quadratic form with symmetric bilinear form f, and suppose that there exists a polarity π such that π is induced by f. Let S_q be the polar space defined by q, and let S_π be the polar space defined by π.*

  (i) *If Char $K \neq 2$, $S_q = S_\pi$.*
  (ii) *If Char $K = 2$, $S_q \neq S_\pi$.*

(b) *Let $f : V \times V \to K$ be a nondegenerate symmetric bilinear form, and suppose that there exists a polarity π such that π is induced by f.*

  (i) *If Char $K \neq 2$, there exists a quadratic form q such that f belongs to q. We have $S_q = S_\pi$.*
  (ii) *If Char $K = 2$ and if there exists an element v of V with $f(v, v) \neq 0$, there is no quadratic form q such that f belongs to q.*

Proof. (a) (i) Let Char $K \neq 2$, and let $x = \langle v \rangle$ be a point of P. Then,

$$f(v, v) = q(v + v) - q(v) - q(v)$$
$$= q(2v) - 2\,q(v)$$
$$= 4\,q(v) - 2\,q(v)$$
$$= 2\,q(v).$$

Hence,

$$q(v) = {}^{1}\!/_{2}\,f(v, v).$$

It follows that

$x = \langle v \rangle$ is a point of S_q
$\Leftrightarrow q(v) = 0$ (definition of S_q)
$\Leftrightarrow f(v, v) = 0$ (since $q(v) = {}^{1}\!/_{2}\,f(v, v)$)
$\Leftrightarrow x \in \pi(x)$ (Theorem 5.2)
$\Leftrightarrow x = \langle v \rangle$ is a point of S_π (definition of S_π).

Let $x = \langle v \rangle$ and $y = \langle w \rangle$ be two points of P, and let $g = xy = \langle v, w \rangle$ be the line of P through x and y. Then,

$g = xy = \langle v, w \rangle$ is a line of S_q
$\Leftrightarrow q(v) = q(w) = 0$ and $f(v, w) = 0$ (Theorem 6.4)
$\Leftrightarrow f(v, v) = f(w, w) = 0$ and f(v, w) $= 0$ (since $q(v) = {}^{1}\!/_{2}\,f(v, v)$)
$\Leftrightarrow x \in \pi(x)$, $y \in \pi(y)$ and $y \in \pi(x)$ (Theorem 5.2)
$\Leftrightarrow g = xy$ is a line of S_π (Theorem 4.6).

It follows that $S_q = S_\pi$.

(ii) Let Char $K = 2$. Let $x = \langle v \rangle$ be a point of P. Then,

$$f(v, v) = q(v + v) - q(v) - q(v) = q(0) - 0 = 0.$$

By Theorem 5.2, every point of P is a point of S_π.[24] On the other hand, there exists a point of P that is not a point of S_q: Otherwise, we would have $q(v) = 0$ for all v of V. Since, by assumption, f is nondegenerate, there exist vectors w_1, w_2 of V such that $f(w_1, w_2) \neq 0$. It follows that

$$0 = q(w_1 + w_2) = q(w_1) + q(w_2) + f(w_1, w_2)$$
$$= f(w_1, w_2) \neq 0, \text{ a contradiction.}$$

It follows that $S_q \neq S_\pi$.

(b) (i) Let Char $K \neq 2$. Set $q(v) := {}^{1}\!/_{2}f(v, v)$ for all v of V. We have

$$q(v\,\lambda) = 1/2\,f(v\,\lambda, v\,\lambda) = \lambda^2\,1/2\,f(v, v) = \lambda^2\,q(v) \text{ and}$$
$$q(v + w) - q(v) - q(w) = 1/2\,f(v + w, v + w) - 1/2\,f(v, v) - 1/2\,f(w, w)$$

---

[24]Note that this fact does not mean that every line of P is a line of S_π.

$$= 1/2 \ f(v, \ w) + 1/2 \ f(w, \ v)$$
$$= 1/2 \ f(v, \ w) + 1/2 \ f(v, \ w).$$
$$= f(v, \ w).$$

It follows that q is a quadratic form with bilinear form f. It follows from Part (a) that $S_q = S_\pi$.

(ii) Let Char $K = 2$, and let v be an element of V with $f(v, \ v) \neq 0$. Assume that there is a quadratic form q such that f is the bilinear form belonging to q. Then,

$$0 \neq f(v, \ v) = q(v+v) - q(v) - q(v) = q(0) - 0 = 0, \text{ a contradiction.} \qquad \square$$

In what follows we shall indicate the motivation for the definition of a pseudo-quadratic form: For, let $f : V \times V \to K$ be a $(\sigma, \ \varepsilon)$-hermitian sesquilinear form. By definition, we have

$$f(w, \ v) = \sigma(f(v, \ w)) \ \varepsilon \text{ for all } v, \ w \text{ of } V.$$

For a quadratic form q and for its symmetric bilinear form f, we have the equation

$$q(v + w) - q(v) - q(w) = f(v, \ w)$$

which is not valid if q is a $(\sigma, \ \varepsilon)$-hermitian sesquilinear form with $(\sigma, \ \varepsilon) \neq (id, \ 1)$, since, otherwise, we would have

$$f(v, \ w) = q(v + w) - q(v) - q(w) = f(w, \ v) = \sigma(f(v, \ w)) \ \varepsilon.$$

Instead of the relation $f(v, \ w) - f(w, \ v) = 0$, we have

$$f(v, \ w) - f(w, \ v) = f(v, \ w) - \sigma(f(v, \ w)) \ \varepsilon \in \{\lambda - \sigma(\lambda) \ \varepsilon \mid \lambda \in K\} = K_{\sigma, \ \varepsilon}.$$

It follows that $f(v, \ w) - f(w, \ v) = \bar{0} \mod K_{\sigma, \ \varepsilon}$. This relation motivates the definition of a pseudo-quadratic form as a transformation $q : V \to K/K_{\sigma, \ \varepsilon}$ with $q(v \ \lambda) = \sigma(\lambda) \ q(v) \ \lambda$ and

$$q(v + w) - q(v) - q(w) = f(v, \ w) + K_{\sigma, \ \varepsilon}$$

for a $(\sigma, \ \varepsilon)$-hermitian sesquilinear form $f : V \times V \to K$. $\qquad \square$

7   The Kleinian Polar Space

Definition. Let P be a 3-dimensional projective space, and let S be the set geometry defined as follows:

(i) The points of S are the lines of P.

(ii) For a point x of P and a plane E of P through x, let E_x be the set of lines of E through x. The lines of S are the sets E_x where x is a point of P and E is a plane of P through x.

S is called **the Kleinian polar space defined by P.**

7.1 Theorem. *Let S be the Kleinian polar space defined by a 3-dimensional projective space P.*

(a) *Two points g and h of S are collinear in S if and only if g and h as lines of P intersect in a point.*

(b) *Let g and h be two collinear points of S, let $x := g \cap h$ be the intersection point of g and h in P, and let E be the plane of P generated by g and h. There is exactly one line of S through g and h, namely the line E_x.*

(c) *S is a nondegenerate polar space.*

(d) *For a point x of P and a plane E of P, let G_x be the set of lines of P through x, and let G_E be the set of lines of P in E. The sets G_x (x point of P) and G_E (E plane of P) are the planes of S.*

(e) *S is a polar space of rank 3.*

(f) *If E_x is a line of S, E_x is incident with exactly the two planes G_x and G_E.*

(g) *Let A and B be two planes of S.*

If $A = G_x$ and $B = G_y$ for two points x and y of P, $A = B$ or $dim_S(A \cap B) = 0$.
If $A = G_x$ and $B = G_E$ for a point x and a plane E of P, $dim_S(A \cap B) \in \{-1, 1\}$.
If $A = G_E$ and $B = G_F$ for two planes E and F of P, $A = B$ or $dim_S(A \cap B) = 0$.

Proof. (a) and (b) are obvious.

(c) Verification of (P_1): Let g be a point of S (that is, a line of P), and let E_x be a line of S such that g is not on E_x.

First case. Let g be contained in E. Let h be a point of E_x, that, is a line of E through x. Since g and h both are contained in E, they meet in P in a point. By (a), they are collinear. It follows that g is collinear with all points of E_x.

Second case. Suppose that x is incident with g. Let h be a point of E_x. Then, the lines g and h meet in P in the point x. By (a), they are collinear. It follows that g is collinear with all points of E_x.

Third case. Suppose that x is not incident with g, and let g be not contained in E. Let y be the intersection point of g and E in P, and let $h := xy$ be the line through x and y in E. By (a),

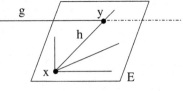

g and h are collinear. All other points on E_x are not collinear with g.

Verification of (P_2): There are at least three points on every line E_x of S since there are at least three lines of P in E through x.

Verification of (P_3): Since, in P, for every line g, there exists a line h skew to g, there exists in S for every point g a non-collinear point h.

(d) Let x and E be a point and a plane of P, and let g and h be two points of G_x or G_E. Then, g and h meet in P in a point s and they generate a plane F. We have

$s = x$ or $F = E$. By (a), g and h are collinear in S. By (b), the line F_s through g and h is contained in G_x and in G_E. It follows that the sets G_x and G_E are subspaces of S.

If g, h and l are three pairwise collinear points in S, there exists in P either a point x such that $x = g \cap h \cap l$ or a plane E such that g, h, l are contained in E. It follows that $\langle g, h, l \rangle_S = G_x$ or $\langle g, h, l \rangle_S = G_E$. It follows that the sets G_x and G_E are the planes of S.

(e) Let E be a plane of P, and let g be a point of S outside of G_E. If s is the intersection point of g and E in P and if h is a line of E not incident with s, the lines g and h are disjoint in P, that is, g and h are non-collinear in S.

It follows that for every point g of S outside of G_E, there exists a point h of G_E non-collinear with g, that is, G_E is a maximal subspace of S. Analogously, for a point x of P, the set G_x is a maximal subspace of S. It follows that S is a polar space of rank 3.

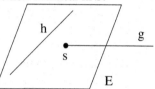

(f) is obvious.

(g)   (i) Let $A = G_x$ and $B = G_y$ for two points x and y of P, and let $A \neq B$. Then, we have $x \neq y$. Let g be the line through x and y. Obviously, we have $\{g\} = G_x \cap G_y$. In S, the planes A and B meet in the point g, hence, $\dim_S(A \cap B) = 0$.

  (ii) Let $A = G_x$ and $B = G_E$ for a point x and a plane E of P. If x is contained in E (in P), $E_x = G_x \cap G_E$. It follows that $\dim_S(A \cap B) = 1$. If x is not contained in E, we have $G_x \cap G_E = \varnothing$, hence, $\dim_S(A \cap B) = -1$.

  (iii) Let $A = G_E$ and $B = G_F$ for two planes E and F of P, and let $A \neq B$. Then, $E \neq F$. In P, $g := E \cap F$ is a line. Hence, $\{g\} = G_E \cap G_F$, and it follows that $\dim_S(A \cap B) = 0$. $\qquad\square$

In Theorem 7.2, we shall see that every polar space S of rank 3 with the property that every line of S is incident with exactly two planes, is a Kleinian polar space.

7.2 Theorem. *Let S be a nondegenerate polar space of rank 3 such that every line of S is incident with exactly two planes. There exists a 3-dimensional projective space P such that S is the Kleinian polar space defined by P.*

Proof. The proof is organised as follows: In Steps 1–3, we shall see that the planes of S split into two equivalence classes T and S where two planes A and B of S are equivalent if $A = B$ or if $\dim_S(A \cap B) = 0$. These two equivalence classes correspond to the sets $\{G_x \mid x \in P\}$ and $\{G_E \mid E \text{ plane of } P\}$ of Theorem 7.1 and hence to the points and planes of P.

In order to construct the projective space P of S, we will define in Step 4 the geometry Γ as follows: The points of Γ are the planes of S of one equivalence class, the planes of Γ are the planes of the other equivalence class. The lines of Γ are the points of S.

In Steps 5 and 6, we shall show that the so-defined geometry Γ is a 3-dimensional projective space.

Step 1. Let A and B be two planes of S through a line l, and let C be a further plane of S. We have $|\dim_S(A \cap C) - \dim_S(B \cap C)| = 1$:

First case. Suppose that l is contained in C. Since there are exactly two planes through l, we have $C = A$ or $C = B$. W.l.o.g., let $C = A$. It follows that $|\dim_S(A \cap C) - \dim_S(B \cap C)| = 2 - 1 = 1$.

Second case. Suppose that $x := l \cap C$ is a point. The residue S_x is a generalized quadrangle. In S_x, l is a point not incident with the line C. By Axiom (V_1), in S_x, there exists exactly one line D through l intersecting C in a point m. In S, D is a plane through l intersecting the plane C in the line m. Since there are exactly two planes through l, we have $D = A$ or $D = B$. W.l.o.g., let $D = A$. Then, $A \cap C = D \cap C = m$ and $B \cap C = x$. It follows that $|\dim_S(A \cap C) - \dim_S(B \cap C)| = 1 - 0 = 1$.

Third case. Let $l \cap C = \emptyset$.

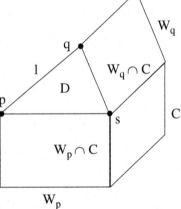

(i) There is a plane D through l intersecting the plane C in a point s: Let p and q be two points on l. Since C is a maximal subspace of S, by the Theorem of Buekenhout–Shult (Theorem 2.18), there exist two planes W_p and W_q through p and q, respectively such that $\dim_S(W_p \cap C) = \dim_S(W_q \cap C) = 1$. Let s be the intersection point of the lines $W_p \cap C$ and $W_q \cap C$ in C.

Since s is contained in W_p and in W_q and since the points p and q are collinear, it follows that there exists a plane D through the points p, q and s containing the line $l = pq$.

(ii) There is at most one plane through l intersecting the plane C in a point: Assume that there are two planes E and F through l both intersecting the plane C in a point. We have $x := C \cap E \neq y := C \cap F$ since, otherwise, $E = \langle x, l \rangle = \langle y, l \rangle = F$. It follows that the points of l and the points x and y are pairwise collinear, hence, there is a subspace $U := \langle l, x, y \rangle$. By construction, the plane E is properly contained in U, in contradiction to the maximality of E.[25]

(iii) We have $|\dim_S(A \cap C) - \dim_S(B \cap C)| = 1$: By (i) and (ii), there exists exactly one plane D through l, intersecting the plane C in a point. Since there are only the two planes A and B through l, w.l.o.g. let $D = A$. It follows that $B \cap C = \emptyset$. Thus

$$|\dim_S(A \cap C) - \dim_S(B \cap C)| = |0 - (-1)| = 1.$$

Step 2. Let A, B, C be three planes of S. Then, the number

$$n := \dim_S(A \cap B) + \dim_S(B \cap C) + \dim_S(A \cap C)$$

---

[25]S is a polar space of rank 3, hence the planes of S are the maximal subspaces of S.

is even: Let $B = X_0, X_1, \ldots, X_r = A$ be a sequence of planes such that $\dim_S(X_{i-1} \cap X_i) = 1$. Such a sequence exists in view of Theorem 7.10 of Chap. 1 since S is residually connected (Theorem 3.2). Furthermore, let $n_i :=$ $\dim_S(X_i \cap B) + \dim_S(B \cap C) + \dim_S(X_i \cap C)$. We have

$$n_0 = \dim_S(X_0 \cap B) + \dim_S(B \cap C) + \dim_S(X_0 \cap C)$$
$$= \dim_S B + \dim_S(B \cap C) + \dim_S(B \cap C)$$
$$= 2 + 2\dim_S(B \cap C).$$

It follows that n_0 is even. By Step 1, for $i = 1, \ldots, r$, we have

$$n_i - n_{i-1} = \dim_S(X_i \cap B) + \dim_S(B \cap C) + \dim_S(X_i \cap C)$$
$$-(\dim_S(X_{i-1} \cap B) + \dim_S(B \cap C) + \dim_S(X_{i-1} \cap C))$$
$$= \dim_S(X_i \cap B) - \dim_S(X_{i-1} \cap B)$$
$$+\dim_S(X_i \cap C) - \dim_S(X_{i-1} \cap C)$$
$$= \pm 1 \pm 1 \in \{-2, 0, 2\}.$$

Since n_0 is even, $n_1, \ldots, n_r = n$ are also even.

Step 3. Two planes A and B of S are called **equivalent** if $A = B$ or if $\dim_S(A \cap B) = 0$. The set of the planes of S splits into two equivalence classes: We first shall show that the relation is an equivalence relation. Obviously, the relation is reflexive and symmetric. In order to verify the transitivity, we consider three planes A, B and C such that A and B as well as B and C are equivalent. Then, $\dim_S(A \cap B)$ and $\dim_S(B \cap C)$ are even, by Step 2, $\dim_S(A \cap C)$ is also even, that is, $A = C$ or $\dim_S(A \cap C) = 0$. It follows that A and C are equivalent.

Let A and B be two planes through a line l. Since $\dim_S(A \cap B) = 1$, A and B are in different equivalence classes. Let C be a further plane of S. If A and C are not in a common equivalence class, $\dim_S(A \cap C) \in \{-1, 1\}$. By Step 1, we have $\dim_S(B \cap C) \in \{0, 2\}$. It follows that B and C are equivalent. Altogether, there exist exactly two equivalence classes.

Step 4. Definition of the geometry Γ : Let $\mathscr{P}$ and $\mathscr{E}$ be the two equivalence classes of the planes of S as described in Step 3, and let $\Gamma = (X, *, \text{type})$ be the geometry over the type set {point, line, plane} defined as follows: The points of Γ are the planes of $\mathscr{P}$. The lines of Γ are the points of S, and the planes of Γ are the planes of $\mathscr{E}$. A point x of Γ or a plane E of Γ is incident with a line g of Γ if, in S, the point g is incident with the plane x or with the plane E. A point x of Γ and a plane E of Γ are incident if, in S, x and E intersect in a line.

We shall use the following notations:

Γ	S
point of Γ (Notation: x_A)	plane of S of $\mathscr{P}$ (Notation: A)
line of Γ (Notation: g_a)	point of S (Notation: a)
plane of Γ (Notation: E_X)	plane of S of $\mathscr{E}$ (Notation: X)

The point x_A of Γ corresponds to the plane A of $\mathscr{P}$, etc. With these notations, we have the following incidence relations:

Γ	S
$x_A * g_a$	$a \in A$
$x_A * E_X$	$\dim_S(A \cap X) = 1$
$g_a * E_X$	$a \in X$

Step 5. Γ is a residually connected geometry with diagram

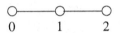

$$\begin{array}{ccc} 0 & 1 & 2 \end{array}$$

(i)   It is easy to see that Γ is a connected geometry.

(ii) Let $x = x_A$ be a point of Γ. The residue Γ_x consists of the lines g_a of Γ and the planes E_x of Γ incident with x_A, that is, Γ_x consists of the points a of S and the planes X of S such that a is contained in A and X and A intersect in a line of S.

Since every line of A is incident with exactly two planes of S (one of them being A itself), the planes of S intersecting A in a line may be identified with the lines of A. Hence, Γ_x is a projective plane. In particular, Γ_x is connected.

(iii) Let $E = E_X$ be a plane of Γ. As in (i), it follows that Γ_E is a projective plane. In particular, Γ_E is connected.

(iv) Finally, let $g = g_a$ be a line of Γ. The residue Γ_g consists of the points x_A and the planes E_X of Γ incident with g_a, that is, Γ_g consists of the planes A of $\mathscr{P}$ and the planes X of $\mathscr{E}$ such that, in S, a is contained in the planes A and X. Since A and X are in different equivalence classes, it follows that $\dim_S(A \cap X) = -1$, or $\dim_S(A \cap X) = 1$. Since the point a is contained in both planes, it follows that $\dim_S(A \cap X) = 1$, hence, A and X are incident. In particular, Γ_g is connected.

Step 6. It follows from Theorem 7.15 of Chap. 1 that Γ is a 3-dimensional projective space. $\square$

We finish this section with the classification theorem of polar spaces of rank at least 3, whose maximal subspaces are Desarguesian projective spaces.

7.3 Theorem (Veldkamp, Tits). *Let S be a polar space of rank at least 3, whose maximal subspaces are Desarguesian projective spaces. Then, S is isomorphic to one of the following geometries Γ:*

(i) *Let **P** be a projective space, and let π be a polarity of **P** such that there exists at least one absolute plane with respect to π. Let Γ be the polar space S_π defined by π.*

(ii) *Let **P** be a projective space over a vector space V, let $K_{\sigma,\varepsilon}$ be a (σ, ε)-subgroup of K such that $K_{\sigma,\varepsilon} \neq K$, and let $q : V \to K/K_{\sigma,\varepsilon}$ be a pseudo-quadratic form such that the pseudo-quadric of **P** defined by q contains at least one plane. Let Γ be the polar space S_q defined by q.*

(iii) *Let **P** be a 3-dimensional projective space. Let Γ be the Kleinian polar space defined by **P**.*

Proof. For a proof see Veldkamp [56], Tits [50] or Buekenhout and Cohen [20]. □

8 The Theorem of Buekenhout and Parmentier

The theorem of Buekenhout [17] and Parmentier [37] characterizes projective spaces as linear spaces admitting a generalization of polarities. Although the theorem is not so well known, it deserves to be called "classical".

Definition. Let L be a linear space with a symmetric and reflexive relation $\sim$ on the point set P of L.

(a) For a point p of L, the set $p^\perp := \{x \in P \mid x \sim p\}$ is called the **polar hyperplane** of p.

(b) For a subset X of the point set of L, let

$$X^\perp := \bigcap_{p \in X} p^\perp.$$

(c) The linear space L (with a symmetric and reflexive relation $\sim$ on its point set) is called a **linear space with polarity** if the following three conditions are fulfilled:

(i) If p is a point and if g is a line of L, either g is contained in $p^\perp$, or g and $p^\perp$ meet in a point.

(ii) For every line g of L, we have $(g^\perp)^\perp = g$.

(iii) For every point p of L, we have $p^\perp \neq P$.

Remark. (a) Condition (i) means that $p^\perp$ either is a hyperplane or the whole point set of L. Condition (iii) excludes the case that $p^\perp$ is the point set of L.

(b) Condition (ii) means that a point p and a line g of L are incident if and only if $g^\perp$ is contained in $p^\perp$ (see Step 8 in the proof of Part (a) of Theorem 8.1).

8.1 Theorem (Buekenhout and Parmentier). *Let **P** be a linear space with polarity such that every line of **P** is incident with at least three points. For a point p of **P**, let $p^\perp := \pi(p)$.*

*(a) **P** is a projective space.*
*(b) If dim **P** < ∞, the transformation π is a polarity of **P**.*[26]

Proof. (a) Step 1. Let X and Y be two subsets of **P** such that X is contained in Y. Then, $Y^\perp$ is contained in $X^\perp$: We have

$$Y^\perp = \bigcap_{y \in Y} y^\perp \subseteq \bigcap_{x \in X} x^\perp = X^\perp.$$

Step 2. For every point p of **P**, the point set $p^\perp$ is a maximal subspace of **P**: The maximality of $p^\perp$ follows from Condition (i) and Theorem 3.4 of Chap. 1.

Step 3. Let p and q be two points of **P**. The point p is contained in $q^\perp$ if and only if the point q is contained in $p^\perp$: We have

$$p \in q^\perp \Leftrightarrow p \sim q \Leftrightarrow q \sim p \Leftrightarrow q \in p^\perp.$$

Step 4. Let p be a point, and let g be a line of **P**. The point p is contained in $g^\perp$ if and only if g is contained in $p^\perp$: We have

$$\begin{aligned} p \in g^\perp &\Leftrightarrow p \in x^\perp \text{ for all } x \text{ on } g \\ &\Leftrightarrow x \in p^\perp \text{ for all } x \text{ on } g \qquad \text{(Step 3)} \\ &\Leftrightarrow g \subseteq p^\perp. \end{aligned}$$

Step 5. Let p and q be two points of **P**, and let g be the line joining p and q. We have $p^\perp \cap q^\perp = g^\perp$:

From $g^\perp := \bigcap_{x \in g} x^\perp$, it follows that $g^\perp$ is contained in $p^\perp \cap q^\perp$.

Conversely, let z be a point of $p^\perp \cap q^\perp$. It follows from Step 3 that p and q are contained in $z^\perp$. Hence, the line $g = pq$ is contained in $z^\perp$. By Step 4, the point z is contained in $g^\perp$. Hence, $p^\perp \cap q^\perp$ is contained in $g^\perp$.

Step 6. Let p and q be two points of **P**, and let x be a point outside of $p^\perp \cap q^\perp$. There exists a unique point y of **P** such that the subspace $\langle x, \ p^\perp \cap q^\perp \rangle$ is contained in $y^\perp$. The point y is incident with the line pq:

Existence of the point y: Let $g = pq$ be the line joining p and q. By Step 5, we have $g^\perp = p^\perp \cap q^\perp$.

Since x is not contained in $p^\perp \cap q^\perp = g^\perp$, it follows from Step 4 that g is not contained in $x^\perp$. By Condition (i), $x^\perp$ and g intersect in a point y.

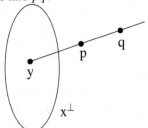

Since the point y is contained in $x^\perp$, the point x is contained in $y^\perp$ (Step 3). Since y is contained in $g = (g^\perp)^\perp$, it follows

---

[26]In particular, $\pi(p) = p^\perp$ is a hyperplane of **P** for all points p of **P**.

from Step 4 that $p^\perp \cap q^\perp = g^\perp$ is contained in $y^\perp$. It follows that $\langle x,\ p^\perp \cap q^\perp \rangle$ is contained in $y^\perp$.

Uniqueness of the point y: Let z be a second point of $\boldsymbol{P}$ such that $\langle x,\ p^\perp \cap q^\perp \rangle$ is contained in $z^\perp$. Since the point x is contained in $z^\perp$, z is contained in $x^\perp$. Since $g^\perp = p^\perp \cap q^\perp$ is contained in $z^\perp$, z is contained in $(g^\perp)^\perp = g = pq$. It follows that $z^\perp = g \cap x^\perp = y$.

Step 7. Let p and q be two distinct points of $\boldsymbol{P}$. We have $p^\perp \neq q^\perp$: Assume that $p^\perp = q^\perp$. Let c be a point of $\boldsymbol{P}$ outside of $p^\perp$. Let $g := pq$ be the line joining p and q. Since c is not contained in $p^\perp$, the point p is not contained in $c^\perp$. It follows that the line g is not contained in $c^\perp$, hence g and $c^\perp$ meet in a point z.

Since z is contained in $c^\perp$, the point c is contained in $z^\perp$. Since z is incident with g, it follows that

$$z^\perp \supseteq \bigcap_{x \in g} x^\perp = g^\perp = p^\perp \cap q^\perp = p^\perp.$$

Hence, $\boldsymbol{P} = \langle p^\perp,\ c \rangle$ is contained in $z^\perp$, in contradiction to Condition (iii).

Step 8. Let p be a point of $\boldsymbol{P}$, and let g be a line of $\boldsymbol{P}$. Then, p and g are incident if and only if $g^\perp$ is contained in $p^\perp$:

(i) Suppose that p and g are incident. Let q be a further point on g. It follows from Step 5 that $g^\perp = p^\perp \cap q^\perp$. In particular, $g^\perp$ is contained in $p^\perp$.

(ii) Suppose that $g^\perp$ is contained in $p^\perp$. Let a and b be two points on g. Since $a^\perp \neq b^\perp$ (Step 7), the subspace $g := a^\perp \cap b^\perp$ is not maximal. Since, by Step 2, the subspace $p^\perp$ is maximal, $p^\perp$ cannot be contained in $g^\perp$. It follows that there exists a point x of $p^\perp$ outside of $g^\perp$. By Step 6, there is a unique point y of $\boldsymbol{P}$ such that the subspace $\langle x,\ a^\perp \cap b^\perp \rangle$ is contained in $y^\perp$. Again by Step 6, the point y is incident with the line $ab = g$. Since $\langle x,\ a^\perp \cap b^\perp \rangle$ is contained in $p^\perp$, it follows that $y = p$. In particular p and g are incident.

Step 9. $\boldsymbol{P}$ is a projective space. It remains to verify the axiom of Veblen–Young. For, let p, x, y, a, b be five points such that the lines xy and ab meet in p.

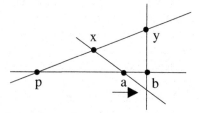

(i) The subspace $x^\perp \cap a^\perp$ is not contained in $y^\perp \cap b^\perp$: Assume that $x^\perp \cap a^\perp$ is contained in $y^\perp \cap b^\perp$. From Step 5, it follows that

$$(xa)^\perp = x^\perp \cap a^\perp \subseteq y^\perp \cap b^\perp = (yb)^\perp.$$

From Step 1, it follows that $xa = ((xa)^\perp)^\perp \supseteq ((yb)^\perp)^\perp = yb$, hence $xa = yb$, a contradiction.

(ii) In view of Step (i), there is a point z of $x^\perp \cap a^\perp$ which is not contained in $y^\perp \cap b^\perp$. By Step 6, there is a unique point r such that the subspace $\langle z,\ y^\perp \cap b^\perp \rangle$ is contained in $r^\perp$. Again by Step 6, the point r is incident with the line yb.

(iii) $x^\perp \cap a^\perp$ is contained in $r^\perp$: Let s be an arbitrary point of $x^\perp \cap a^\perp$. Since z is contained in $r^\perp$, we can assume w.l.o.g. that $s \neq z$. Since z is contained in $x^\perp \cap a^\perp$, it follows that the whole line sz is contained in $x^\perp \cap a^\perp$.

By Condition (i), the line sz and the subspace $p^\perp$ have at least one point u in common. It follows that u is contained in $p^\perp \cap x^\perp = (px)^\perp = (py)^\perp = p^\perp \cap y^\perp$ and that u is contained in $p^\perp \cap a^\perp = (pa)^\perp = (pb)^\perp = p^\perp \cap b^\perp$. In particular, u is contained in $y^\perp \cap b^\perp$.

By Step (ii), the line $zs = zu$ is contained in $r^\perp$. Since s is an arbitrary point of $x^\perp \cap a^\perp$, it follows that $x^\perp \cap a^\perp$ is contained in $r^\perp$.

(iv) The lines xa and yb meet in r: By Step (ii), the point r is incident with the line yb. By Step (iii), the subspace $(xa)^\perp = x^\perp \cap a^\perp$ is contained in $r^\perp$. It follows from Step 8 that the point r is incident with the line xa.

(b) Let P be the point set of $\boldsymbol{P}$. By Step 2 of (a), $\pi(x) := x^\perp$ is a maximal subspace of $\boldsymbol{P}$, hence a hyperplane for all points x of $\boldsymbol{P}$.

Step 1. Let z be a point of $\boldsymbol{P}$, and let H be a hyperplane of $\boldsymbol{P}$. We have $z^\perp = H$ if and only if z is contained in $H^\perp$:

$$H \subseteq z^\perp \Leftrightarrow x \in z^\perp \text{ for all } x \text{ of } H$$

$$\Leftrightarrow z \in x^\perp \text{ for all } x \text{ of } H \text{ (Step 3 of (a))}$$

$$\Leftrightarrow z \in \bigcap_{x \in H} x^\perp$$

$$\Leftrightarrow z \in H^\perp.$$

As H is a hyperplane of $\boldsymbol{P}$, the inclusion $H \subseteq z^\perp$ implies $H = z^\perp$.

Step 2. Let z be a point of $\boldsymbol{P}$, and let H be a hyperplane of $\boldsymbol{P}$. We have $z^\perp = H$ if and only if $z = H^\perp$: In view of Step 1, we need to show that $z = H^\perp$ if and only if $z \in H^\perp$. Assume that there is a second point y in $H^\perp$. By Step 3 of Part (a), we have $y^\perp \neq z^\perp$, a contradiction.

Step 3. Let z be a point of $\boldsymbol{P}$. Then, we have $z = (z^\perp)^\perp$: Let $H := z^\perp$. It follows from Step 2 that $(z^\perp)^\perp = H^\perp = z$.

Step 4. Let $\boldsymbol{P}^*$ be the dual projective space of $\boldsymbol{P}$, and denote by $\mathrm{Im}(\pi) := \{\pi(x) \mid x \in P\}$ the image of $\boldsymbol{P}$ under π. Then, $\mathrm{Im}(\pi)$ is a subspace of $\boldsymbol{P}^*$: For, let $p^\perp$ and $q^\perp$ be two points of $\mathrm{Im}(\pi)$, and let H be a further point of $\boldsymbol{P}^*$ on the line of $\boldsymbol{P}^*$ through $p^\perp$ and $q^\perp$. By definition of $\boldsymbol{P}^*$, H is a hyperplane of $\boldsymbol{P}$ through $p^\perp \cap q^\perp$. Let x be a point of H outside of $p^\perp \cap q^\perp$. We have $H = \langle x,\ p^\perp \cap q^\perp \rangle$.

By Step 6 of the proof of Part (a), there exists a (unique) point y of $\boldsymbol{P}$ such that the subspace $\langle x,\ p^\perp \cap q^\perp \rangle$ is contained in $y^\perp$. Since the subspaces $\langle x,\ p^\perp \cap q^\perp \rangle$ and $y^\perp$ both are hyperplanes of $\boldsymbol{P}$, it follows that $H = \langle x,\ p^\perp \cap q^\perp \rangle = y^\perp = \pi(y)$. Hence, H is a point of $\mathrm{Im}(\pi)$.

Step 5. π is a bijective transformation of the set of points of P into the projective space $\text{Im}(\pi)$: It follows from Step 3 that $(z^\perp)^\perp = z$ for all points z of P. Hence, π is an injective transformation with $\pi^2 = 1$.

Step 6. The transformation $\pi : P \to \text{Im}(\pi)$ is a collineation: For, let x, y and z be three points of P, and let $H := x^\perp$, $L := y^\perp$ and $M := z^\perp$.

(i) Suppose that the points x, y and z are collinear. It follows that the point z is contained in the line xy implying that

$$z^\perp \supseteq (xy)^\perp = x^\perp \cap y^\perp.$$

Hence, the subspace $H \cap L$ is contained in M, that is, H, L and M are collinear in P^*.

(ii) Suppose that the subspaces H, L and M are collinear in P^*. It follows that the subspace $H \cap L$ is contained in M, that is, $z^\perp \supseteq x^\perp \cap y^\perp$. It follows that

$$z = (z^\perp)^\perp \in (x^\perp \cap y^\perp)^\perp = ((xy)^\perp)^\perp = (g^\perp)^\perp = g.$$

The equation $z = (z^\perp)^\perp$ follows from Step 3, the relation $(z^\perp)^\perp \in (x^\perp \cap y^\perp)^\perp$ follows from Step 1 of Part (a), the equation $(x^\perp \cap y^\perp)^\perp = ((xy)^\perp)^\perp$ follows from Step 5 of Part (a), and the equation $(g^\perp)^\perp = g$ follows from Condition (ii).

From (i) and (ii), it follows that the points x, y and z are collinear if and only if $x^\perp$, $y^\perp$ and $z^\perp$ are collinear. It follows together with Step 5 that $\pi : P \to \text{Im}(\pi)$ is a collineation.

Step 7. If P is of finite dimension d, the transformation π is a polarity: By Theorem 4.1, the dual projective space P^* is also of dimension d. By Step 6, $\pi : P \to \text{Im}(\pi)$ is a collineation. By Theorem 4.2 of Chap. 2, we have $\dim(P) = \dim(\text{Im}(\pi))$. It follows that $\text{Im}(\pi) = P^*$. Hence, π is a bijective transformation of the set of points into the set of hyperplanes of P with the property that for any two points x and y of P, the relation "$x \in \pi(y)$" implies the relation "$y \in \pi(x)$" (Step 2 of the proof of Part (a)). It follows from Theorem 4.5 that π is a polarity. $\square$

Chapter 5
Quadrics and Quadratic Sets

1 Introduction

The present chapter is devoted to the study of quadrics and quadratic sets of a projective space. The main result is the Theorem of Buekenhout saying that every nondegenerate quadratic set is either an ovoid or a quadric (Theorems 6.4 and Corollary 6.5).

In Sect. 2, we shall introduce the notion of a quadratic set and we shall see that a quadratic set defines a polar space. Furthermore, we shall prove some elementary properties of quadratic sets.

Quadrics and quadratic forms, which have already been introduced in Sect. 6 of Chap. IV as special cases of pseudo-quadrics and pseudo-quadratic forms, are the subject of Sect. 3. We shall discuss the relation between quadratic forms and homogeneous quadratic polynomials, and we shall see that every quadric is a quadratic set.

In Sect. 4, we will consider the quadratic sets of a 3-dimensional projective space. Mainly, we shall investigate the so-called hyperbolic quadrics. We will see that a nondegenerate quadratic set of a 3-dimensional projective space is either an ovoid or a hyperbolic quadric.

For the investigation of quadratic sets and of quadrics, central collineations play a crucial role. In Sect. 5, we will see that nondegenerate quadratic sets containing at least one line, are perspective, that is, that there are "many" central collineations fixing the quadratic set.

Section 6 contains the main result of the present chapter: Every nondegenerate quadratic set is either an ovoid or a quadric.

In Sect. 7, we will introduce the Kleinian quadric. This quadric is of particular interest, since it provides a proper family of polar spaces which has been introduced in Sect. 7 of Chap. IV.

In Sect. 8, we will turn to one of the most famous results of finite geometries, namely the Theorem of Segre [43] saying that every oval of a finite Desarguesian projective plane of odd order is a conic (that is, a quadric in a projective plane).

J. Ueberberg, *Foundations of Incidence Geometry*, Springer Monographs in Mathematics, DOI 10.1007/978-3-642-20972-7_5, © Springer-Verlag Berlin Heidelberg 2011

2   Quadratic Sets

In the present section, we shall introduce quadratic sets and we shall prove some of their elementary properties.

Definition. Let P be a projective space, let U be a subspace of P, and let Q be a set of points of U.

(a) Let x be a point of Q. A line g of P is called a **tangent of** Q **at** x if g intersects the set Q in the point x or if all points of g are contained in Q.

(b) Let x be a point of Q. The **tangent set** Q_x consists of the point x and all points y of P such that the line xy is a tangent of Q at x.[1]

(c) The set Q is called a **quadratic set** of U if Q fulfils the following two conditions:

   (Q_1) Every line of U has at most two points in common with Q, or it is completely contained in Q.
   (Q_2) For any point x of Q, the tangent set Q_x is the point set of a hyperplane of U or it is the whole point set of U. The set Q_x is called the **tangent space of** Q **at** x **with respect to** U. If $U = P$ the set Q_x is just called the **tangent space of** Q **at** x.

(d) A point x of Q with the property that $Q_x = P$ is called a **double point**. The set of all double points of Q is called the **radical** of Q. It is denoted by $\mathrm{Rad}(Q)$.[2]

(e) A non-empty quadratic set Q which does not contain any double points (that is, $\mathrm{Rad}(Q) = \varnothing$) is called **nondegenerate**.[3]

The definition of a quadratic set is due to Buekenhout [11].

2.1 Theorem. *Let P be a projective space, and let Q be a quadratic set. Furthermore, let x and y be two points of Q. The point y is contained in the tangent space Q_x if and only if the line xy is contained in Q. In particular the point y is contained in Q_x if and only if the point x is contained in Q_y.*

Proof. The point y is contained in Q_x if and only if the line xy is a tangent of Q. Since the line xy contains the points x and y of Q, the line xy is a tangent of Q if and only if all points of the line xy are contained in Q.   □

2.2 Theorem. *Let P be a projective space, and let Q be a quadratic set. Let S be the geometry over the type set {point, line} defined as follows:*

(i) *The points of S are the points of Q.*
(ii) *The lines of S are the lines of P contained in Q.*
(iii) *The incidence in S is induced by the incidence in P.*

---

[1] The tangent set Q_x is a subset of the point set of P, but, in general, not a subset of Q.

[2] Note that a quadratic set may be empty. In this case, the radical $\mathrm{Rad}(Q)$ is empty as well.

[3] A nondegenerate quadratic set is a quadratic set such that for any point x of Q, the tangent space Q_x is a hyperplane of P.

Then, we have:

(a) If Q contains at least one line, the geometry S is a polar space.
(b) If Q is nondegenerate, S is nondegenerate.

Proof. (a) Verification of (P_1): Let x be a point of Q, and let g be a line in Q not containing x. Let Q_x be the tangent space at Q through x. By definition, Q_x is a hyperplane or P.

In any case, the line g contains at least one point y of $Q_x \cap Q$. By Theorem 2.1, the line xy is contained in Q. It follows that there exists at least one line of S through x intersecting the line g in a point.

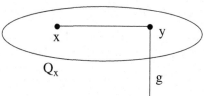

Next, we consider the case that there are at least two points y and z on g such that the lines xy and xz are contained in Q. It follows that y, z are contained in Q_x. Hence, the plane generated by x, y and z is contained in Q_x, that is, every line of P through x intersecting g in a point, is a tangent of Q through x. Any of these lines contains at least two points of Q (the point x and the intersection point with g), thus, any of these lines is contained in Q.
It follows that either exactly one point or all points of g are joined with x by a line of S.
Verification of (P_2): Since on every line of P, there are at least three points, on any line of S, there are also at least three points.

(b) Let Q be a nondegenerate quadratic set, and let x be a point of Q. Furthermore, let Q_x be the tangent space at Q through x. Since Q_x is a hyperplane, there exists a line g through x not contained in Q_x. Since g is not a tangent and contains a point of Q (namely x), there is exactly one further point y of Q on g. It follows that the points x and y are non-collinear in S.   □

Definition. Let P be a projective space, and let Q be a quadratic set in P. Let t be the maximal dimension of a subspace of P contained in Q. Then, $r := t + 1$ is called the **rank** of Q. (A possible value for t is $t = \infty$.)
The rank of a quadratic set Q is, by definition, the rank of the polar space defined by Q.

2.3 Theorem. *Let P be a d-dimensional projective space, and let Q be a nondegenerate quadratic set. If U is a subspace of P contained in Q, we have $\dim U < \tfrac{1}{2} d$.*

Proof. By Theorem 2.2, the subspaces of P contained in Q define a nondegenerate polar space S. Let U be a subspace of S, and let M be a maximal subspace of S through U. By Theorem 2.15 of Chap. IV, there exists a maximal subspace W of S disjoint to M. Since M and W are disjoint subspaces of P, it follows that $\dim U \leq \dim M < \tfrac{1}{2} d$.   □

2.4 Theorem. *Let P be a projective space, and let Q be a quadratic set. Furthermore, let E be a plane of P which has a point p in common with Q. Then, either E*

contains exactly one tangent of Q through p, or all lines of E through p are tangents of Q.

Proof. Since the tangent space Q_p is either a hyperplane of P or P, either the intersection $E \cap Q_p$ is a line, or we have $E \cap Q_p = E$.   □

2.5 Theorem. *Let P be a projective space, and let Q be a quadratic set. If U is a subspace of P, the set $Q \cap U$ is a quadratic set of U. Furthermore, for any point p of $Q \cap U$, the relation $Q_p \cap U = (Q \cap U)_p$ holds.*

Proof. Step 1. We first shall prove the relation $Q_p \cap U = (Q \cap U)_p$: For, let x be a point of U distinct from p. Then, we have

$$x \in Q_p \cap U \Leftrightarrow x \in U \text{ and } px \text{ is a tangent of } Q \text{ at } p$$

$$\Leftrightarrow px \subseteq U \text{ and } px \text{ is a tangent of } Q \text{ at } p$$

$$\Leftrightarrow px \text{ is a tangent of } Q \cap U \text{ at } p$$

$$\Leftrightarrow x \in (Q \cap U)_p.$$

Step 2. $Q \cap U$ is a quadratic set of U: Axiom (Q_1) follows from the fact that every line of P intersects the set Q in at most two points or is contained in Q. Hence, this is true for every line of U.

In order to verify Axiom (Q_2), we consider the set $(Q \cap U)_p = Q_p \cap U$ (Step 1). Since Q_p is a hyperplane of P, or $Q_p = P$, it follows that $Q_p \cap U$ is a hyperplane of U, or $Q_p \cap U = U$.   □

The rest of this section is devoted to the investigation of the radical Rad(Q) (the set of double points) of a quadratic set Q.

2.6 Theorem. *Let Q be a nondegenerate quadratic set of a projective space P, and let H be a hyperplane of P. Furthermore, let p be a point of $Q \cap H$.*
The point p is a double point of the quadratic set $Q \cap H$ if and only if $Q_p = H$.

Proof. Let $Q' := Q \cap H$. By Theorem 2.5, we have $Q'_p = Q_p \cap H$. Since Q is a nondegenerate quadratic set, Q_p is a hyperplane of P. It follows that

p is a double point of $Q' := Q \cap H \Leftrightarrow Q'_p = H \Leftrightarrow Q_p \cap H = H \Leftrightarrow Q_p = H.$

□

2.7 Theorem. *Let P be a projective space, and let Q be a quadratic set. Then, the radical Rad(Q) of Q is a subspace of P.*

Proof. Let x and y be two double points of Q, and let $g := xy$ be the line through x and y. Let z be a point on g. Since x is a double point of Q, the line g is a tangent of Q. Since g contains the two points x, y of Q, it follows that g is contained in Q. In particular, the point z is contained in Q.

Let h be a line of P through z.
Assume that h is not a tangent. On h,
there is exactly one further point s of Q
different from z. Since s is contained
in $Q \cap Q_x \cap Q_y$, it follows from
Theorem 2.1 that the lines sx and sy
are contained in Q. Since the points
and lines of Q define a polar space,
every line through s intersecting the line
$g = xy$, is contained in S.

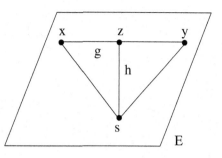

It follows that $h = sz$ is a tangent of Q, in contradiction to the assumption. Thus,
every line through z is a tangent, hence, z is a double point. $\qquad\square$

2.8 Theorem. *Let Q be a quadratic set of a projective space P.*

(a) *If x is a point of Q, every line xy, where y is a point of $Rad(Q)$, is contained in Q.*

(b) *For every point x of Q, the subspace $\langle x, \ Rad(Q)\rangle$ is contained in Q.*

(c) *For every point x of Q, the subspace $Rad(Q)$ is contained in Q_x.*

Proof. (a) Let y be a point of $Rad(Q)$. Every line through y is a tangent of Q. Since the line xy contains two points of Q (namely x and y), it follows that xy is contained in Q.

(b) and (c) follow from (a). $\qquad\square$

2.9 Theorem. *Let Q be a quadratic set of a projective space P with $Q \neq Rad(Q) \neq \emptyset$. Furthermore, let U be a complement[4] of $Rad(Q)$ in P.*

(a) *$Q' := Q \cap U$ is a nondegenerate quadratic set of U.*

(b) *Let $Q' := Q \cap U$. Q consists of all points incident with a line joining a point of Q' with a point of $Rad(Q)$, that is,*

$$Q = \{x \in P \mid \exists \, a \in Q' \, \exists \, b \in Rad(Q) \ with \ x \in ab\}.$$

Proof. (a) Let x be a point of Q'. Assume that x is a double point of Q'. Then, $U = Q'_x = Q_x \cap U$ is contained in Q_x. By Theorem 2.8, it follows that Q_x contains $\langle U, Rad(Q)\rangle = P$. It follows that x is a double point of Q, in contradiction to the assumption that U is a complement of $Rad(Q)$.

(b) Let $X := \{x \in P \mid \exists \, a \in Q' \, \exists \, b \in Rad(Q) \ with \ x \in ab\}$. We need to show that $X = Q$.

(i) X is contained in Q: Let x be a point of X. Then, there exists a point a of $Q' \subseteq Q$ and a point b of $Rad(Q)$ such that x is incident with the line ab. By Theorem 2.8, we have

$$x \in ab \subseteq \langle a, \ Rad(Q)\rangle \subseteq Q.$$

---

[4]That is, U is a subspace of P with $U \cap Rad(Q) = \emptyset$ and $\langle U, Rad(Q)\rangle = P$.

(ii) Q is contained in X: Let x be a point of Q. If x is contained in $\mathrm{Rad}(Q)$, x is contained in X. W.l.o.g., let x be a point outside of $\mathrm{Rad}(Q)$.

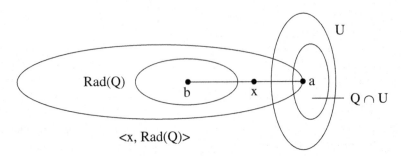

By Theorem 2.8, the subspace $\langle x, \mathrm{Rad}(Q)\rangle$ is contained in Q. Since the subspaces $\mathrm{Rad}(Q)$ and U are complementary, the subspaces $\langle x, \mathrm{Rad}(Q)\rangle$ and U meet in a point a of $U \cap Q = Q'$. Since the line ax is contained in the subspace $\langle x, \mathrm{Rad}(Q)\rangle$, the line ax intersects the subspace $\mathrm{Rad}(Q)$ in a point b.

It follows that x is incident with the line ab where a is contained in Q' and b is contained in $\mathrm{Rad}(Q)$, that is, x is contained in X.   □

In view of Theorem 2.9, the study of quadratic sets can be restricted to the study of nondegenerate quadratic sets.

2.10 Theorem. *Let Q be a quadratic set of a projective space P. Q is a subspace of P if and only if $Q = \mathrm{Rad}(Q)$.*

Proof. (i) If $Q = \mathrm{Rad}(Q)$, by Theorem 2.7, Q is a subspace of P.

(ii) Conversely, let Q be a subspace of P. Since every line of P is either contained in Q or has at most one point common with Q, all lines through a point x of Q are tangents of Q. It follows that every point of Q is a double point, that is, $Q = \mathrm{Rad}(Q)$.   □

2.11 Theorem. *Let Q be a quadratic set of a projective space P with $Q \neq \mathrm{Rad}(Q)$. Then, $\langle Q\rangle = P$.*

Proof. Since $Q \neq \mathrm{Rad}(Q)$, there exists a point x of Q which is not a double point. Then, $H := Q_x$ is a hyperplane of P. For every line g through x not contained in H, there is point of Q on g distinct from x (otherwise, g would be a tangent of Q). It follows that $\langle Q\rangle \supseteq \langle P \setminus H\rangle = P$.   □

Next, we shall determine the quadratic sets of a projective plane.

Definition. Let P be a projective plane, and let $\mathcal{O}$ be a set of points of P satisfying the following properties:

(i) On every line of P, there are at most two points of $\mathcal{O}$.

(ii) Through any point of $\mathcal{O}$, there is exactly one tangent, that is, there is exactly one line of $\boldsymbol{P}$ intersecting $\mathcal{O}$ in exactly one point.

The set $\mathcal{O}$ is called an **oval** of $\boldsymbol{P}$.

2.12 Theorem. *Let $\boldsymbol{P}$ be a projective plane, and let Q be a quadratic set of $\boldsymbol{P}$. Then, Q is one of the following sets:*

(i) Q is a subspace of $\boldsymbol{P}$. In this case, we have $\mathrm{Rad}(Q) = Q$.
(ii) Q consists of the set of the points of two lines, intersecting in a point x. In this case, we have $\mathrm{Rad}(Q) = x$.
(iii) Q is an oval. In this case, we have $\mathrm{Rad}(Q) = \varnothing$.

Proof. First case: Let $\mathrm{Rad}(Q) = \varnothing$, and let $Q \neq \varnothing$.

Step 1. Any point of Q is incident with exactly one tangent: Assume that there is a point p of Q incident with two tangents t_1 and t_2. Then, $Q_p \supseteq \langle t_1, t_2 \rangle = \boldsymbol{P}$, in contradiction to $\mathrm{Rad}(Q) = \varnothing$.

Step 2. Q does not contain a line: Assume on the contrary that Q contains a line g. Since $\langle Q \rangle = \boldsymbol{P}$ (Theorem 2.11), Q contains a point x not on g. The point x is incident with exactly one tangent t intersecting the line g in a point z. Hence, the point z of Q is incident with two tangents, a contradiction.

It follows from Steps 1 and 2 that Q is an oval.

Second case. Let $\mathrm{Rad}(Q) = x$ be a point. Let g be a line of $\boldsymbol{P}$ not incident with x. Then, g is a subspace of $\boldsymbol{P}$ complementary to x. Hence, we can apply Theorem 2.9:

If $g \cap Q = \varnothing$, we have $Q = \mathrm{Rad}(Q) = x$.

If g and Q meet in a point x, we have $Q = xy$ and $xy = \mathrm{Rad}(Q)$, in contradiction to $\mathrm{Rad}(Q) = x$.

If g and Q meet in two points y and z, we have $Q = xy \cup xz$, and we have $x = \mathrm{Rad}(Q)$.

If g is contained in Q, we have $Q = \langle x, g \rangle = \boldsymbol{P}$ and $\boldsymbol{P} = \mathrm{Rad}(Q)$, a contradiction to $\mathrm{Rad}(Q) = x$.

Third case. Let $\mathrm{Rad}(Q) = g$ be a line. Let p be a point outside of g. We shall, once more, apply Theorem 2.9.

If p is not contained in Q, we have $Q = g$ and $g = \mathrm{Rad}(Q)$.

If p is contained in Q, we have $Q = \langle p, g \rangle = \boldsymbol{P}$ and $\boldsymbol{P} = \mathrm{Rad}(Q)$, a contradiction to $\mathrm{Rad}(Q) = g$.

Fourth case. We have $Q = \boldsymbol{P}$ and $\boldsymbol{P} = \mathrm{Rad}(Q)$. □

3 Quadrics

Quadrics have been introduced in Sect. 6 of Chap. IV as special cases of pseudo-quadrics. In the present chapter we shall consider the relation between quadrics and homogeneous quadratic polynomials, and we shall show that every quadric is a quadratic set. Since the main results about quadrics and quadratic sets presented in the present chapter are about quadrics and quadratic sets in finite-dimensional projective spaces, we generally restrict ourselves to finite-dimensional projective spaces.

Every quadratic form defines a homogenous quadratic polynomial and vice versa. This relation is the subject of Theorems 3.1 and 3.2.

Definition. Let V be a $(d + 1)$-dimensional vector space over a commutative field K, and let $P = PG(d, K)$ be the corresponding d-dimensional projective space.

For $i, j = 0, 1, \ldots, d$, let q_{ij} be some elements of K. Define the transformation $q : V \to K$ by

$$q(x_0, x_1, \ldots, x_d) := \sum_{i,j=0}^{d} q_{ij} \, x_i \, x_j.$$

Then, q is called a **homogenous quadratic polynomial**.

3.1 Theorem. *Let V be a $(d + 1)$-dimensional vector space over a commutative field K, and let $q : V \to K$ be a quadratic polynomial. Then, q is a quadratic form.*

Proof. Since q is a quadratic homogenous polynomial, for $i, j = 0, 1 \ldots, d$, there exist elements q_{ij} of K such that

$$q(x_0, x_1, \ldots, x_d) = \sum_{i,j=0}^{d} q_{ij} \, x_i \, x_j.$$

(i) We have $q(\lambda \, v) = \lambda^2 \, q(v)$ for all v of V and for all λ of K: For a vector $v = (x_0, x_1, \ldots, x_d)^t$ of V and an element λ of K, we have[5]

$$q(\lambda \, v) = q(\lambda \, x_0, \lambda \, x_1, \ldots, \lambda \, x_d)$$

$$= \sum_{i,j=0}^{d} q_{ij} \lambda x_i \lambda x_j$$

$$= \lambda^2 \sum_{i,j=0}^{d} q_{ij} x_i x_j = \lambda^2 q(v).$$

(ii) The transformation $f : V \times V \to K$ defined by

$$f(v, w) := q(v + w) - q(v) - q(w)$$

is a symmetric bilinear form: For all vectors v, w of V with $v = (x_0, x_1, \ldots, x_d)^t$ and $w = (y_0, y_1, \ldots, y_d)^t$, we have

---

[5]Note that, by assumption, K is commutative.

$$f(v, w) = q(v + w) - q(v) - q(w)$$

$$= \sum_{i,j=0}^{d} q_{ij} (x_i + y_i)(x_j + y_j) - q_{ij} x_i x_j - q_{ij} y_i y_j$$

$$= \sum_{i,j=0}^{d} q_{ij} (x_i y_j + x_j y_i).$$

A simple computation yields for u, v, w of V and λ of K:

$$f(v + u, w) = f(v, w) + f(u, w)$$

$$f(\lambda v, w) = \lambda f(v, w)$$

$$f(v, w + u) = f(v, w) + f(v, u)$$

$$f(v, \lambda w) = \lambda f(v, w) \text{ and}$$

$$f(v, w) = f(w, v).$$

□

Remark. If $q : V \to K$ is a quadratic homogenous polynomial with

$$q(x_0, x_1, \ldots, x_d) = \sum_{i,j=0}^{d} q_{ij} x_i x_j,$$

by Theorem 3.1, q is a quadratic form and hence defines a quadric Q. One says that the quadric Q is defined by the quadratic equation $\sum_{i,j=0}^{d} q_{ij} x_i x_j = 0$.

By Theorem 3.1, a quadratic homogenous polynomial is a quadratic form. As we shall see in Theorem 3.2, conversely, every quadratic form of a finite-dimensional vector space defines a homogenous quadratic polynomial. For finite-dimensional vector spaces, the two notions are equivalent.

3.2 Theorem. *Let V be a $(d + 1)$-dimensional vector space over a commutative field K, and let $q : V \to K$ be a quadratic form with symmetric bilinear form f.*
   Then, q is a quadratic homogenous polynomial.

Proof. Let $\{e_0, e_1, \ldots, e_d\}$ be a basis of V. For i, $j = 0, 1, \ldots, d$, set

$$q_{ii} := q(e_i)$$

$$q_{ij} := f(e_i, e_j) = q(e_i + e_j) - q(e_i) - q(e_j) \text{ for } i < j$$

$$q_{ij} := 0 \text{ for } i > j.$$

Let $v = \sum_{r=0}^{d} x_r e_r$ be an element of V. Furthermore, let

$$v_k := \sum_{r=0}^{k} x_r e_r \text{ for } k = 0, \ldots, d.$$

Finally, let $g : V \rightarrow K$ be the transformation defined by $g(v) := \sum\limits_{i,k=0}^{d} q_{ik} x_i x_k$.
We shall prove by induction on k that

$$q(v_k) = g(v_k) \text{ for all } k = 0, \ldots, d.$$

$k = 0$: We have $q(v_0) = q(x_0 e_0) = x_0^2 q(e_0) = q_{00} x_0^2 = g(v_0)$.
$k - 1 \rightarrow k$: We have

$$
\begin{aligned}
q(v_k) &= q(v_{k-1} + x_k e_k) \\
&= f(v_{k-1}, x_k e_k) + q(v_{k-1}) + q(x_k e_k) \\
&= \sum_{i=0}^{k-1} x_i x_k f(e_i, e_k) + g(v_{k-1}) + x_k^2 q(e_k) \\
&= \sum_{i=0}^{k-1} q_{ik} x_i x_k + \sum_{i,j=0}^{k-1} q_{ij} x_i x_j + q_{kk} x_k^2 \\
&= \sum_{i=0}^{k} q_{ik} x_i x_k = g(v_k).
\end{aligned}
$$

It follows from $q(v_d) = g(v_d)$ that q is a homogenous quadratic polynomial.   □

In Theorem 3.7, we shall see that every quadric is a quadratic set. Theorems 3.3 to 3.6 prepare the proof of this assertion.

3.3 Theorem. *Let $P = PG(d, K)$ be a d-dimensional projective space over a vector space V over a commutative field K. Furthermore, let Q be a quadric of P.*

Every line of P intersects the quadric Q in at most two points, or all points of the line are contained in Q.

Proof. Let $q : V \rightarrow K$ be the quadratic form defining the quadric Q. Let g be a line of P which has three points $x = \langle v \rangle$, $y = \langle w \rangle$ and $z = \langle v + \lambda w \rangle$ for some element $0 \neq \lambda$ of K in common with Q. We need to show that all points of g are contained in Q.

For, let $a := \langle v + \mu w \rangle$ with $0, \lambda \neq \mu \in K$ be a further point on g. Since x, y, z are contained in Q, it follows that $q(v) = q(w) = q(v + \lambda w) = 0$. In particular, we have

$$
\begin{aligned}
0 = q(v + \lambda w) &= f(v, \lambda w) + q(v) + q(\lambda w) \\
&= \lambda f(v, w) + q(v) + \lambda^2 q(w) \\
&= \lambda f(v, w).
\end{aligned}
$$

Since $\lambda \neq 0$, we have $f(v, w) = 0$. It follows that

$$
\begin{aligned}
q(v + \mu w) &= f(v, \mu w) + q(v) + q(\mu w) \\
&= \mu f(v, w) + q(v) + \mu^2 q(w) \\
&= 0,
\end{aligned}
$$

that is, $a = \langle v + \mu w \rangle$ is contained in Q.   □

3.4 Theorem. *Let $P = PG(d, K)$ be a d-dimensional projective space over a vector space V over a commutative field K. Let Q be a quadric of P with quadratic form $q : V \to K$ and bilinear form f. Let $x = \langle v \rangle$ and $w = \langle y \rangle$ be two points of Q. The line xy is contained in Q if and only if $f(v, w) = 0$.*

Proof. (i) Suppose that the line xy is contained in Q. In particular, the point $\langle v + w \rangle$ is contained in Q. It follows that $f(v, w) = q(v + w) - q(v) - q(w) = 0$.

(ii) Suppose that $f(v, w) = 0$. Let $\langle v + \lambda w \rangle$ be an arbitrary point on xy. It follows that

$$q(v + \lambda w) = f(v, \lambda w) + q(v) + q(\lambda w) = \lambda\, f(v, w) + q(v) + \lambda^2\, q(w) = 0.$$

Hence, the point $\langle v + \lambda w \rangle$ is contained in Q.   □

Definition. Let V be a $(d + 1)$-dimensional vector space over a commutative field K, and let $q : V \to K$ be a quadratic form with bilinear form f. For an element $0 \neq v$ of V, set

$$\langle v \rangle^\perp := \{w \in V \mid f(v, w) = 0\}.$$

Note that $\langle v \rangle^\perp$ is a subspace of V and hence a subspace of $P = PG(d, K)$.

3.5 Theorem. *Let $P = PG(d, K)$ be a d-dimensional projective space over a vector space V over a commutative field K. Let Q be a quadric of P with quadratic form $q : V \to K$.*

(a) *For every point $\langle v \rangle$ of P, we have: If $\langle v \rangle$ is contained in Q, the point $\langle v \rangle$ is contained in $\langle v \rangle^\perp$. If Char $K \neq 2$, the converse is also true, that is, if $\langle v \rangle$ is not contained in Q, the point $\langle v \rangle$ is not contained in $\langle v \rangle^\perp$.*

(b) *For an element $0 \neq v$ of V, the subspace $\langle v \rangle^\perp$ is a hyperplane of P or $\langle v \rangle^\perp$ equals P.*

(c) *For every point $\langle v \rangle$ of Q, the subspace $\langle v \rangle^\perp$ is the tangent set at Q through $\langle v \rangle$.*

(d) *For every two points $\langle v \rangle$ and $\langle w \rangle$ of Q, we have: The line $\langle v, w \rangle$ is contained in Q if and only if $\langle w \rangle$ is contained in $Q \cap \langle v \rangle^\perp$, or $\langle v \rangle$ is contained in $Q \cap \langle w \rangle^\perp$.*

Proof. (a) Let $\langle v \rangle$ be a point of Q. We have $q(v) = 0$, and it follows that

$$f(v, v) = q(v + v) - q(v) - q(v) = 2^2\, q(v) - q(v) - q(v) = 0,$$

thus, v is contained in $\langle v \rangle^\perp$.

Let Char $K \neq 2$, and let $\langle v \rangle \notin Q$. Then, $q(v) \neq 0$, and it follows that

$$f(v, v) = q(v + v) - q(v) - q(v) = 2^2\, q(v) - q(v) - q(v) = 2\, q(v) \neq 0,$$

hence, $v \notin \langle v \rangle^\perp$.

(b) The transformation $f(v, \cdot) : V \to K$ is a linear transformation with Kern $f(v, \cdot) = \langle v \rangle^\perp$. It follows that $\langle v \rangle^\perp$ is a hyperplane, or $\langle v \rangle^\perp$ equals V.

(c) Step 1. Every line of $\langle v \rangle^\perp$ through $\langle v \rangle$ is a tangent of Q through $\langle v \rangle$: Let g be a line of $\langle v \rangle^\perp$ through $\langle v \rangle$ containing besides $\langle v \rangle$ a further point $\langle w \rangle$ of Q. Since $g = \langle v, w \rangle$ is contained in $\langle v \rangle^\perp$, we have $f(v, w) = 0$. Let $x = \langle v + \lambda\, w \rangle$, $0 \neq \lambda \in K$, be a further point on g.

Then,

$$q(v + \lambda\, w) = f(v, \lambda\, w) + q(v) + q(\lambda\, w)$$
$$= \lambda\, f(v,\, w) + q(v) + \lambda^2\, q(w)$$
$$= 0.$$

It follows that x is contained in Q, that is, the line g is contained in Q.

Step 2. Let $\langle v \rangle$ be a point of Q. Then, every tangent through $\langle v \rangle$ at Q is contained in $\langle v \rangle^{\perp}$: By Theorem 3.3, every line through $\langle v \rangle$ intersects the quadric Q in one or in two points or it is completely contained in Q. It follows that we need to show that every line of $\boldsymbol{P}$ through $\langle v \rangle$ not contained in $\langle v \rangle^{\perp}$, intersects the quadric Q in exactly two points.

For, let $g = \langle v, w \rangle$ be a line through $\langle v \rangle$ not contained in $\langle v \rangle^{\perp}$. Then, w is not contained in $\langle v \rangle^{\perp}$, hence, $f(v, w) \neq 0$. Hence, for any point $\langle v \rangle \neq x = \langle \lambda\, v + w \rangle$ on g, we have

$$x \in Q \Leftrightarrow 0 = q(\lambda\, v + w)$$
$$= f(\lambda\, v,\, w) + q(\lambda\, v) + q(w)$$
$$= \lambda\, f(v,\, w) + q(w)$$
$$\Leftrightarrow \lambda = -\frac{q(w)}{f(v, w)}.$$

Besides the point x, there is exactly one further point of Q on g, namely the point $y = \langle \mu\, v + w \rangle$ with $\mu = -q(w)/f(v, w)$.

(d) The proof is obvious.   □

3.6 Theorem. *Let $\boldsymbol{P} = PG(d, K)$ be a d-dimensional projective space over a vector space V over a commutative field K. Furthermore, let Q be a quadric of $\boldsymbol{P}$ defined by the equation*

$$\sum_{i,j=0}^{d} q_{ij}\, x_i\, x_j = 0.$$

The tangent space Q_p of a point $p = (p_0,\, p_1,\, \ldots,\, p_d)^t$ of Q is defined by the equation

$$\sum_{i,j=0}^{d} q_{ij}\, (p_i\, x_j + x_i\, p_j) = 0.$$

Proof. Let $p = \langle v \rangle$ with $0 \neq v = (p_0,\, p_1,\, \ldots,\, p_d)^t \in V$. By Theorem 3.5, we have

$$Q_p = \langle v \rangle^{\perp} = \{w \in \mathbf{V} \mid f(v,\, w) = 0\}.$$

If $w = (x_0,\, x_1,\, \ldots,\, x_d)^t$, we have

$$f(v, w) = \sum_{i,j=0}^{d} q_{ij} \left(p_i x_j + x_i p_j \right).$$

The assertion follows.   □

3.7 Theorem. *Let* $P = PG(d, K)$ *be a d-dimensional projective space over a vector space V over a commutative field K. Furthermore, let Q be a quadric of P with quadratic form q.*

(a) Q is a quadratic set.
(b) The quadratic set Q is nondegenerate if and only if the quadratic form q is nondegenerate.

Proof. (a) Verification of (Q_1): By Theorem 3.3, every line of P intersects the quadric Q in at most two points or is completely contained in Q.

Verification of (Q_2): By Theorem 3.5, for any point $\langle v \rangle$ of Q, the set $\langle v \rangle^{\perp}$ is the tangent set at Q through $\langle v \rangle$. Again by Theorem 3.5, $\langle v \rangle^{\perp}$ is a hyperplane of P or $\langle v \rangle^{\perp}$ equals P.

(b) By Theorem 3.5, for any point $\langle v \rangle$ of Q, the tangent space $Q_{\langle v \rangle}$ at Q through $\langle v \rangle$ equals $\langle v \rangle^{\perp}$. It follows that
q is nondegenerate is the first part of the equivalences

$\Leftrightarrow$ For any point $\langle v \rangle$ of Q, there exists an element $0 \neq w$ of $\mathbf{V}$
with $f(v, w) = 0$.
$\Leftrightarrow$ We have $\langle v \rangle^{\perp} \neq P$ for all $\langle v \rangle$ of Q.
$\Leftrightarrow$ We have $Q_{\langle v \rangle} \neq P$ for all $\langle v \rangle$ of Q.
$\Leftrightarrow$ Q is nondegenerate.

□

The previous theorems imply a construction method for quadratic sets: Every homogeneous quadratic polynomial of the form

$$q(x_0, x_1, \ldots, x_d) := \sum_{i,j=0}^{d} q_{ij} x_i x_j$$

with $q_{ij} \in K$ defines a quadric in $P = PG(d, K)$ over a commutative field K. By Theorem 3.7, this quadric is a quadratic set.

4   Quadratic Sets in $PG(3, K)$

The aim of the present section is the classification of the quadratic sets of a 3-dimensional projective space $P = PG(3, K)$. This includes a detailed investigation of the so-called hyperbolic quadrics. The main result of this section (Theorem 4.13) says that a quadratic set in $P = PG(3, K)$ is either a subspace of P, the point set of two planes, a cone, an ovoid or a hyperbolic quadric.

At the end of this section, we shall see that there are central collineations of P fixing a hyperbolic quadric (Theorem 4.15).

Definition. Let P be a d-dimensional projective space, and let U and W be two subspaces of P. The subspaces U and W are called **skew** if U and W are disjoint.

4.1 Theorem. *Let P be a 3-dimensional projective space, and let g and h be two skew lines of P. Furthermore, let p be a point of P which is neither incident with g nor with h. There exists exactly one line l through p, intersecting each of the lines g and h in a point.*

Proof. (i) Existence:

Let $E := \langle p, g \rangle$ be the plane generated by p and g. Since dim $P = 3$, the plane E meets the line h in a point x. Since the lines g and $l := px$ are contained in the plane E, the lines g and l meet in a point y.

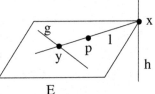

(ii) Uniqueness: Assume that there are two lines l_1 and l_2 through p intersecting each of the lines g and h in a point. Then, the plane $E := \langle l_1, l_2 \rangle$ contains the lines g and h, in contradiction to the assumption that g and h are disjoint. $\square$

Definition. Let P be a d-dimensional projective space, and let $\mathcal{R}$ be a set of pairwise skew subspaces of P. A line t of P is called a **transversal of** $\mathcal{R}$ if the line t intersects each of the subspaces of $\mathcal{R}$ in exactly one point.

4.2 Theorem. *Let P be a 3-dimensional projective space, and let G be a set of three pairwise disjoint lines. For the set $\mathcal{T}$ of the transversals of G, we have:*

(a) The lines of $\mathcal{T}$ are pairwise disjoint.
(b) Any point on a line of G is incident with exactly one transversal.
(c) The lines of G are transversals of $\mathcal{T}$.

Proof. The proof of the assertion follows from Theorem 4.1. $\square$

Definition. Let P be a 3-dimensional projective space, and let $\mathcal{R}$ be a set of pairwise disjoint lines. The set $\mathcal{R}$ is called a **regulus** of P if the set $\mathcal{R}$ fulfils the following conditions:

(R_1) Each point on a line of $\mathcal{R}$ is incident with a transversal of $\mathcal{R}$.
(R_2) Each point on a transversal of $\mathcal{R}$ is incident with a line of $\mathcal{R}$.

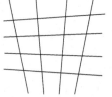

4.3 Theorem. *Let P be a 3-dimensional projective space, and let $\mathcal{R}$ be a regulus of P.*

(a) $\mathscr{R}$ *contains at least three lines.*

(b) *Let x be a point on a line of* $\mathscr{R}$*. Then, x is incident with exactly one transversal.*

(c) *Any two transversals of* $\mathscr{R}$ *are skew.*

(d) *Let* $\mathscr{T}$ *be the set of the transversals of* $\mathscr{R}$*. Then,* $\mathscr{T}$ *is a regulus. The set of the transversals of* $\mathscr{T}$ *is* $\mathscr{R}$*.*

Proof. (a) Let t be a transversal of $\mathscr{R}$. Since each point of t is incident with a line of $\mathscr{R}$, it follows that $|\mathscr{R}| \geq 3$.

(b) By Axiom (R_1), any point on a line of $\mathscr{R}$ is incident with at least one transversal. The uniqueness follows from Theorem 4.2 together with (a).

(c) Since $|\mathscr{R}| \geq 3$, the assertion follows from Theorem 4.2.

(d) Step 1. By (c), the set $\mathscr{T}$ consists of pairwise disjoint lines.

Step 2. Let g be a line of $\mathscr{R}$. Then, g is a transversal of $\mathscr{T}$: For, let t be a line of $\mathscr{T}$. Since t intersects every line of $\mathscr{R}$,[6] in particular, t and g meet in a point. It follows that g meets every line of $\mathscr{T}$, that is, g is a transversal of $\mathscr{T}$.

Step 3. The lines of $\mathscr{R}$ are exactly the transversals of $\mathscr{T}$: By Step 2, every line of $\mathscr{R}$ is a transversal of $\mathscr{T}$. Conversely, let h be a transversal of $\mathscr{T}$. Let t_1, t_2 and t_3 be three lines of $\mathscr{T}$, and let $x := h \cap t_1$ be the intersection point of h and t_1. By Axiom (R_2), the point x is incident with a line g of $\mathscr{R}$. Since g and h are incident with the point x and since g and h meet the lines t_2 and t_3, it follows from Theorem 4.1 that $h = g$ is contained in $\mathscr{R}$.

Step 4. $\mathscr{T}$ is a regulus: By Step 3, the lines of $\mathscr{R}$ are exactly the transversals of $\mathscr{T}$. Since Axioms (R_1) and (R_2) are symmetric with respect to the sets $\mathscr{R}$ and $\mathscr{T}$, it follows from the fact that $\mathscr{R}$ a regulus, that $\mathscr{T}$ is a regulus as well.   □

Definition. Let *P* be a 3-dimensional projective space, and let $\mathscr{R}$ be a regulus of *P*. Let $\mathscr{T}$ be the set of the transversals of $\mathscr{R}$. Then, $\mathscr{T}$ is called the **opposite regulus**[7] of $\mathscr{R}$.

4.4 Theorem. *Let **P** be a 3-dimensional projective space, and let* $\mathscr{R}$ *be a regulus of* **P***. Furthermore, let R be the set of the points on the lines of* $\mathscr{R}$*. Finally, let* $\mathscr{T}$ *be the opposite regulus of* $\mathscr{R}$*.*

(a) *Each line of* **P** *intersects the set R in at most two points or is completely contained in R.*

(b) *The lines completely contained in R are exactly the lines of* $\mathscr{R}$ *and* $\mathscr{T}$*.*

(c) *Each point of R is incident with exactly one line of* $\mathscr{R}$ *and with exactly one line of* $\mathscr{T}$*.*

---

[6]By assumption, t is a transversal of $\mathscr{R}$.

[7]Since $\mathscr{T}$ is a regulus, $\mathscr{R}$ is the opposite regulus of $\mathscr{T}$. In this sense, regulus and opposite regulus are two symmetric notions.

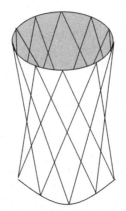

Regulus and Opposite Regulus

Proof. (a) and (b) Let h be a line intersecting the set R in at least three points. Then, either h is a line of $\mathscr{R}$, or h meets at least three lines g_1, g_2, g_3 of $\mathscr{R}$ in a point. If h is a line of $\mathscr{R}$, the assertion is shown. Let us suppose that $x := h \cap g_1$ is a point. By Theorem 4.2, h is the uniquely determined transversal of these three lines through the point x. It follows that h is a transversal of $\mathscr{R}$ through x. By Axiom (R_2), h is contained in R.

(c) Since each of the sets $\mathscr{R}$ and $\mathscr{T}$ consists of pairwise skew lines, any point of R is incident with exactly one line of $\mathscr{R}$ and with exactly one line of $\mathscr{T}$. □

4.5 Theorem. *Let **P** be a 3-dimensional projective space, and let $\mathscr{R}$ be a regulus of **P**. Let $\mathscr{T}$ be the opposite regulus of $\mathscr{R}$. If R is the set of the points on the lines of $\mathscr{R}$, R and the lines of $\mathscr{R} \cup \mathscr{T}$ form a generalized quadrangle.*

Proof. We shall verify Axioms (V_1), (V_2) and (V_3):

Verification of (V_1): Obviously, any two points of R are incident with at most one line of $\mathscr{R} \cup \mathscr{T}$.

Verification of (V_2): Let g be a line of $\mathscr{R} \cup \mathscr{T}$, and let x be a point of R which is not on g. If g is contained in $\mathscr{R}$, there exists exactly one line of $\mathscr{T}$ (and no line of $\mathscr{R}$) through x intersecting the line g. If g is contained in $\mathscr{T}$, there exists exactly one line of $\mathscr{R}$ (and no line of $\mathscr{T}$) through x intersecting the line g.

Verification of (V_3): Obviously, on each line of $\mathscr{R} \cup \mathscr{T}$, there are at least two points. By Theorem 4.4, any point of R is incident with two lines of $\mathscr{R} \cup \mathscr{T}$. □

4.6 Theorem. *Let **P** be a 3-dimensional projective space over a 4-dimensional vector space **V**. Let g, g_0 and g_1 be three pairwise disjoint lines, and let t, t_0 and t_1 be three transversals of $\{g, g_0, g_1\}$. There exists a basis $\{v_0, v_1, v_2, v_3\}$ of **V** with the following properties:*

(a) We have

$$g_0 \cap t_0 = \langle v_0 \rangle$$
$$g_0 \cap t = \langle v_1 \rangle$$
$$g_0 \cap t_1 = \langle v_0 + v_1 \rangle$$
$$g \cap t_0 = \langle v_2 \rangle$$
$$g \cap t = \langle v_3 \rangle$$
$$g \cap t_1 = \langle v_2 + v_3 \rangle$$
$$g_1 \cap t_0 = \langle v_0 + v_2 \rangle$$
$$g_1 \cap t = \langle v_1 + v_3 \rangle$$
$$g_1 \cap t_1 = \langle v_0 + v_1 + v_2 + v_3 \rangle.$$

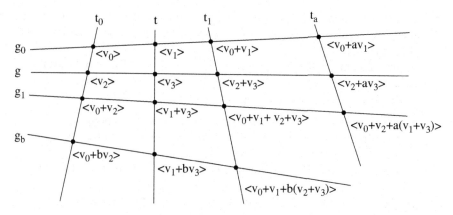

(b) For an element a of K, let t_a be the line $\langle v_0 + a\, v_1,\ v_2 + a\, v_3 \rangle$. The lines t and t_a $(a \in K)$ are exactly the transversals of $\{g, g_0, g_1\}$.

(c) For an element b of K, let g_b be the line $\langle v_0 + b\, v_2,\ v_1 + b\, v_3 \rangle$. The lines g and g_b $(b \in K)$ are exactly the transversals of $\{t, t_0, t_1\}$.

Proof. (a) Since the lines g_0 and g are skew, the points $g_0 \cap t_0$, $g_0 \cap t$, $g \cap t_0$ and $g \cap t$ generate $\boldsymbol{P}$. For $\langle v_0 \rangle := g_0 \cap t_0$, $\langle w_1 \rangle := g_0 \cap t$, $\langle w_2 \rangle := g \cap t_0$ and $\langle w_3 \rangle := g \cap t$, the set $\{v_0, w_1, w_2, w_3\}$ is a basis of V.

For the point $g_0 \cap t_1$, there exists an element $0 \neq \lambda$ of K such that $g_0 \cap t_1 = \langle v_0 + \lambda\, w_1 \rangle$. Set $v_1 := \lambda\, w_1$.

Similarly, there exist for the points $g_1 \cap t_0$ and $g_1 \cap t$ elements $0 \neq \mu$, ν of K such that $g_1 \cap t_0 = \langle v_0 + \mu\, w_2 \rangle$ and $g_1 \cap t = \langle v_1 + \mu\, w_3 \rangle$. Set $v_2 := \mu\, w_2$ and $v_3 = \nu\, w_3$.

Let h be the line $\langle v_0 + v_1,\ v_2 + v_3 \rangle$. Then, h meets the lines g_0, g and g_1 in the points $\langle v_0 + v_1 \rangle$, $\langle v_2 + v_3 \rangle$ and $\langle v_0 + v_1 + v_2 + v_3 \rangle$, respectively. It follows that h is a transversal of $\{g_0, g, g_1\}$ through the point $g_0 \cap t_1 = \langle v_0 + v_1 \rangle$. By Theorem 4.1, we have $h = t_1$. It follows that

$$g \cap t_1 = g \cap h = \langle v_2 + v_3 \rangle \text{ and}$$

$$g_1 \cap t_1 = g_1 \cap h = \langle v_0 + v_1 + v_2 + v_3 \rangle.$$

(b) The points on g_0 are exactly the points $\langle v_1 \rangle$ and $\langle v_0 + a v_1 \rangle$, $a \in K$. Obviously, for an element a of K, the line t_a meets the lines g_0, g and g_1 in the points $g_0 \cap t_a = \langle v_0 + a\, v_1 \rangle$, $g \cap t_a = \langle v_2 + a\, v_3 \rangle$ and $g_1 \cap t_a = \langle v_0 + v_2 + a\, (v_1 + v_3) \rangle$, respectively.

Hence, the set $\{t\} \cup \{t_a \mid a \in K\}$ is the set of the transversals of $\{g, g_0, g_1\}$.

(c) follows analogously to (b). $\square$

4.7 Theorem. *Let* $P = PG(3, K)$ *be a 3-dimensional projective space over a vector space* V *and a skew field K.*

(a) *In* P, *there exists a regulus if and only if K is commutative.*

(b) *If K is commutative, any three pairwise skew lines are contained in exactly one regulus.*

Proof. (a) In P, there exists a regulus if and only if there exist three pairwise skew lines g, g_0, g_1 which are contained in a regulus. Let t, t_0, t_1 be three transversals of $\{g, g_0, g_1\}$.

By Theorem 4.6, there exists a basis $\{v_0, v_1, v_2, v_3\}$ of V with the properties described in Theorem 4.6. The following assertions are equivalent:

The lines g, g_0, g_1 are contained in a regulus.

↔ For all elements a, b of K, the lines t_a and g_b meet in a point s_{ab}.

↔ For all elements a, b of K, there exist λ, μ of K such that $v_0 + b\, v_2 + \lambda(v_1 + b\, v_3) = s_{ab} = v_0 + a\, v_1 + \mu(v_2 + a\, v_3)$.

↔ For all elements a, b of K, there exist λ, μ of K such that $0 = (a - \lambda)\, v_1 + (\mu - b)\, v_2 + (\mu\, a - \lambda\, b)\, v_3$.

↔ For all elements a, b of K, there exist λ, μ of K such that $\lambda = a$, $\mu = b$, $\mu\, a = \lambda b$.[8]

↔ For all elements a, b of K, we have $ab = ba$.

↔ K is commutative.

(b) Since every three pairwise disjoint lines g, g_0, g_1 and every three transversals t, t_0, t_1 of $\{g, g_0, g_1\}$ can be transformed into the form described in Theorem 4.6, the assertion follows from (a). $\square$

4.8 Theorem. *Let* $P = PG(3, K)$ *be a 3-dimensional projective space over a vector space* V *over a commutative field K, and let* $\mathcal{R}$ *be a regulus of* P. *Furthermore, let R be the set of the points on the lines of* $\mathcal{R}$. *There exists a basis* $\{v_0, v_1, v_2, v_3\}$ *of V such that the set R consists exactly of the following points:*

$$R = \{(0, 0, 0 1)^t\} \cup \{(0, 0, 1, a)^t \mid a \in K\} \cup \{(0, 1, 0, a)^t \mid a \in K\}$$

$$\cup \{(1, a, b, ab)^t \mid a, b \in K\}.$$

---

[8]Note that the vectors v_1, v_2, v_3 are linearly independent.

Proof. We choose the notations and the basis $\{v_0, v_1, v_2, v_3\}$ as in Theorem 4.6. The points of R are exactly the points on the lines g, g_b ($b \in K$). The line g consists of the point $(0, 0, 0, 1)^t$ and the points $(0, 0, 1, a)^t$, $a \in K$.

For an element b of K, the line g_b consists of the point $(0, 1, 0, b)^t$ and the points $(1, a, b, ab)^t$, $a \in K$. The assertion follows. □

4.9 Theorem. *Let $P = PG(3, K)$ be a 3-dimensional projective space over a vector space V over a commutative field K, and let $\mathscr{R}$ be a regulus of P. Furthermore, let R be the set of the points on the lines of $\mathscr{R}$.*

(a) The set R is a quadric of P.
(b) There exists a basis $\{v_0, v_1, v_2, v_3\}$ of V such that the quadric R is defined by the equation

$$x_0\, x_3 - x_1\, x_2 = 0.$$

Proof. We shall prove Parts (a) and (b) in common. By Theorem 4.8, there exists a basis $\{v_0, v_1, v_2, v_3\}$ such that

$$R = \{(0, 0, 0, 1)^t\} \cup \{(0, 0, 1, a)^t \mid a \in K\} \cup \{(0, 1, 0, a)^t \mid a \in K\}$$

$$\cup \{(1, a, b, ab)^t \mid a, b \in K\}.$$

Let Q be the quadric of P defined by the equation $x_0\, x_3 - x_1\, x_2 = 0$. Since each point of R fulfils the equation $x_0\, x_3 - x_1\, x_2 = 0$, it follows that R is contained in Q.

Conversely, let $z = (z_0, z_1, z_2, z_3)$ be a point of Q. If $z_0 = z_1 = z_2 = 0$, $z = (0, 0, 0, 1)^t$ is an element of R. If $z_0 = z_1 = 0$ and $z_2 \neq 0$, we have $z = (0, 0, 1, a)^t$ for an element a of K. Again, it follows that z is a point of R.

If $z_0 = 0$ and $z_1 \neq 0$, we have $z = (0, 1, a, b)^t$ for some elements a, b of K. Since z is an element of Q, it fulfils the equation $z_0\, z_3 - z_1 z_2 = 0$. Thus, we get $a = 0$. Hence, we have $z = (0, 1, 0, b)^t \in R$.

Finally, let $z_0 \neq 0$. Then, we have $z = (1, a, b, c)^t$ for some a, b, c of K. From $z_0\, z_3 - z_1\, z_2 = 0$, it follows that $c = ab$, that is, $z = (1, a, b, ab)^t$ is contained in R. Altogether, it follows that Q is contained in R. □

Definition. Let $P = PG(3, K)$ be a 3-dimensional projective space, and let $\mathscr{R}$ be a regulus of P. Furthermore, let R be the set of the points on the lines of $\mathscr{R}$. Then, R is called a **hyperbolic quadric**.

4.10 Theorem. *Let P be a 3-dimensional projective space, let $\mathscr{R}$ be a regulus of P, and let R be the hyperbolic quadric defined by the regulus $\mathscr{R}$. Let E be a plane of P. Then, E and R meet in one of the following sets:*

(i) The set $E \cap R$ is an oval in E.
(ii) The set $E \cap R$ consists of a line of $\mathscr{R}$ and a transversal of $\mathscr{R}$.

Proof. Since R is a quadratic set (Theorem 4.9), $E \cap R$ is a quadratic set in E. By Theorem 2.12, $E \cap R$ is a subspace of E, the set of the points of two lines intersecting in a point, or R is an oval.

Since E has at least one point in common with each line of $\mathscr{R}$, we have $|E \cap R| \geq 3$. It follows that $E \cap R$ is an oval, or $E \cap R$ contains a line.

We consider the case that $E \cap R$ contains a line g. Since the lines contained in R are either lines of $\mathscr{R}$ or transversals of $\mathscr{R}$, we can assume w.l.o.g. that g is a line of $\mathscr{R}$. Each of the other lines of $\mathscr{R}$ intersect E in a point. It follows that $E \cap R$ contains besides the line g at least one further point, that is, $E \cap R$ consists of the set of the points of two lines, or we have $E \cap R = E$. Since no planes are contained in R, the assertion follows.   □

Definition. Let P be a d-dimensional projective space, and let $\mathscr{O}$ be a non-empty set of points of P. The set $\mathscr{O}$ is called an **ovoid** if $\mathscr{O}$ satisfies the following conditions:

(i) No three points of $\mathscr{O}$ are collinear.
(ii) For any point x of $\mathscr{O}$, the tangent set $\mathscr{O}_x$ is the point set of a hyperplane of P.

For $d = 2$, an ovoid is an oval.

4.11 Theorem. *Let P be a d-dimensional projective space, and let $\mathscr{O}$ be an ovoid of P. Then, $\mathscr{O}$ is a nondegenerate quadratic set.*

Proof. The assertion follows from the definition of an ovoid.   □

Definition. Let P be a 3-dimensional projective space, and let E be a plane of P. Furthermore, let s be a point of P outside of E. Finally, let $\mathscr{O}$ be an oval in E. The set of the points on the lines sx, where x is contained in $\mathscr{O}$, is called a **cone** with **vertex** s.

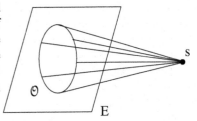

4.12 Theorem. *Let P be a 3-dimensional projective space, and let $\mathscr{K}$ be a cone of P with vertex s. Then, $\mathscr{K}$ is a degenerate quadratic set with $Rad(\mathscr{K}) = s$.*

Proof. Let $\mathscr{O}$ be an oval in a plane E of P such that $\mathscr{K}$ consists of the points on the lines sx where x is contained in $\mathscr{O}$. Furthermore, let G be the set of the lines of P through s which intersect the plane E in a point of $\mathscr{O}$.

Step 1. By construction of a cone, all lines of P through s are either contained in $\mathscr{K}$ or they intersect $\mathscr{K}$ in the point s. In other words, all lines through s are tangents of $\mathscr{K}$.

Step 2. Each line of P, which is not incident with s, has at most two points in common with $\mathscr{K}$:

Assume that there is a line g of P not containing s such that g has at least three points x_1, x_2, x_3 in common with $\mathcal{H}$. Then, the lines sx_1, sx_2 and sx_3 intersect the plane E in three points y_1, y_2 and y_3 of $\mathcal{O}$, respectively. If h is the intersection line of the two planes E and $\langle s, g \rangle$, it follows that y_1, y_2, y_3 are incident with h. In particular,

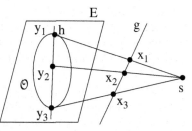

the line h contains three points of $\mathcal{O}$. This contradicts the definition of an oval.

Step 3. Let $s \neq x$ be a point of $\mathcal{H}$, and let $y := sx \cap E$ be contained in $\mathcal{O}$. Let t be the tangent of $\mathcal{O}$ in E through y. Then, $F := \langle s, t \rangle$ is the tangent plane of $\mathcal{H}$ at x:

Since $t \cap \mathcal{O} = y$, we have $F \cap \mathcal{H} = sy$. It follows that all lines through x in F are tangents of $\mathcal{H}$.

Conversely, if g is a line through x not contained in E, $\langle s, g \rangle \cap E$ is a line l through y distinct from t. Since t is the only tangent through y at $\mathcal{O}$ in E, on the line l, there is exactly one further point z of $\mathcal{O}$. It

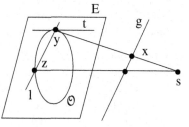

follows that the line g is incident with the points x and $sz \cap g$ of $\mathcal{O}$.

From Steps 1 and 2, it follows that every line of P has at most two points in common with $\mathcal{H}$ or is completely contained in $\mathcal{H}$. From Steps 1 and 3, it follows that the tangent space at $\mathcal{H}$ through a point x of $\mathcal{H} \setminus \{s\}$ is a plane and that the tangent space at $\mathcal{H}$ through s equals P. It follows that $\mathcal{H}$ is a quadratic set with $\mathrm{Rad}(\mathcal{H}) = s$. $\qquad\square$

The following theorem classifies all quadratic sets of a 3-dimensional projective space:

4.13 Theorem. *Let $P = PG(3, K)$ be a 3-dimensional projective space over a skew field K, and let Q be a quadratic set in P. Then, one of the following cases occurs:*

(a) *Q is a subspace U of P. We have $U = \mathrm{Rad}(Q)$.*
(b) *Q is the set of the points of two planes E_1 and E_2. If $g = E_1 \cap E_2$, $g = \mathrm{Rad}(Q)$.*
(c) *Q is a cone with vertex s. We have $s = \mathrm{Rad}(Q)$.*
(d) *Q is an ovoid. We have $\varnothing = \mathrm{Rad}(Q)$.*
(e) *Q is a hyperbolic quadric. We have $\varnothing = \mathrm{Rad}(Q)$. Furthermore, K is commutative.*

Proof. By Theorem 2.10, every subspace U of P is a quadratic set with $U = \mathrm{Rad}(Q)$. From now on, let us assume that Q is not a subspace of P.

Step 1. We have dim $\mathrm{Rad}(Q) \leq 1$: Assume that $E := \mathrm{Rad}(Q)$ is a plane. By assumption, we have $E \neq Q$, that is, there exists a point x of $Q \setminus E$. By Theorem 2.8, it follows that $P = \langle x, E \rangle$ is contained in Q, hence, $Q = P$, a contradiction.

Step 2. Let $g := \mathrm{Rad}(Q)$ be a line. Then, Q consists of the set of the points of two planes E_1 and E_2 intersecting in g: Since $Q \neq \mathrm{Rad}(Q)$, there exists a point x of Q not on g. Let h be a line through x disjoint to g. Then, by Theorem 2.9, Q and h meet in a nondegenerate quadratic set of h, that is, $h \cap Q$ consists of two points x and y. By Theorem 2.9, we have $Q = \langle x, g \rangle \cup \langle y, g \rangle$.

Step 3. Let $s := \mathrm{Rad}(Q)$ be a point. Then, Q is a cone with vertex s: Let E be a plane of P which does not contain the point s and which is not contained in Q. By Theorem 2.9, $Q \cap E$ is a nondegenerate quadratic set of E. By Theorem 2.12, it follows that $\mathcal{O} := Q \cap E$ is an oval. It follows from Theorem 2.9 that Q is a cone with vertex s.

Step 4. Let $\varnothing = \mathrm{Rad}(Q)$. If Q does not contain any line of P, Q is an ovoid: By definition, an ovoid is a nondegenerate quadratic set of a 3-dimensional projective space which does not contain any lines.

Step 5. Let $\varnothing = \mathrm{Rad}(Q)$. If Q contains a line of P, Q is a hyperbolic quadric:

(i)   By Theorem 2.3, no plane of P is contained in Q.
(ii) Let g be a line in Q, and let x be a point of Q outside of g. Then, there is exactly one line of Q through x intersecting the line g: By Theorem 2.2, the subspaces contained in Q define a nondegenerate polar space S. Since S does not contain any planes, by Theorem 2.20 of Chap. IV, S is a generalized quadrangle. The assertion follows.
(iii) There are two disjoint lines in Q: Since the subspaces contained in Q define a polar space (Theorem 2.2), the assertion follows from Theorem 2.15 of Chap. IV.
(iv) Let g and h be two disjoint lines in Q. For any point x on g, by Part (ii), there exists exactly one line l_x in Q through x intersecting the line h in a point. For two points x and y on g, the lines l_x and l_y are disjoint:

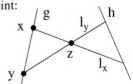

Assume that the lines l_x and l_y have a point z in common. If z is not contained in h, the lines g and h are contained in $\langle l_x, l_y \rangle$, in contradiction to the assumption that g and h are disjoint.

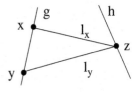

If z is incident with h, the point z of Q is incident with two lines l_x and l_y of Q intersecting the line g. This yields a contradiction to Part (ii).

(v)   From Part (iv), it follows in particular that Q contains at least three pairwise disjoint lines.

(vi) Q is a hyperbolic quadric: Let g_1, g_2, g_3 be three pairwise disjoint lines in Q. Since every transversal of $\{g_1, g_2, g_3\}$ contains three points of Q, every transversal is contained in Q. Let $\mathcal{T}$ be the set of the transversals of $\{g_1, g_2, g_3\}$.

First case. If P contains a regulus $\mathcal{R}$ through $\{g_1, g_2, g_3\}$, Q is a hyperbolic quadric: Let R be the set of the points on the lines of $\mathcal{R}$. Since $\mathcal{R}$ is a regulus, R is the set of the points on the lines of $\mathcal{T}$. It follows that R is contained in Q. If $R = Q$, Q is a hyperbolic quadric.

Assume that there exists a point x of $Q \setminus R$. By Theorem 4.1, there exists a line h through x meeting the lines g_1 and g_2 in two different points y and z. It follows that h is contained in Q. Let t be the transversal of $\mathscr{R}$ through y. Since t is contained in R, we have $t \neq h$. It follows that there are two lines h and t of Q through y intersecting the line g_2, in contradiction to Part (ii).

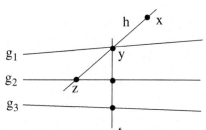

Second case. The case, that there does not exist a regulus through $\{g_1, g_2, g_3\}$, yields a contradiction: For, let R be the set of the points on the lines of $\mathscr{T}$. Since $\mathscr{T}$ is not a regulus (otherwise, by Theorem 4.2, the opposite regulus of $\mathscr{T}$ would be a regulus through $\{g_1, g_2, g_3\}$), one of the following two cases occurs:

(α) There exists a line h which intersects three of the lines of $\mathscr{T}$ without being contained in R.

(β) There exists a line l which intersects three of the lines of $\mathscr{T}$, which is contained in R, but which misses at least one line of $\mathscr{T}$.

The case (α) cannot occur: Since such a line h contains three points of $R \subseteq Q$, it follows that h is contained in Q. It follows that there exists a point x of $Q \setminus R$. This fact yields, as in Case 1, a contradiction.

It remains to show that the Case (β) cannot occur. For, let l be a line intersecting three of the lines of $\mathscr{T}$ which is contained in R, but which misses at least one line t of $\mathscr{T}$.

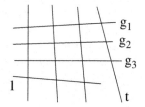

Since $\mathscr{T}$ consists of the transversals of $\{g_1, g_2, g_3\}$ and since l intersects at least three of the lines of $\mathscr{T}$, it follows from Theorem 4.2 that either l is one of the lines of $\{g_1, g_2, g_3\}$ or l is disjoint to the lines g_1, g_2 and g_3. Since t is a transversal of $\{g_1, g_2, g_3\}$ and since l is disjoint to t, it follows that l is different from g_1, g_2 and g_3. Hence, l and g_1, g_2, g_3 are disjoint.

Obviously, t is different from g_1. Let E be the plane generated by t and g_1. Since P is of dimension 3, the line l and the plane E meet in a point p. Since l is disjoint to g_1 and to t, the point p is neither on g_1 nor on t.

Since the line l is contained in R, there is a transversal t' of $\mathscr{T}$ through p. The transversal t' meets the line g_1 in a point $p' \neq p$. Hence, $t' = pp'$ is contained in E.

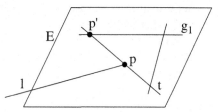

It follows that the lines t and t' intersect in a point, a contradiction.

□

Let Q be a hyperbolic quadric of a 3-dimensional projective space P. In
Theorem 4.15, we shall see that for every point z of $P \setminus Q$, there exists a central
collineation σ with centre z such that $\sigma(Q) = Q$. The following lemma prepares
the proof of this theorem.

4.14 Lemma. *Let $P = PG(3, K)$ be a 3-dimensional projective space, and let Q
be a hyperbolic quadric of P.*

(a) *Let z be a point of $P \setminus Q$. There exists a plane E of P through z such that $Q \cap E$
consists of the set of the points of two lines.*

(b) *Let a and b be two points of Q which are not
joined by a line of Q. There exist exactly two
points r and s such that r and s are joined with a
and b by some lines of Q. Furthermore, we have
$Q_a \cap Q_b = rs$.*

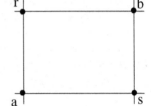

(c) *Let E be a plane of P such that $E \cap Q =
g \cup h$ for two lines g and h of E. Let $e :=
g \cap h$ be the intersection point of g and h.
Let a and c be two points on g distinct from
e, and let b and d be two points on h distinct
from e. Finally, let r (respectively s) be the, in
view of (b), uniquely determined points of P
different from e which are joined with a and
with b (respectively with c and with d) by a
line of Q.*

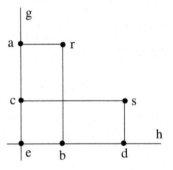

Then, $H := \langle Q_a \cap Q_b, Q_c \cap Q_d \rangle = \langle e, r, s \rangle$. In particular, H is a plane of P.[9]

Proof. (a) Let g be a line of Q. Since z is not contained in Q, the point z is not on
g. It follows that $E := \langle z, g \rangle$ is a plane. Since E contains the line g of Q, by
Theorem 4.10, $E \cap Q$ consists of the set of the points of two intersecting lines.

(b) By Theorem 4.4, there are exactly two lines g and
h of Q through the point a. Since the line ab is
not contained in Q, the point b is neither incident
with g nor with h.

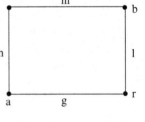

By Theorem 4.5, there exists exactly one line l
of Q through b intersecting the line g in a point r
and exactly one line m of Q through b intersecting
the line h in a point s.

---

[9] Note that the points r and s are not contained in the plane $E = \langle g, h \rangle$.

Obviously, r and s are contained in $Q_a \cap Q_b$, hence, rs is contained in $Q_a \cap Q_b$. Since b is not contained in Q_a, we have $Q_a \neq Q_b$. From dim $Q_a =$ dim $Q_b = 2$, it follows that dim $(Q_a \cap Q_b) = 1$. Thus, $rs = Q_a \cap Q_b$.

(c) In view of (b), we have $er = Q_a \cap Q_b$ and $es = Q_c \cap Q_d$. It follows that

$$\langle Q_a \cap Q_b, \ Q_c \cap Q_d \rangle = \langle e, \ r, \ s \rangle.$$

Assume that $l := \langle e, r, s \rangle$ is a line. Then, l intersects the lines ar, cs and bd, and it follows that l is a transversal of $\{ar, cs, bd\}$. From the uniqueness of this transversal (Theorem 4.1), it follows that $l = ac$, in contradiction to the fact that r, s are not incident with ac. $\square$

4.15 Theorem. *Let $\boldsymbol{P} = PG(3, \ K)$ be a 3-dimensional projective space over a vector space $\boldsymbol{V}$, and let Q be a hyperbolic quadric of $\boldsymbol{P}$. Furthermore, let z be a point of $\boldsymbol{P}$ outside of Q.*

(a) There exists a central collineation $\sigma : \boldsymbol{P} \to \boldsymbol{P}$ with centre z such that $\sigma(Q) = Q$.

(b) Let E be a plane of $\boldsymbol{P}$ through z such that $Q \cap E$ consists of the set of the points of two lines g and h of E. [10] *Let $e := g \cap h$, and let a and c be two points on the line g distinct from e. Furthermore, let $b := za \cap h$ and $d := zc \cap h$. Finally, let $H := \langle Q_a \cap Q_b, \ Q_c \cap Q_d \rangle$, and let σ be the central collineation of $\boldsymbol{P}$ with centre z, axis H and $\sigma(a) = b$.* [11]

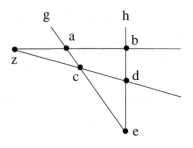

Then, $\sigma(Q) = Q$ and σ is an automorphism of order 2. In particular, we have $\sigma(b) = a$.

Proof. (a) Obviously, (a) follows from (b).
(b) By Theorem 4.14, there exist two points r and s such that r is joined with the points a and b by some lines of Q and such that s is joined with the points c and d by some lines of Q. Furthermore, again by Theorem 4.14, $H := \langle e, r, s \rangle$ is a plane of $\boldsymbol{P}$.

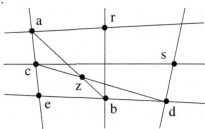

---

[10]Such a plane E exists by Theorem 4.14.
[11]By Theorem 4.14, H is a plane of $\boldsymbol{P}$.

By Theorem 4.6, there exists a basis $B = \{v_0, v_1, v_2, v_3\}$ of V such that $a = \langle v_0 \rangle$, $b = \langle v_1 + v_3 \rangle$, $c = \langle v_2 \rangle$, $d = \langle v_0 + v_1 + v_2 + v_3 \rangle$, $e = \langle v_0 + v_2 \rangle$, $r = \langle v_1 \rangle$ and $s = \langle v_2 + v_3 \rangle$. Since $z = ab \cap cd$, one easily verifies that $z = \langle v_0 + v_1 + v_3 \rangle$.

Let $A : V \to V$ be the linear transformation with the following matrix with respect to the basis B:

$$A = \begin{pmatrix} 0 & 0 & 1 & -1 \\ -1 & 1 & 1 & -1 \\ 0 & 0 & 1 & 0 \\ -1 & 0 & 1 & 0 \end{pmatrix},$$

and let $\sigma : P \to P$ be the projective collineation defined by A. Since $H = \langle e, r, s \rangle$, we have $H = \langle v_0 + v_2, v_1, v_2 + v_3 \rangle$. From $A(v_0 + v_2) = v_0 + v_2$, $A(v_1) = v_1$ and $A(v_2 + v_3) = v_2 + v_3$, it follows that $A|_H = id$, hence, $\sigma|_H = id$. By Theorem 5.3 of Chap. II, σ is a central collineation with centre x.

From $A(v_0) = -v_1 - v_3$ and $A(v_2) = v_0 + v_1 + v_2 + v_3$, it follows that $\sigma(a) = b$ and $\sigma(c) = d$.

From $A(v_1 + v_3) = -v_0$ and $A(v_0 + v_1 + v_2 + v_3) = v_2$, it follows that $\sigma(b) = a$ and $\sigma(d) = c$.

It follows that σ is a central collineation with axis H, centre $x = ab \cap cd = z$ and $\sigma(a) = b$. Since σ^2 is also a central collineation with axis H and centre z and since $\sigma^2(a) = \sigma(b) = a$, it follows that $\sigma^2 = id$, that is, σ is of order 2.

In order to verify the assertion $\sigma(Q) = Q$, we shall consider a point $x = (x_0, x_1, x_2, x_3)^t$ of Q with homogeneous coordinates with respect to the basis B. By Theorem 4.9, Q is defined by the equation $x_0 x_3 - x_1 x_2 = 0$. We have

$$\sigma(x) = A \begin{pmatrix} x_0 \\ x_1 \\ x_2 \\ x_3 \end{pmatrix} = \begin{pmatrix} 0 & 0 & 1 & -1 \\ -1 & 1 & 1 & -1 \\ 0 & 0 & 1 & 0 \\ -1 & 0 & 1 & 0 \end{pmatrix} \begin{pmatrix} x_0 \\ x_1 \\ x_2 \\ x_3 \end{pmatrix} = \begin{pmatrix} x_2 - x_3 \\ -x_0 + x_1 + x_2 - x_3 \\ x_2 \\ -x_0 + x_2 \end{pmatrix}.$$

From $(x_2 - x_3)(-x_0 + x_2) - (-x_0 + x_1 + x_2 - x_3) x_2 = x_3 x_0 - x_1 x_2 = 0$, it follows that $\sigma(x)$ is contained in Q. It follows that $\sigma(Q)$ is contained in Q.

It remains to show that Q is contained in $\sigma(Q)$: For, let y be an element of Q. Then, $x := \sigma(y)$ is an element of $\sigma(Q) \subseteq Q$. It follows that $\sigma(x) = \sigma^2(y) = y$. Hence, y is an element of $\sigma(Q)$. $\square$

5   Perspective Quadratic Sets

Definition. Let $P = PG(d, K)$ be a d-dimensional projective space, and let Q be a quadratic set of P. Q is called **perspective** if for every point z outside of Q, such that z is not contained in $\bigcap_{x \in Q} Q_x$, there exists a central collineation σ_z with centre z such that $\sigma_z(Q) = Q$.

By Theorem 4.15, a hyperbolic quadric of a 3-dimensional projective space is perspective. In the present section, we shall see that every nondegenerate quadratic set Q of rank $r \geq 2$ is perspective (Theorem 5.4). In Theorem 5.5, we shall see that the central collineation σ_z is uniquely determined.

5.1 Theorem. *Let* $P = PG(d, K)$ *be a d-dimensional projective space. Let* Q *be a nondegenerate quadratic set of rank* $r \geq 2$.

(a) *We have* $d \geq 3$.
(b) *Let* x *and* y *be two points of* Q *such that the line* xy *is contained in* Q. *For any point* z *incident with the line* xy, *the set* $Q_x \cap Q_y$ *is contained in* Q_z.
(c) *For any two points* x *and* y *of* Q, *we have* $Q_x \neq Q_y$.
(d) *For any two points* x *and* y *of* Q, *we have* $\dim (Q_x \cap Q_y) = d - 2$.

Proof. (a) Since the subspaces contained in Q define a nondegenerate polar space of rank r (Theorem 2.2), there exist two disjoint subspaces of dimension $r - 1$ in Q. It follows that $d \geq 2r - 1 \geq 3$.

(b) Let a be a point of $Q_x \cap Q_y$.

If a is a point of Q, ax and ay are lines of Q through a intersecting the line $xy \subseteq Q$ in the two points x and y. By Theorem 2.2, all lines through a intersecting the line xy are contained in Q. In particular, the line az is contained in Q. Therefore, the point a is contained in Q_z.

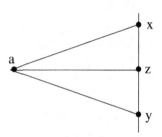

Next, we shall consider the case that a is a point outside of Q. If za is a tangent of Q, a is contained in Q_z. If za is not a tangent, there exists a point b of $za \cap Q$ different from a and z. Since b is contained in $Q_x \cap Q_y$,[12] the lines bx and by are contained in Q. As above, it follows that bz is contained in Q. In particular, the line $za = bz$ is a tangent, in contradiction to the assumption.

(c) Assume that $Q_x = Q_y$. Then, x is contained in Q_y, and y is contained in Q_x, that is, the line xy is contained in Q.

Step 1. For any point z on xy, we have $Q_x = Q_z$: By assumption, we have $Q_x = Q_y$. Hence, it follows from (b) that $Q_x = Q_x \cap Q_y$ is contained in Q_z. From $\dim Q_x = \dim Q_z = d - 1$, it follows that $Q_x = Q_z$.

---

[12]Since the line xy is contained in Q and since the point z is incident with xy, z is contained in $Q_x \cap Q_y$. Since the point a is also contained in $Q_x \cap Q_y$, it follows that the line za is contained in $Q_x \cap Q_y$. Finally, it follows that b is contained in $Q_x \cap Q_y$ since b is a point on za incident with Q.

Step 2. The assumption $Q_x = Q_y$ yields a contradiction: Let r be a point of Q outside of Q_x, and let $s := Q_r \cap xy$. Since s is contained in $Q \cap Q_r$, the line sr is contained in Q. It follows that r is contained in Q_s. By Step 1, the point r is contained in $Q_s = Q_x$, in contradiction to the choice of r.

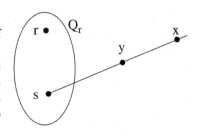

(d) follows from (c).   □

Definition. Let $P = PG(d, K)$ be a projective space, and let Q be a nondegenerate quadratic set. For two points x, y of Q, we denote by $\boldsymbol{Q_{xy}}$ the subspace $Q_{xy} := Q_x \cap Q_y$.

5.2 Theorem. *Let* $\boldsymbol{P} = PG(d, K)$ *be a projective space, and let* Q *be a nondegenerate quadratic set of rank* $r \geq 2$. *Let* x *and* y *be two points of* Q *such that the line* xy *is not contained in* Q.

(a) *We have* $\dim Q_{xy} = d - 2$.
(b) *We have* $Q \cap Q_{xy} \neq \varnothing$.
(c) *The set* $Q \cap Q_{xy}$ *is a nondegenerate quadratic set in* Q_{xy}.
(d) *We have* $Q_{xy} = \langle Q \cap Q_{xy} \rangle$.

Proof. (a) follows from Theorem 5.1.
(b) Since Q is a quadratic set of rank $r \geq 2$, the point x is incident with a line g of Q. In particular, g is contained in Q_x. Let $z := Q_y \cap g$. Then, z is contained in $Q_x \cap Q_y = Q_{xy}$.
(c) By Theorem 2.5, $Q \cap Q_{xy}$ is a quadratic set in Q_{xy}. Assume that $Q \cap Q_{xy}$ admits a double point p. Then, all lines of Q_{xy} through p are tangents of Q, and it follows that Q_{xy} is contained in Q_p. On the other hand, it follows from the fact that p is contained in Q_{xy} and that Q_{xy} is contained in Q_x that x is contained in Q_p. Analogously, it follows that y is contained in Q_p.

Since the line xy is not contained in Q, the point x is not contained in Q_{xy}. It follows that $\dim \langle x, Q_{xy} \rangle = d - 1$, hence $\langle x, Q_{xy} \rangle = Q_x$. Similarly, it follows that $\langle y, Q_{xy} \rangle = Q_y$. By Theorem 5.1, we have $Q_x \neq Q_y$. It follows that $\langle x, y, Q_{xy} \rangle = \langle Q_x, Q_y \rangle = P$.

Altogether, we have $P = \langle x, y, Q_{xy} \rangle \subseteq Q_p$, that is, $Q_p = P$. It follows that p is a double point of Q, in contradiction to the assumption that Q is nondegenerate.
(d) follows from (b) and (c) in view of Theorem 2.11.   □

5.3 Theorem. *Let* $\boldsymbol{P} = PG(d, K)$ *be a d-dimensional projective space, and let* Q *be a nondegenerate quadratic set of* $\boldsymbol{P}$ *of rank* $r \geq 2$. *Let* z *be a point of* $\boldsymbol{P}$ *outside of* Q *and outside of* $\bigcap_{x \in Q} Q_x$.

(a) *There exists a plane E of **P** through z such that $Q \cap E = g \cup h$ for two lines g and h.*

(b) *Let $e := g \cap h$ be the intersection point of the lines g and h. Furthermore, let a and c be two points on the line g distinct from e, and let b and d be two points on the line h distinct from e. Then, the subspace $\langle Q_{ab}, Q_{cd} \rangle$ is a hyperplane of **P**.*

Proof. (a) By assumption, there exists a point a in Q such that z is not contained in Q_a. It follows that on the line az, there is exactly one further point b of Q.

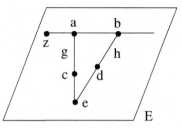

By Theorem 5.2, we have $Q \cap Q_{ab} \neq \varnothing$, that is, there exists a point e of $Q \cap Q_{ab}$. It follows that the lines $g := ae$ and $h := be$ are contained in Q. Let $E := \langle e, a, b \rangle$ be the plane generated by the points e, a and b. Then, $g \cup h$ is contained in E.

Since the line ab is not contained in Q, we have $Q \cap E \neq E$. From Theorem 2.12, it follows that $Q \cap E = g \cup h$.

(b) By Theorem 5.1, the hyperplanes Q_a, Q_b, Q_c, Q_d, Q_e are pairwise distinct.

Step 1. We have $Q_{ab} \neq Q_{cd}$: By Theorem 5.1, we have dim Q_{ab} = dim $Q_{cd} = d - 2$. Assume that $Q_{ab} = Q_{cd}$. Then, $U := Q_{ab} = Q_{ab} \cap Q_{cd} = Q_a \cap Q_b \cap Q_c \cap Q_d$ and dim $U = d - 2$. In particular, we have $U = Q_b \cap Q_d = Q_{bd}$. Since the line bd is contained in Q, the point b is contained in $Q_b \cap Q_d = U$. It follows that b is contained in $U = Q_{ab} \subseteq Q_a$, in contradiction to the assumption that the line ab is not a tangent of Q.

Step 2. The subspace $\langle Q_{ab}, Q_{cd} \rangle$ is a hyperplane of **P**: By Theorem 5.1, $Q_a \cap Q_c$ is contained in Q_e. In particular, we have $Q_{ac} = Q_a \cap Q_c \subseteq Q_a \cap Q_e = Q_{ae}$. From dim $Q_{ac} = d - 2 =$ dim Q_{ae} (Theorem 5.1), it follows that $Q_{ac} = Q_{ae}$. Analogously, we have $Q_{bd} = Q_{be}$. It follows that

$$Q_{ab} \cap Q_{cd} = Q_a \cap Q_b \cap Q_c \cap Q_d$$
$$= Q_a \cap Q_c \cap Q_b \cap Q_d$$
$$= Q_a \cap Q_e \cap Q_b \cap Q_e$$
$$= Q_a \cap Q_b \cap Q_e.$$

Since Q_a, Q_b, Q_e are pairwise distinct hyperplanes of **P**, it follows that

$$\dim(Q_{ab} \cap Q_{cd}) = \dim(Q_a \cap Q_b \cap Q_e) \in \{d - 2, d - 3\}.$$

If dim $(Q_{ab} \cap Q_{cd}) = d - 2$, $Q_{ab} = Q_{cd}$, in contradiction to Step 1. It follows that $\dim(Q_{ab} \cap Q_{cd}) = d - 3$, implying that $\langle Q_{ab}, Q_{cd} \rangle$ is a hyperplane of **P**. □

5.4 Theorem (**Buekenhout**). *Let* $P = PG(d, K)$ *be a d-dimensional projective space, and let Q be a nondegenerate quadratic set of rank $r \geq 2$.*

(a) *The quadratic set Q is perspective.*

(b) *Let z be a point of P outside of Q and outside of $\bigcap_{x \in Q} Q_x$. Furthermore, let E be a plane of P through z such that $Q \cap E = g \cup h$ for two lines g and h.[13] Let $e := g \cap h$ be the intersection point of the lines g and h, and let a and c be two points on the line g, both distinct from e. Let $b := za \cap h$ and $d := zc \cap h$.*

    Finally, let σ be the central collineation of P with centre z, axis $\langle Q_{ab}, Q_{cd} \rangle$ and $\sigma(a) = b$.[14] Then, $\sigma(Q) = Q$, $\sigma(b) = a$, $\sigma(c) = d$ and $\sigma(d) = c$. In particular, σ is of order 2.

Proof. (a) follows from (b).

(b) The proof of (b) is organized as follows: In Step 1, we shall show that $\sigma(g) = h$. In particular, $\sigma(x)$ is contained in Q for all points x on g.

    In Steps 2 and 3, we shall see that the lines of Q through a (respectively through c) are mapped under σ on lines of Q through b (respectively through d) and that we have $\sigma(Q_a) = Q_b$ and $\sigma(Q_c) = Q_d$.

For a point x of $Q \setminus E$, there exists at least one line l contained in Q such that the lines l and $g = ae$ meet in a point f.[15] In Step 4, we shall see that $\sigma(x)$ is contained in Q if $f \neq e$.

Step 5 is devoted to the proof of the equations $\sigma(b) = a$ and $\sigma(Q_b) = Q_a$. In particular, we have $\sigma(h) = g$, that is, $\sigma(x)$ is contained in Q for all x of $Q \cap E$ (Step 1).

In Step 6, we shall show that $\sigma(x)$ is contained in Q for all points x of $Q \setminus E$ such that the line $l := xe$ is contained in Q.

From Steps 1 to 6, it follows that $\sigma(Q)$ is contained in Q. It remains to show that $\sigma(Q) = Q$. This verification is the subject of the final Step 7.

    In what follows, let $H := \langle Q_{ab}, Q_{cd} \rangle$.

Step 1. We have $\sigma(e) = e$, $\sigma(c) = d$ and $\sigma(g) = h$: Since the lines g and h are tangents of Q through a and b, respectively, it follows that e is contained in $Q_{ab} \subseteq H$. Since H is the axis of σ, it follows that $\sigma(e) = e$. From $\sigma(a) = b$ and $\sigma(e) = e$, it follows that $\sigma(g) = h$. Since z is the centre of σ, the point $\sigma(c)$ is on the line zc. It follows that $\sigma(c) = \sigma(zc \cap g) = \sigma(zc) \cap \sigma(g) = zc \cap h = d$.

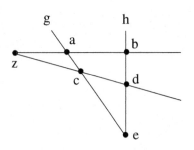

---

[13] Such a plane exists by Theorem 5.3.

[14] By Theorem 5.3, the subspace $\langle Q_{ab}, Q_{cd} \rangle$ is a hyperplane of P.

[15] The existence of the point f follows from the fact that the points and lines of Q form a polar space.

Step 2. σ maps the lines of Q through a (respectively through c) onto the lines of Q through b (respectively through d):

For, let l be a line of Q through a, and let $s := l \cap Q_b$. Then, s is contained in $Q_a \cap Q_b = Q_{ab} \subseteq H$, and it follows that $\sigma(s) = s$.

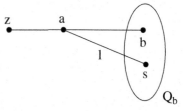

Since $\sigma(a) = b$, $\sigma(s) = s$ and since s is contained in Q_b, it follows that $\sigma(l) = bs$ is a line through b in Q_b, that is, a tangent of Q. Since the points s, b are contained in $Q \cap \sigma(l)$, it follows that $\sigma(l)$ is contained in Q.

Step 3. We have $\sigma(Q_a) = Q_b$ and $\sigma(Q_c) = Q_d$: By Theorem 5.2, we have $\dim Q_{ab} = d - 2$, and $Q_{ab} = \langle Q \cap Q_{ab} \rangle$. Since a is not contained in Q_b, it follows that a is not contained in Q_{ab}. Hence, we have $Q_a = \langle a, Q_{ab} \rangle$. Analogously, it follows that $Q_b = \langle b, Q_{ab} \rangle$. Finally, we have $\sigma(Q_{ab}) = Q_{ab}$, since Q_{ab} is contained in H. Altogether, we have

$$\sigma(Q_a) = \sigma(\langle a, Q_{ab} \rangle) = \langle \sigma(a), \sigma(Q_{ab}) \rangle = \langle b, Q_{ab} \rangle = Q_b.$$

Step 4. Let x be a point of $Q \backslash E$, and let l be a line of Q through x intersecting the line $g = ac$ in a point f. If f and e are distinct, $\sigma(x)$ is contained in Q:

If $f = a$ or $f = c$, the assertion follows from Step 2. Let $f \neq a, c$. Let $U := \langle E, l \rangle$ be the 3-dimensional subspace of P generated by E and l. Since a, b, e, x are contained in $Q \cap U$, we have $\langle Q \cap U \rangle = U$. Since $Q \cap U$ contains the lines $g = ac$, $h = bd$ and l, it follows that $Q \cap U$ is neither an ovoid nor a cone. Since z is contained in $U \backslash Q$, we have $Q \cap U \neq U$. By Theorem 4.13, either $Q \cap U$ is the set of the points of two planes or $Q \cap U$ is a hyperbolic quadric.

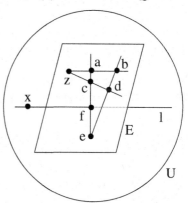

First case. If $Q \cap U$ is the set of the points of two planes, $\sigma(x)$ is contained in Q:

Let F_1 and F_2 be two planes of U such that $Q \cap U = F_1 \cup F_2$. Since the plane E contains the point z outside of Q, we have $E \neq F_1, F_2$. It follows that g is contained in F_1 and h is contained in F_2 (or g is contained in F_2 and h is contained in F_1). Since l is contained in Q and since $l \cap g = f \neq e$, the line l is contained in the plane F_1. It follows that the points a and x are contained in F_1, that is, the line

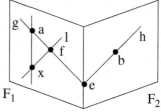

ax is contained in F_1 and hence contained in Q. From Step 2, it follows that $\sigma(x)$ is contained in Q.

Second case. If $Q \cap U$ is a hyperbolic quadric, $\sigma(x)$ is contained in Q, and we have $\sigma(b) = a$: Let $H' := H \cap U$. Then, H' is a plane of U not containing the points a and b. From $\sigma(H') = H'^{16}$ and $\sigma(a) = b$, it follows that $\sigma(U) = \sigma(\langle a, H' \rangle) = \langle b, H' \rangle = U$.

Let $\sigma' := \sigma|_U$. Then, $\sigma' : U \to U$ is a central collineation with centre z and axis $H' := H \cap U$. Let $Q' := Q \cap U$. By Theorem 2.5, the set Q' is a quadratic set with the property that $Q_x \cap U = (Q \cap U)_x = Q'_x$ for all points x of $Q \cap U = Q'$. Using Theorem 2.5, we obtain

$$
\begin{aligned}
H' &= H \cap U \\
&= \langle Q_{ab}, Q_{cd} \rangle \cap U \\
&\supseteq \langle Q_{ab} \cap U, Q_{cd} \cap U \rangle \\
&= \langle Q_a \cap Q_b \cap U, Q_c \cap Q_d \cap U \rangle \\
&= \langle (Q_a \cap U) \cap (Q_b \cap U), (Q_c \cap U) \cap (Q_d \cap U) \rangle \\
&= \langle (Q \cap U)_a \cap (Q \cap U)_b, (Q \cap U)_c \cap (Q \cap U)_d \rangle \\
&= \langle Q'_a \cap Q'_b, Q'_c \cap Q'_d \rangle.
\end{aligned}
$$

By Lemma 4.14, we have $\dim \langle Q'_a \cap Q'_b, Q'_c \cap Q'_d \rangle = 2$. Since $\dim H' = 2$, it follows that $H' = \langle Q'_a \cap Q'_b, Q'_c \cap Q'_d \rangle$.

By Theorem 4.15, we have $\sigma'(Q') = Q'$ and $\sigma'(b) = a$. From $\sigma' = \sigma|_U$, it follows that $\sigma(b) = a$ and that $\sigma(x)$ is contained in Q.

Step 5. We have $\sigma(b) = a$ and $\sigma(Q_b) = Q_a$. Furthermore, σ is of order 2: We first shall show that there is a 3-dimensional subspace U through the plane $E = \langle a, b, e \rangle$ such that $Q \cap U$ is a hyperbolic quadric:

By Theorem 5.2, $Q \cap Q_{ab}$ is a nondegenerate quadratic set in Q_{ab}. It follows that there is a line l contained in Q_{ab} with $|l \cap Q| = 2$ through the point e of Q_{ab}. Let x be the point of Q on l distinct from e. Since x is contained in $Q \cap Q_{ab}$, the lines ax and bx are contained in Q. Since $Q \cap E = g \cup h$, x is not contained in E.

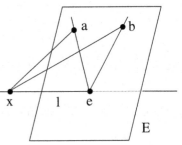

Let $U := \langle x, E \rangle$ be the 3-dimensional subspace of P generated by x and E. Then, $Q \cap U$ is a hyperbolic quadric: Since $U = \langle x, a, b, e \rangle$ and since $Q \cap U$

---

[16] H' is a subspace of H, and H is the axis of σ.

contains the lines xa, xb, $g = ae$ and $h = be$, by Theorem 4.13, $Q \cap U$ is the set of the points of two planes of P or a hyperbolic quadric.

Assume that $Q \cap U$ is the set of the points of two planes F_1 and F_2 of P. Since E is not contained in Q, we have $E \neq F_1$, F_2. It follows that w.l.o.g. the line g is contained in F_1 and the line h is contained in F_2. For the line $m := F_1 \cap F_2$, we get

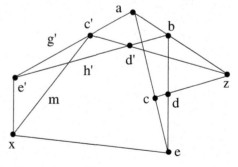

$$m = F_1 \cap F_2 \supseteq g \cap h = e.$$

Since x is contained in $Q \cap U = F_1 \cup F_2$, the line $l = xe$ is contained in Q, a contradiction. It follows that $Q \cap U$ is a hyperbolic quadric. From Step 4 (Case 2), it follows that $\sigma(b) = a$. Thus, σ^2 is a central collineation with centre z, axis H and $\sigma^2(a) = \sigma(\sigma(a)) = \sigma(b) = a$. It follows that $\sigma^2 = id$.

It remains to show that $\sigma(Q_b) = Q_a$. Since $Q_b = \langle b, Q_{ab} \rangle$, $\sigma(b) = a$ and $\sigma(Q_{ab}) = Q_{ab}$ (since Q_{ab} is contained in H), we have $\sigma(Q_b) = \langle \sigma(b), \sigma(Q_{ab}) \rangle = \langle a, Q_{ab} \rangle = Q_a$.

Step 6. Let x be a point of Q such that the line $l := xe$ is contained in Q. Then, $\sigma(x)$ is contained in Q: If x is contained in $H = \langle Q_{ab}, Q_{cd} \rangle$, $\sigma(x) = x$ is contained in Q. W.l.o.g., we shall suppose that x is not contained in H.

(i) There is a point e' of $Q \cap (Q_{ab} \cup Q_{cd})$ such that the line $e'x$ is not contained in Q: Otherwise, for any point y of $(Q \cap Q_{ab}) \cup (Q \cap Q_{cd})$, the line yx is contained in Q. It follows that $Q \cap Q_{ab}$ is contained in Q_x and hence, $Q_{ab} = \langle Q \cap Q_{ab} \rangle$ is contained in Q_x. Analogously, we can see that Q_{cd} is contained in Q_x. Thus, $Q_x = \langle Q_{ab}, Q_{cd} \rangle = H$, in contradiction to the assumption that x is not contained in H.

(ii) W.l.o.g., we shall assume that e' is contained in Q_{ab}. Let m be a line of Q through x intersecting the line $e'a$ in a point c'. Since the line $e'x$ is not contained in Q, we have $c' \neq e'$. If $c' = a$, it follows from Step 2 that $\sigma(x)$ is contained in Q. Hence, we may assume that $c' \neq a$. Let $d' := zc' \cap e'b$.[17]

If $g' := ac'$ and $h' := bd'$, the lines g' and h' are lines of Q since e' is contained in $Q \cap Q_{ab}$.

Let $E' := \langle z, a, c' \rangle$. Then, $Q \cap E' = g' \cup h'$.

---

[17]Note that the plane $\langle e', c', b \rangle$ contains the point a and therefore the line ab. Hence, the point z is also contained in this plane. It follows that the lines zc' and $e'b$ meet in a point.

(iii) Let σ' be the central collineation of P with centre z and axis $H' :=$ $\langle Q_{ab}, Q_{c'd'} \rangle$ such that $\sigma'(a) = b$. Then, $\sigma'(b) = a$. Furthermore, $\sigma'(x)$ is contained in Q: Steps 1 to 5 can be applied to the central collineation σ' with the following replacements:

$$z \to z,\, a \to a,\, b \to b,\, c \to c',\, d \to d',\, e \to e',\, E \to E',\, H \to H'.$$

From Step 5, it follows that $\sigma'(b) = a$. Since the point x is on the line xc' of Q, it follows from Step 2 that $\sigma'(x)$ is contained in Q.

(iv) We have $\sigma = \sigma'$. In particular, $\sigma(x)$ is contained in Q: Since the point z is the centre of σ and of σ', the transformation $\tau := \sigma \sigma'$ admits also the point z as centre. By Theorem 5.6 of Chap. II, τ is a central collineation with centre z. Let L be the axis of τ.

Since Q_{ab} is contained in $H \cap H'$, Q_{ab} is contained in L. From $\tau(a) = \sigma(\sigma'(a)) = \sigma(b) = a$ and $\tau(b) = \sigma(\sigma'(b)) = \sigma(a) = b$, it follows that the points a, b are contained in L. It follows that L contains $\langle Q_{ab}, a, b \rangle = P$. Hence, we have $\tau = id$. Since σ is a collineation of order 2 (Step 5), we have $\sigma = \sigma^{-1} = \sigma'$.

Step 7. We have $\sigma(Q) = Q$: It follows from Steps 1 to 6 that $\sigma(Q)$ is contained in Q. Let y be a point of Q, and let $x := \sigma(y)$. Then, x is an element of Q, and it follows that $\sigma(x) = \sigma^2(y) = y$. Hence, $\sigma(Q) = Q$. $\square$

5.5 Theorem. *Let $P = PG(d, K)$ be a d-dimensional projective space, and let Q be a nondegenerate quadratic set of rank $r \geq 2$. Let z be a point of P outside of Q and outside of $\bigcap\limits_{x \in Q} Q_x$.*

There exists exactly one central collineation $id \neq \sigma$ with centre z and $\sigma(Q) = Q$.

Proof. By Theorem 5.4, there exists a central collineation σ with centre z and $\sigma(Q) = Q$. Let H be the axis of σ.

By Theorem 5.3, there exists a plane E of P through z such that $Q \cap E = g \cup h$ for two lines g and h of Q. Let $e := g \cap h$ be the intersection point of g and h, and let a be a point on g distinct from e.

Step 1. If $b := za \cap h$ is the intersection point of the lines za and h, Q_{ab} is contained in H: Let x be a point of $Q \cap Q_{ab}$. Then, the lines xa and xb are contained in Q. Let $E := \langle a, b, x \rangle$ be the plane generated by the points a, b and x.

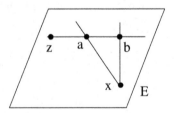

Since E contains the lines ax and bx of Q, we have $Q \cap E = ax \cup bx$ or $Q \cap E = E$ (Theorem 2.12). Since the point z is contained in E, it follows that $Q \cap E \neq E$, hence, $Q \cap E = ax \cup bx$. In particular, we have $zx \cap Q = x$. Since z is the centre of σ and since $\sigma(Q) = Q$, it follows that

$$\sigma(x) = \sigma(zx \cap Q) = \sigma(zx) \cap \sigma(Q) = zx \cap Q = x.$$

Thus, x is contained in H. By Theorem 5.2, $Q_{ab} = \langle Q \cap Q_{ab}\rangle$ is contained in H.

Step 2. Let c be a point on the line g distinct from e and a, and let $d := zc \cap h$. Then, Q_{cd} is contained in H: The assertion follows as in Step 1.

Step 3. The transformation σ is uniquely determined: It follows from Steps 1 and 2 that H contains $\langle Q_{ab}, Q_{cd}\rangle$. By Theorem 5.3, $\langle Q_{ab}, Q_{cd}\rangle$ is a hyperplane of P, it follows that $H = \langle Q_{ab}, Q_{cd}\rangle$. Thus, for every central collineation $id \neq \sigma$ with centre z, axis H and the property that $\sigma(Q) = Q$, we have $H = \langle Q_{ab}, Q_{cd}\rangle$ and $\sigma(a) = b$. It follows that σ is uniquely determined.   □

6   Classification of the Quadratic Sets

In this section we shall present the main result of the present chapter, namely the classification of the quadratic sets. The main results are as follows:

- An oval of the projective plane $P = PG(2, K)$, which is perspective, is a quadric. In particular, the field K is commutative (Corollary 6.1).
- If Q is a nondegenerate quadratic set of rank $r \geq 2$ of the projective space $P = PG(d, K)$, Q is perspective, Q is a quadric, and K is commutative (Theorem 6.4).
- If Q is a nondegenerate quadratic set of the projective space $P = PG(d, K)$, Q is either an ovoid or a quadric, and K is commutative (Corollary 6.5).

6.1 Theorem. *Let $P = P(V)$ be a projective plane over a right vector space over a skew field K, and let Q be an oval of P. If Q is perspective, we have:*

(a) Q is a quadric.
(b) The field K is commutative.

Proof. (a) Step 1. Let a, b and c be three points on Q. W.l.o.g., let $a = (0, 1, 0)^t$, $b = (1, 0, 0)^t$ and $c = (1, 1, 1)^t$. Let t_a and t_b be the tangents of Q through a and b, respectively, and let $s := t_a \cap t_b$ be the intersection point of the tangents t_a and t_b. Since the points a, b, c, s form a frame, we can choose w.l.o.g. the point s as $s = (0, 0, 1)^t$.

Step 2. Let $g := ab$ be the line joining the two points a and b. Then, for every point z on g different from a and b, it follows that z is not contained in Q and that z is not contained in $\cap \{Q_x \mid x \in Q\}$:

Since Q is an oval and since a, b are contained in Q, it follows that $Q \cap g = \{a, b\}$. In particular,

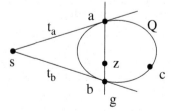

z is not contained in Q. Since $Q_a = t_a$ and $Q_b = t_b$, it follows from $z \neq t_a \cap t_b$ that z is not contained in $\underset{x \in Q}{\cap} Q_x$.

Step 3. The points on g different from a and b are exactly the points $z = (1, m, 0)^t$ with $0 \neq m \in K$. Since Q is perspective, for every point $z_m = (1, m, 0)^t$ with $0 \neq m \in K$, there exists a central collineation $id \neq \sigma_m$ with centre z_m such that $\sigma_m(Q) = Q$.

Step 4. We have $\sigma_m(s) = s$: Since the line $g = ab$ is incident with the centre z_m of σ_m, we have $\sigma_m(g) = g$. It follows from $\sigma_m(Q) = Q$ that $\sigma_m(a) = a$ and $\sigma_m(b) = b$ or $\sigma_m(a) = b$ and $\sigma_m(a) = b$. It follows that for the tangents t_a and t_b, we obtain: $\sigma_m(t_a) = t_a$ and $\sigma_m(t_b) = t_b$ or $\sigma_m(t_a) = t_b$ and $\sigma_m(t_a) = t_b$. Since $s = t_a \cap t_b$, it follows that $\sigma_m(s) = s$.

Step 5. Let h be a line through z_m intersecting the oval Q in two points x and y. Then, $\sigma_m(x) = y$ and $\sigma_m(y) = x$:

Since the line h is incident with the centre z_m of σ_m, we have $\sigma_m(h) = h$. From $\sigma_m(Q) = Q$, it follows that $\sigma_m(x) = x$ and $\sigma_m(y) = y$ or $\sigma_m(x) = y$ and $\sigma_m(y) = x$. Assume that $\sigma_m(x) = x$ and $\sigma_m(y) = y$. Then, the line h is incident with the three fixed points x, y and z_m, and it follows that h is the axis of σ_m. Since the centre z_m of σ_m is incident with h, h is the only fixed line of σ_m through x. Since $\sigma_m(x) = x$ and $\sigma_m(Q) = Q$, the collineation σ_m fixes the tangent t_x through x at Q. It follows that $t_x = h$, in contradiction to $h \cap Q = \{x, y\}$.

Step 6. We have $\sigma_m(a) = b$ and $\sigma_m(b) = a$: The line g meets the oval Q in the points a and b. Furthermore, z_m is incident with g, and s is not incident with g. From Step 5, it follows that $\sigma_m(a) = b$ and $\sigma_m(b) = a$.

Step 7. Let A_m be the matrix representation of σ_m. Then, we have

$$A_m = \begin{pmatrix} 0 & m^{-1} & 0 \\ m & 0 & 0 \\ 0 & 0 & -1 \end{pmatrix} :$$

From $a = (0, 1, 0)^t$, $b = (1, 0, 0)^t$, $s = (0, 0, 1)^t$ and $\sigma_m(a) = b$, $\sigma_m(b) = a$, $\sigma_m(s) = s$, it follows that there exist non-zero elements α, β and γ of K such that

$$A_m = \begin{pmatrix} 0 & \alpha & 0 \\ \beta & 0 & 0 \\ 0 & 0 & \gamma \end{pmatrix}.$$

Since $z_m = (1, m, 0)^t$ and $A_m(z_m) = z_m$, it follows that there exists an element $0 \neq \lambda$ of K such that

$$\begin{pmatrix} 1 \\ m \\ 0 \end{pmatrix} \lambda = A_m \begin{pmatrix} 1 \\ m \\ 0 \end{pmatrix} = \begin{pmatrix} 0 & \alpha & 0 \\ \beta & 0 & 0 \\ 0 & 0 & \gamma \end{pmatrix} \begin{pmatrix} 1 \\ m \\ 0 \end{pmatrix} = \begin{pmatrix} \alpha m \\ \beta \\ 0 \end{pmatrix}.$$

It follows that $\lambda = \alpha m$ and $m\lambda = \beta$. Hence, $\alpha = m^{-1}\beta m^{-1}$.

The lines g_k through the point $z_m = (1, m, 0)^t$ are of the form

$$-kmx_0 + kx_1 + x_2 = 0$$

for some element k of K. In other words, the lines through the point z_m are exactly the lines $g_k = [-km, \ k, \ 1]$. Since the point z_m is the centre of σ_m, it follows that $\sigma_m(g_k) = g_k$ for all k of K. On the other hand, we have

$$\sigma_m(g_k) = [-km, \ k, \ 1] \begin{pmatrix} 0 & \alpha & 0 \\ \beta & 0 & 0 \\ 0 & 0 & \gamma \end{pmatrix} = [k\beta, \ -km\alpha, \ \gamma].$$

Hence, there exists an element $0 \neq \mu \in K$ such that $\mu \ [-km, \ k, \ 1] = [k\beta, \ -km\alpha, \ \gamma]$. It follows that $\mu = \gamma$ and that

$$k\beta = -\gamma k \, m \text{ for all } k \text{ of } K. \tag{1}$$

If we consider the case $k = 1$, it follows from (1) that

$$\beta = -\gamma \, m. \tag{2}$$

From (1) and (2), it follows that

$$k\gamma = \gamma k \text{ for all } k \text{ of } K. \tag{3}$$

Hence, γ is an element of the centre $Z(K)$ of K. W.l.o.g. we may suppose that $\gamma = -1$ (otherwise, by Theorem 4.4 of Chap. III, we may replace the matrix A_m by $-\gamma^{-1} A_m$). It follows from (2) that $\beta = m$. Hence, $\alpha = m^{-1}\beta m^{-1} = m^{-1}$. Altogether, we have

$$A_m = \begin{pmatrix} 0 & m^{-1} & 0 \\ m & 0 & 0 \\ 0 & 0 & -1 \end{pmatrix}.$$

Step 8. We have $Q = \{(m^{-1}, \ m, \ -1)^t \mid 0 \neq m \in K\} \cup \{(1, 0, 0)^t, \ (0, 1, 0)^t\}$: Let $M := \{(m^{-1}, \ m, \ -1)^t \mid 0 \neq m \in K\} \cup \{(1, 0, 0)^t, \ (0, 1, 0)^t\}$.

(i) M is contained in Q: Let $0 \neq m$ be an element of K. By Step 1, the point $c = (1, 1, 1)^t$ is contained in Q. Let σ_m be the central collineation with centre $z_m = (1, m, 0)^t$ and $\sigma_m(Q) = Q$ (see Step 3). By Step 7, it follows that

$$\sigma_m(c) = A_m \begin{pmatrix} 1 \\ 1 \\ 1 \end{pmatrix} = \begin{pmatrix} 0 & m^{-1} & 0 \\ m & 0 & 0 \\ 0 & 0 & -1 \end{pmatrix} \begin{pmatrix} 1 \\ 1 \\ 1 \end{pmatrix} = \begin{pmatrix} m^{-1} \\ m \\ -1 \end{pmatrix}.$$

Since $\sigma_m(Q) = Q$, $(m^{-1}, \ m, \ -1)^t = \sigma_m(c)$ is contained in Q.

(ii) Q is contained in M: Obviously, the points a and b are contained in M. Let x be a point on Q distinct from a and b. If $x = c$, the point x is contained in M. Hence, we may assume that $x \neq c$. Let $z := ab \cap xc$ be the intersection point of the lines ab and xc.

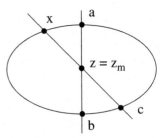

By Step 3, there exists an element $0 \neq m$ of K such that $z = z_m = (1, m, 0)^t$. Let σ_m be the central collineation with centre z_m such that $\sigma_m(Q) = Q$. By Step 7, we have $\sigma_m(c) = (m^{-1}, m, -1)^t$. From Step 5, it follows that $\sigma_m(c) = x$. Altogether, the point $x = (m^{-1}, m, -1)^t$ is contained in M.

Step 9. The field K is commutative: For, let m and n be two elements of $K \setminus \{0\}$. By Step 8, the point $x := (n^{-1}, n, -1)^t$ is contained in Q. Let σ_m be the central collineation defined in Step 3. By Step 7, σ_m has the matrix representation

$$A_m = \begin{pmatrix} 0 & m^{-1} & 0 \\ m & 0 & 0 \\ 0 & 0 & -1 \end{pmatrix}.$$

It follows that

$$\sigma_m(x) = A_m \begin{pmatrix} n^{-1} \\ n \\ -1 \end{pmatrix} = \begin{pmatrix} 0 & m^{-1} & 0 \\ m & 0 & 0 \\ 0 & 0 & -1 \end{pmatrix} \begin{pmatrix} n^{-1} \\ n \\ -1 \end{pmatrix} = \begin{pmatrix} m^{-1}n \\ mn^{-1} \\ 1 \end{pmatrix}$$

is contained in Q. Since m, $n \neq 0$, it follows from Step 8 that $\sigma_m(x)$ is of the form $(\lambda^{-1}, \lambda, -1)$ for some $0 \neq \lambda$ of K. Hence, $\sigma_m(x) = (-m^{-1}n, -mn^{-1}, -1)^t$ with

$$1 = (-m^{-1}n)(-mn^{-1}) = m^{-1}nmn^{-1}.$$

It follows that $mn = nm$, hence, K is commutative.

Step 10. Let Q' be the quadric of P with the equation $x_0 x_1 - x_2{}^2 = 0$. Then, $Q = Q'$. In particular, Q is a quadric:

Let $M := \{(m^{-1}, m, -1)^t \mid 0 \neq m \in K\} \cup \{(1, 0, 0)^t, (0, 1, 0)^t\}$. By Step 8, we have $M = Q$.

(i) M is contained in Q': Obviously, the points $(1, 0, 0)^t$ and $(0, 1, 0)^t$ are contained in Q'. Since $m^{-1}m - 1 = 0$, every point $(m^{-1}, m, -1)^t$ with $0 \neq m \in K$ is contained in Q.

(ii) Q' is contained in M: Let $x = (x_0, x_1, x_2)^t$ be a point of Q. If $x_2 = 0$, it follows that $x_0 x_1 = 0$, that is, $x = (1, 0, 0)^t$ or $x = (0, 1, 0)^t$. It follows that x is contained in M. If $x_2 \neq 0$, we may assume w.l.o.g. that $x_2 = -1$. It follows that $x_0x_1 - 1 = 0$, that is, $x_0x_1 = 1$. Hence, x_0, $x_1 \neq 0$ and $x_1 = x_0^{-1}$. Thus, the point $x = (x_0, x_0^{-1}, -1)^t$ is contained in M.

(b) The commutativity of K has been shown in Step 9 of the proof of Part (a). □

6.2 Theorem. *Let* $P = PG(2, K)$ *be a projective plane over a commutative field* K, *and let* $\mathcal{O}_1$ *and* $\mathcal{O}_2$ *be two conics with the following properties:*

(i) We have $\mathcal{O}_1$, $\mathcal{O}_2 \neq P$.
(ii) The set $\mathcal{O}_1 \cap \mathcal{O}_2$ *contains three mutually non-collinear points x, y and z.*
(iii) The tangents of $\mathcal{O}_1$ *at the points x and y equal the tangents of* $\mathcal{O}_2$ *at the points x and y.*

Then, $\mathcal{O}_1 = \mathcal{O}_2$.

Proof. Let t_x and t_y be the tangents of $\mathcal{O}_1$ at the points x and y, respectively, and let $s := t_x \cap t_y$ be the intersection point of the tangents t_x and t_y. Then, the points s, x and y are not collinear. Since, by Theorem 4.6 of Chap. III, the automorphism group of P operates transitively on the frames of P, we can assume w.l.o.g. that $x = (1, 0, 0)^{t, y} = (0, 1, 0)^t$ and $s = (0, 0, 1)^t$. It follows that the lines $t_x = xs$ and $t_y = ys$ are defined by the equations

$$t_x : x_1 = 0$$

$$t_y : x_0 = 0.$$

Since the points x, y and z are mutually non-collinear, the point z is neither incident with the tangent t_x nor with the tangent t_y.

Since the points x, y, and z are non-collinear and since z is neither incident with t_x nor with t_y, the points x, y, z and s form a frame. Hence, we may assume w.l.o.g. that $z = (1, 1, 1)^t$.

The conic $\mathcal{O}_1$ is defined by an equation of the form

$$ax_0^2 + bx_1^2 + cx_2^2 + dx_0x_1 + ex_0x_2 + fx_1x_2 = 0$$

for some a, b, c, d, e, f of K. Since x is contained in $\mathcal{O}_1$, it follows that $a = 0$. Since y is contained in $\mathcal{O}_1$, it follows that $b = 0$.

By Theorem 3.6, the tangents t_x and t_y are defined by the equations

$$t_x : dx_1 + ex_2 = 0$$

$$t_y : dx_0 + fx_2 = 0.$$

From $t_x : x_1 = 0$, it and $t_y : x_0 = 0$, it follows that $e = f = 0$. Finally, since z is contained in $\mathcal{O}_1$, it follows that $c + d = 0$. Altogether the conic $\mathcal{O}_1$ satisfies the equation

$$cx_2^2 - cx_0x_1 = 0.$$

Since $\mathcal{O}_1 \neq P$, we have $c \neq 0$. Hence, $\mathcal{O}_1$ is defined by the equation $x_2^2 - x_0x_1 = 0$. Similarly, it follows that $\mathcal{O}_2$ is defined by the equation $x_2^2 - x_0x_1 = 0$. Hence, $\mathcal{O}_1 = \mathcal{O}_2$.   □

6.3 Theorem. *Let $P = PG(d, K)$ be a d-dimensional projective space over a commutative field K. Furthermore, let Q be a quadratic set of P which is not a subspace of P. Assume that $E \cap Q$ is a conic for all planes E such that $E \cap Q$ is an oval. Then, Q is a quadric.*

Proof. We shall prove the assertion by induction on d. For $d = 2$, the assertion is obvious. Hence, we may assume that $d \geq 3$.

Step 1. Since Q is not a subspace of P, there exist two points a and b of Q such that the line $g := ab$ is not contained in Q. In particular, g is not a tangent of Q. Let H be a hyperplane of P through a and b. Then, $H \cap Q$ is not a subspace of P. By induction, $Q_H := H \cap Q$ is a quadric of H.

Step 2. There is a point x of $Q \setminus (Q \cap H)$ which is not a double point of Q: Since Q is not a subspace of P, by Theorem 2.11, we have $\langle Q \rangle = P$. It follows that there exists a point x of Q which is not contained in $Q \cap H$.

Assume that each point of $Q \setminus (Q \cap H)$ is a double point. In particular, the point x is a double point of Q. It follows that the lines xa and xb are tangents of Q. Hence, they are contained in Q. Let z be a point on the line xa distinct from x and a. Since xa is contained in Q, z is a point of $Q \setminus (Q \cap H)$.

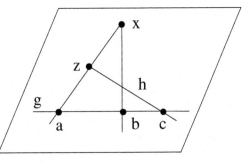

By assumption, z is a double point of Q, that is, all lines of P through z are tangents of Q. Let h be a line through z intersecting the line $g = ab$ in a point $c \neq a, b$. Since the line h is contained in the plane generated by the points x, a and b, the lines h and xb intersect in a point. It follows that there are two points (z and $h \cap xb$) of Q on h. Since h is a tangent of Q, it follows that h is contained in Q. Hence, c is contained in Q, in contradiction to Step 1.

Step 3. Let p be a point of $Q \setminus (Q \cap H)$ which is not a double point. In P, there exist coordinates such that $p = (1, 0, \ldots, 0)^t$ and such that the hyperplanes H and Q_p are defined by the equations $H : x_0 = 0$ and $Q_p : x_d = 0$.

Step 4. There is a point q of $Q \cap H$ such that the line pq is not contained in Q: Since the point p is not a double point, the tangent space Q_p is a hyperplane of P.

Since p is not contained in H, the subspace $Q_p \cap H$ is a $(d-2)$-dimensional subspace of H.

Since $\langle Q \cap H \rangle = H$ (Theorem 2.11), there exists a point q of $Q \cap H$ such that q is not contained in Q_p. Since q is not contained in Q_p, the line pq is not contained in Q.

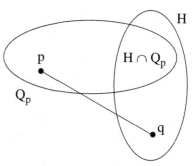

Step 5. Since, by Step 1, $Q_H = Q \cap H$ is a quadric in H and since the subspace H is defined by the equation $x_0 = 0$, there exist elements a_{ij} of K for $i, j = 1, \ldots, d$ such that the quadric Q_H is defined by the equation

$$Q_H : \sum_{i,j=1}^{d} a_{ij}\, x_i\, x_j = 0.$$

Step 6. For every element $0 \neq \lambda$ of K, let Q_λ be the quadric of P defined by the equation

$$Q_\lambda : \lambda x_0 x_d + \sum_{i,j=1}^{d} a_{ij}\, x_i\, x_j = 0.$$

Then:

(i) We have $Q_\lambda \cap H = Q \cap H = Q_H$.
(ii) The point p is contained in Q_λ.
(iii) The tangent space Q_p of Q at p is the tangent space $(Q_\lambda)_p$ of Q_λ at p.

The statements (i) and (ii) are obvious.

By Theorem 3.6, the tangent space $(Q_\lambda)_p$ of Q_λ at $p = (p_0, p_1, \ldots, p_d)^t = (1, 0, \ldots, 0)^t$ is defined by the equation

$$0 = \lambda\,(p_0 x_d + p_d x_0) + \sum_{i,j=1}^{d} a_{ij}\,(p_i x_j + p_j x_i) = \lambda x_d.$$

Since $\lambda \neq 0$, the tangent space $(Q_\lambda)_p$ is defined by the equation $x_d = 0$. From Step 3, it follows that $(Q_\lambda)_p = Q_p$.

Step 7. There is an element $0 \neq \lambda$ of K with $(Q_\lambda)_q = Q_q$ (see Step 4; in the following, this quadric Q_λ is denoted by Q'):

Since the point q is contained in H, q is of the form $q = (0, \mu_1, \ldots, \mu_d)^t$. Since q is not contained in Q_p, it follows that $\mu_d \neq 0$. By Theorem 3.6, the tangent space $(Q_\lambda)_q$ of Q_λ at q is given by the equation

$$0 = \lambda\,(\mu_0 x_d + \mu_d x_0) + \sum_{i,j=1}^{d} a_{ij}\,(\mu_i x_j + \mu_j x_i) = \lambda \mu_d x_0 + \sum_{i,j=1}^{d} a_{ij}\,(\mu_i x_j + \mu_j x_i)$$

(note that $\mu_0 = 0$). By Step 5 and Theorem 3.6, the tangent space $(Q_H)_q$ of Q_H at q in H is given by the equation

$$(Q_H)_q : \sum_{i,j=1}^{d} a_{ij} \left(\mu_i x_j + \mu_j x_i \right) = 0.$$

Since, by Theorem 2.5, we have $Q_q \cap H = (Q \cap H)_q = (Q_H)_q$, there exists an element α of K such that Q_q is defined in $\mathbf{P}$ by the equation

$$Q_q : \alpha x_0 + \sum_{i,j=1}^{d} a_{ij} \left(\mu_i x_j + \mu_j x_i \right) = 0.$$

Since $p = (p_0, \ p_1, \ \ldots, \ p_d)^t = (1, \ 0, \ \ldots, \ 0)^t$ is not contained in Q_q, we have

$$0 \neq \alpha p_0 + \sum_{i,j=1}^{d} a_{ij} \left(\mu_i \ p_j + \mu_j \ p_i \right) = \alpha.$$

It follows that there exists (exactly) one element $0 \neq \lambda$ of K, namely $\lambda := \alpha \mu_d^{-1}$, such that $(Q_\lambda)_q = Q_q$. Set $Q' := Q_\lambda$ for $\lambda := \alpha \mu_d^{-1}$.

   In what follows, we shall show that $Q = Q'$. In particular, this implies that Q is a quadric. The purpose of Steps 8 to 10 is to verify that Q is contained in Q'.

   Step 8. Let x be a point of $Q \cap H$. Then, by Step 6, the point x is contained in Q' as well.

   Step 9. Let x be a point of Q such that x is not contained in H and such that x is not contained in $\langle p, (Q \cap H)_q \rangle$. Then, x is contained in Q':

   By construction, the points p, q and x are non-collinear. We consider the plane $E := \langle x, \ p, \ q \rangle$.

   The line $l := E \cap H$ is not contained in the subspace $\langle p, (Q \cap H)_q \rangle$: Otherwise, $E = \langle p, l \rangle$ would be contained in $\langle p, (Q \cap H)_q \rangle$, in contradiction to the fact that x is not a point of $\langle p, (Q \cap H)_q \rangle$.

   It follows that l is not a tangent of Q. Hence, there is a point y of Q on l different from q.

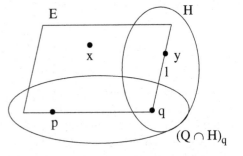

Since the line pq is not a tangent of Q, by Theorem 2.4, the plane E contains exactly one tangent t_p of Q through p. Since $Q_p = Q'_p$ (Step 6), t_p is a tangent of Q' at p. Analogously, it follows that there is exactly one tangent t_q of Q at q in E. The tangent t_q is a tangent of Q' at q as well. Since pq is not a tangent of Q, we have $t_q \neq t_p$. Finally, the plane E is neither contained in Q nor in Q', since the line pq is neither contained in Q nor in Q'.[18]

Since the points p, q and x are non-collinear, $E \cap Q$ is either the set of the points of two intersecting lines or it is an oval.

If $E \cap Q$ is the set of two intersecting lines g and h with $s := g \cap h$, each of the points p and q is incident with exactly one of the two lines g and h, and we have p, $q \neq s$. We may assume w.l.o.g. that p is incident with g and that q is incident with h. Hence, $g = t_p$ and $h = t_q$.

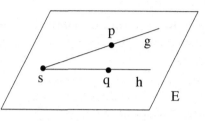

It follows that $E \cap Q = t_p \cup t_q = E \cap Q'$. Hence, x is contained in Q'. Analogously, it follows that x is contained in Q' if $E \cap Q'$ is the point set of two intersecting lines.

Hence, we may assume that $\mathscr{O} := E \cap Q$ and $\mathscr{O}' := E \cap Q'$ are ovals of E. By assumption, $\mathscr{O}$ is a conic. Since Q' is a quadric, $\mathscr{O}'$ is a conic as well. Since the conics $\mathscr{O}$ and $\mathscr{O}'$ have the points p, q and y and the tangents t_p and t_q in common, it follows from Theorem 6.2 that $\mathscr{O} = \mathscr{O}'$. In particular, x is a point of Q'.

Step 10. Let x be a point of Q such that x is not contained in H and such that x is contained in $\langle p, (Q \cap H)_q \rangle$. Then, x is contained in Q':

If the line px is contained in Q, it follows from $Q_p = Q'_p$ that the line px is contained in Q'. Hence, x is contained in Q'. Thus, we may assume w.l.o.g. that the line px is not contained in Q.

Let E be a plane through the points p and x which is not contained in $\langle p, (Q \cap H)_q \rangle$. Since the line px is not a tangent of Q, there exists exactly one tangent t_p of Q in E at p. It follows that there exists a point z of Q in E which is not on the line px and hence not contained in $\langle p, (Q \cap H)_q \rangle$.

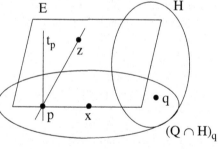

Since the points p, x and z are non-collinear, $E \cap Q$ is either the set of the points of two intersecting lines or of an oval. In the latter case, the oval is by assumption a conic. Since $Q_p = Q'_p$, there is a point of Q' distinct from p on the line px. Hence, $E \cap Q'$ is the set of the points of two intersecting lines or of a conic (note that Q' is a quadric). By Step 9, all points of $E \cap Q$, which are not incident with the line px, are contained in $E \cap Q'$. It follows that $E \cap Q = E \cap Q'$. In particular, the point x is contained in Q'.

---

[18]$t_p \neq t_q$ is both in Q and in Q' the only tangent in E through p.

Step 11. From Steps 8 to 10, it follows that Q is contained in Q'. Analogously, it follows that Q' is contained in Q. Altogether, $Q = Q'$ is a quadric.   □

The following two theorems of Buekenhout [11] are the main results about quadratic sets.

6.4 Theorem (Buekenhout). *Let* $P = PG(d, K)$ *be a d-dimensional projective space over a skew field* K, *and let* Q *be a nondegenerate quadratic set of rank* $r \geq 2$.

(a) *The quadratic set* Q *is perspective.*
(b) *The field* K *is commutative.*
(c) Q *is a quadric.*

Proof. (a) follows from Theorem 5.4.

(b) Let p be an arbitrary point of Q, and let q be a point of Q not contained in the tangent space Q_p. There is a point r of Q which is neither contained in Q_p nor in Q_q: Since Q defines a polar space of rank $r \geq 2$, there exists a line g of Q through p. There is a line of Q through q intersecting the line g in a point a. By construction, the point a belongs to the set $p^\perp \cap q^\perp = Q_p \cap Q_q \cap Q$. It follows from Theorem 2.7 of Chap. IV that $p^\perp \cap q^\perp$ is a nondegenerate polar space. In particular, there exists a point b of $p^\perp \cap q^\perp$ which is not collinear with a.

Let c be a point on the line pa different from p and a, and let d be a point on the line bq collinear with c. We have $d \neq q$ and $d \neq b$ (otherwise, the point d would be collinear with all points on the line pa implying that q (resp. b) would be collinear with p (resp. a)).

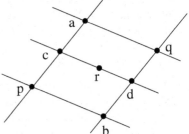

Hence, the point c is not collinear with q, and the point d is not collinear with p. Let r be a point on the line cd different from c and d. Then, r is neither collinear with p nor with q.

Let $E := \langle p, q, r \rangle$ be the plane generated by the points p, q, and r. Since the points p, q and r are non-collinear, $E \cap Q$ is either the point set of two intersecting lines or the point set of an oval.

Assume that $E \cap Q$ is the point set of two intersecting lines g and h. Then, at least two of the points p, q, r would be incident with one of the lines g or h. We may assume w.l.o.g. that the points p and q are incident with g. It follows that $pq = g$ is contained in Q, in contradiction to the fact that q is not contained in Q_p.

It follows that $E \cap Q$ is an oval. Since Q is perspective, $E \cap Q$ is a perspective quadratic set in E. By Theorem 6.1, it follows that K is commutative and that $E \cap Q$ is a conic.

(c) Let E be a plane of Q such that $E \cap Q$ is an oval. Since Q is perspective, by Theorem 6.1, $E \cap Q$ is a conic. In view of Theorem 6.3, Q is a quadric.   □

6.5 Corollary (Buekenhout). *Let* $P = PG(d, K)$ *be a d-dimensional projective space over a skew field K, and let Q be a nondegenerate quadratic set in* P. *Then, one of the following cases occurs:*

(i) Q is an ovoid.
(ii) Q is a quadric, and K is commutative.

Proof. A quadratic set Q is of rank $r \geq 2$ if and only if Q contains at least one line. Hence, the assertion follows from Theorem 6.4.   □

7   The Kleinian Quadric

In Sect. 7 of Chap. IV, the Kleinian polar space S has been introduced whose points are the lines of a 3-dimensional projective space $P = PG(3, K)$. In the present section, we shall see that if the field K is commutative, there exists a quadric Q (the so-called Kleinian quadric) in $PG(5, K)$ such that the Kleinian polar space S and the polar space S_Q defined by Q are isomorphic.

In other words, if K is commutative, the Kleinian polar space S stems from a quadric. If K is not commutative, the Kleinian polar space S forms its proper class of polar spaces which is not related to quadrics.

7.1 Theorem. *Let $P_3 = PG(3, K)$ and $P_5 = PG(5, K)$ be two projective spaces over the vector spaces V_4 and V_6 over the same commutative field K of dimension 3 and 5, respectively.*[19] *Let L_3 be the set of the lines of P_3.*

(a) For two points $x = \langle (x_0, x_1, x_2, x_3)^t \rangle$ and $y = \langle (y_0, y_1, y_2, y_3)^t \rangle$ of P_3, let

$$\pi : L_3 \to P_5$$

$$\pi : g = xy \to a = \langle (a_{01}, a_{02}, a_{03}, a_{23}, a_{31}, a_{12})^t \rangle$$

where $a_{ij} := x_i y_j - x_j y_i$.
   $\pi : L_3 \to P_5$ is a well-defined transformation of the set L_3 of the lines of P_3 into the set of the points of P_5.
(b) The image of π is the quadric of P_5 with the equation

$$p_0 p_3 + p_1 p_4 + p_2 p_5 = 0.$$

Proof. (a) Let $g = xy$ be a line of P_3 with $x = \langle (x_0, x_1, x_2, x_3)^t \rangle$ and $y = \langle (y_0, y_1, y_2, y_3)^t \rangle$. Obviously, $\pi(xy) = a = \langle (a_{01}, a_{02}, a_{03}, a_{23}, a_{31}, a_{12})^t \rangle$ is a point of P_5.

---

[19] The vector spaces V_4 and V_6 are of dimension 4 and 6, respectively.

In order to show that the transformation π is well-defined, we will consider two points $r := \lambda x + \mu y$ and $s := \alpha x + \beta y$ on g with λ, μ, α, $\beta \in K$ such that $\lambda \beta - \mu \alpha \neq 0$.

Let $b = \langle (b_{01},\ b_{02},\ b_{03},\ b_{23},\ b_{31},\ b_{12})^t \rangle := \pi(rs)$. Then, we have for $ij = 01, 02, 03, 23, 31, 12$:

$$
\begin{aligned}
b_{ij} &= (\lambda\, x_i + \mu y_i)(\alpha\, x_j + \beta\, y_j) - (\lambda\, x_j + \mu y_j)(\alpha x_i + \beta y_i) \\
&= \lambda\,\alpha\, x_i\, x_j + \lambda\,\beta x_i\, y_j + \mu\alpha\, x_j\, y_i + \mu\,\beta\, y_i\, y_j - \lambda\,\alpha\, x_i x_j - \lambda\,\beta x_j\, y_i \\
&\quad - \mu\,\alpha x_i y_j - \mu\,\beta\, y_i\, y_j \\
&= (\lambda\,\beta - \mu\,\alpha)(x_i\, y_j - x_j\, y_i) \\
&= (\lambda\,\beta - \mu\alpha)a_{ij}.
\end{aligned}
$$

Since $\lambda\beta - \mu\alpha \neq 0$, it follows that $b = a$. Hence, $\pi : L_3 \to P_5$ is well-defined.

(b) Let Q be the quadric of P_5 with the equation $p_0 p_3 + p_1 p_4 + p_2 p_5 = 0$.

Step 1. The image of π is contained in Q: Let $g = xy$ be a line of P_3 with $x = \langle (x_0,\ x_1,\ x_2,\ x_3)^t \rangle$ and $y = \langle (y_0,\ y_1,\ y_2,\ y_3)^t \rangle$, and let $\pi(xy) = \langle (a_{01},\ a_{02},\ a_{03},\ a_{23}, a_{31},\ a_{12})^t \rangle$. Then,

$$
\begin{aligned}
a_{01}\, &a_{23} + a_{02}\, a_{31} + a_{03}\, a_{12} \\
&= (x_0\, y_1 - x_1\, y_0)\, (x_2\, y_3 - x_3\, y_2) \\
&\quad + (x_0\, y_2 - x_2 y_0)(x_3\, y_1 - x_1\, y_3) + (x_0\, y_3 - x_3\, y_0)(x_1\, y_2 - x_2\, y_1) \\
&= x_0\, x_2\, y_1\, y_3 - x_0\, x_3\, y_1\, y_2 - x_1\, x_2\, y_0\, y_3 \\
&\quad + x_1\, x_3\, y_0\, y_2 + x_0\, x_3\, y_1\, y_2 - x_0\, x_1\, y_2\, y_3 - x_2\, x_3\, y_0\, y_1 \\
&\quad + x_1\, x_2\, y_0\, y_3 + x_0\, x_1\, y_2\, y_3 - x_0\, x_2\, y_1 y_3 - x_1 x_3 y_0 y_2 + x_2 x_3 y_0 y_1 \\
&= 0.
\end{aligned}
$$

It follows that $\pi(xy)$ is a point of Q.

Step 2. Q is contained in the image of π:

(i) Let $c = (c_0, c_1, c_2, c_3, c_4, c_5)$ be a point of Q with $c_0 \neq 0$. Then, $c = \pi(xy)$ with $x = \langle (0,\ c_0,\ c_1,\ c_2)^t \rangle$ and $y = \langle (-c_0,\ 0,\ c_5,\ -c_4)^t \rangle$: For, let $\pi(xy) = \langle (a_{01},\ a_{02},\ a_{03},\ a_{23},\ a_{31},\ a_{12})^t \rangle$. Then,

$$
a_{01} = x_0 y_1 - x_1 y_0 = c_0 c_0
$$

$$
a_{02} = x_0 y_2 - x_2 y_0 = c_0 c_1
$$

$$
a_{03} = x_0 y_3 - x_3 y_0 = c_0 c_2
$$

$$a_{23} = x_2 y_3 - x_3 y_2 = -c_1 c_4 - c_2 c_5 = c_0 c_3 \text{ (since } c \in Q)$$

$$a_{31} = x_3 y_1 - x_1 y_3 = c_0 c_4$$

$$a_{12} = x_1 y_2 - x_2 y_1 = c_0 c_5.$$

Since $c_0 \neq 0$, we have $\pi(xy) = \langle c_0 \cdot (c_0, \ c_1, \ c_2, \ c_3, \ c_4, \ c_5)^t \rangle = \langle (c_0, \ c_1, \ c_2, \ c_3, \ c_4, \ c_5)^t \rangle = c$.

(ii) Analogously, one easily verifies the relation $\pi(xy) = c_i$ with $c_i \neq 0$ ($i = 1$, $2, \ldots, 5$) as indicated in the following table.

$c_i \neq 0$	x	y
$c_1 \neq 0$	$\langle (0, \ c_0, \ c_1, \ c_2)^t \rangle$	$\langle (-c_1, -c_5, \ 0, \ c_3)^t \rangle$
$c_2 \neq 0$	$\langle (0, -c_0, -c_1, -c_2)^t \rangle$	$\langle (c_2, -c_4, \ c_3, \ 0)^t \rangle$
$c_3 \neq 0$	$\langle (c_1, \ c_5, \ 0, -c_3)^t \rangle$	$\langle (c_2, -c_4, \ c_3, \ 0)^t \rangle$
$c_4 \neq 0$	$\langle (c_0, \ 0, -c_5, \ c_4)^t \rangle$	$\langle (-c_2, \ c_4, -c_3, \ 0)^t \rangle$
$c_5 \neq 0$	$\langle (c_0, \ 0, -c_5, \ c_4)^t \rangle$	$\langle (c_1, \ c_5, \ 0, -c_3)^t \rangle$

□

Definition. Let $P_3 = PG(3, \ K)$ and $P_5 = PG(5, \ K)$ be two projective spaces over the same commutative field K of dimension 3 and 5, respectively, and let L_3 be the set of the lines of P_3.

(a) The transformation $\pi : L_3 \to P_5$ defined in Theorem 7.1 is called a **Plücker transformation**.

(b) The quadric Q of P_5 with the equation $p_0 p_3 + p_1 p_4 + p_2 p_5 = 0$ and every quadric which is projectively equivalent to Q, is called a **Kleinian quadric** of P_5.

In Theorem 7.2, the properties of a Kleinian quadric and of a Plücker transformation will be investigated. First, we recall the following fact: Let V_6 be a 6-dimensional vector space over a commutative field K, and let $q : V_6 \to K$ be the transformation defined by

$$q(p_0, \ p_1, \ p_2, \ p_3, \ p_4, \ p_5) := p_0 p_3 + p_1 p_4 + p_2 p_5.$$

Then, q is the quadratic form belonging to the Kleinian quadric Q. For two elements $a := (a_0, \ a_1, \ a_2, \ a_3, \ a_4, \ a_5)^t$ and $b := (b_0, \ b_1, \ b_2, \ b_3, \ b_4, \ b_5)^t$ of V_6, let

$$f(a, \ b) := q(a + b) - q(a) - q(b)$$

$$= (a_0 + b_0)(a_3 + b_3) + (a_1 + b_1)(a_4 + b_4) + (a_2 + b_2)(a_5 + b_5)$$

$$\quad -a_0 a_3 - a_1 a_4 - a_2 a_5 - b_0 b_3 - b_1 b_4 - b_2 b_5$$

$$= a_0 b_3 + b_0 a_3 + a_1 b_4 + a_4 b_1 + a_2 b_5 + a_5 b_2.$$

Then, f is the symmetric bilinear form belonging to q, and for any two points a and b of Q, the line $g := ab$ is contained in Q if and only if $f(a, b) = 0$ (Theorem 3.4).

7.2 Theorem. *Let $P_3 = PG(3, K)$ and $P_5 = PG(5, K)$ be two projective spaces over the same commutative field K of dimension 3 and 5, respectively, and let L_3 be the set of the lines of P_3. Furthermore, let $\pi : L_3 \to P_5$ be the Plücker transformation, and let $Q := \mathrm{Im}\,\pi$ be the Kleinian quadric defined by π.*

(a) *For two lines $g = xy$ and $h = rs$ of P_3 with $x = \langle(x_0, x_1, x_2, x_3)^t\rangle$, $y = \langle(y_0, y_1, y_2, y_3)^t\rangle$, $r = \langle(r_0, r_1, r_2, r_3)^t\rangle$ and $s = \langle(s_0, s_1, s_2, s_3)^t\rangle$, let $a := \pi(g)$ and $b := \pi(h)$. Then,*

$$f(a,b) = \det \begin{pmatrix} x_0 & x_1 & x_2 & x_3 \\ y_0 & y_1 & y_2 & y_3 \\ r_0 & r_1 & r_2 & r_3 \\ s_0 & s_1 & s_2 & s_3 \end{pmatrix}.$$

(b) *For two lines g and h of P_3, the line of P_5 through $\pi(g)$ and $\pi(h)$ is contained in Q if and only if $g \cap h \neq \emptyset$.*

(c) *The Plücker transformation $\pi : L_3 \to P_5$ is injective.*

(d) *Let $l := rs$ be a line of P_3, and let x be a point of P_3 not on l. Furthermore, let $t := \lambda r + \mu s$ be an arbitrary point on l, and let $g := xr$, $h := xs$ and $m := xt$ be the lines joining x with r, s and t, respectively. Then,*

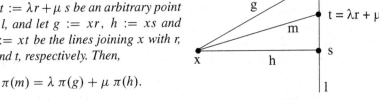

$$\pi(m) = \lambda\,\pi(g) + \mu\,\pi(h).$$

(e) *For a plane E of P_3 and a point x of P_3 such that x is contained in E, let E_x be the set of the lines of E through x. Then, the set*

$$\pi(E_x) := \{\pi(g) \mid g \in E_x\}$$

is the set of the points of a line of Q.

(f) *Conversely, there exists for every line l of Q a plane E and a point x of E such that $l = \pi(E_x)$.*

(g) *Let x be a point of P_3, and let G_x be the set of the lines of P_3 through x. Then, the set*

$$\pi(G_x) := \{\pi(g) \mid g \in G_x\}$$

is the set of the points of a plane of Q.

(h) *Let E be a plane of P_3, and let G_E be the set of the lines of P_3 in E. Then, the set*

$$\pi(G_E) := \{\pi(g) \mid g \in G_E\}$$

is the set of the points of a plane of Q.

(i) *Conversely, to every plane X of Q, there exists a point x or a plane E of P_3 such that $X = \pi(G_x)$ or $X = \pi(G_E)$.*

(j) *Let $R =: \{\pi(G_x) \mid x \in P_3\}$ and let $S =: \{\pi(G_E) \mid E$ plane of $P_3\}$. Then, every line of Q is incident with exactly one plane of R and with exactly one plane of S.*

(k) *Let x and y be two points, and let E and F be two planes of P_3. Then, we have:*

    (α) *If $x \neq y$, $\pi(G_x) \cap \pi(G_y) = \pi(xy)$ is a point of Q.*
    (β) *If x is a point of E, $\pi(G_x) \cap \pi(G_E) = \pi(E_x)$ is a line of Q.*
    (γ) *If x is not contained in E, $\pi(G_x) \cap \pi(G_E) = \emptyset$.*
    (δ) *If $E \neq F$, $\pi(G_E) \cap \pi(G_F) = \pi(E \cap F)$ is a point of Q.*

Proof. (a) The assertion follows by computing the determinant.

(b) Let $g = xy$ and $h = rs$ with $x = \langle(x_0, x_1, x_2, x_3)^t\rangle$, $y = \langle(y_0, y_1, y_2, y_3)^t\rangle$, $r = \langle(r_0, r_1, r_2, r_3)^t\rangle$ and $s = \langle(s_0, s_1, s_2, s_3)^t\rangle$. For $a := \pi(g)$ and $b := \pi(h)$, by (a), we have:

$$g \cap h \neq \emptyset \Leftrightarrow \langle x, y, r, s\rangle \neq P_3$$

$$\Leftrightarrow f(a,b) = \det \begin{pmatrix} x_0 & x_1 & x_2 & x_3 \\ y_0 & y_1 & y_2 & y_3 \\ r_0 & r_1 & r_2 & r_3 \\ s_0 & s_1 & s_2 & s_3 \end{pmatrix} = 0.$$

(c) We consider two different lines g and h of L_3, and we set $a := \pi(g)$ and $b := \pi(h)$. If $g \cap h = \emptyset$, by (a) and (b), we have

$$0 \neq f(a, b) = f(a, a) = 0,$$

hence, $a \neq b$.

    Suppose that $x := g \cap h$ is a point. Let m be a line of P_3 intersecting the line g in a point and being skew to the line h, and let $c := \pi(h)$. It follows from Part (b) and Theorem 3.4 that $f(a, c) = 0$ and $f(b, c) \neq 0$. Hence, $a \neq b$.

(d) Let $\pi(g) = \langle(a_{01}, a_{02}, a_{03}, a_{23}, a_{31}, a_{12})^t\rangle$, $\pi(h) = \langle(b_{01}, b_{02}, b_{03}, b_{23}, b_{31}, b_{12})^t\rangle$ and $\pi(m) = \langle(c_{01}, c_{02}, c_{03}, c_{23}, c_{31}, c_{12})^t\rangle$. Then, for $ij = 01, 02, 03, 23, 31, 12$, we get

$$c_{ij} = x_i(\lambda r_j + \mu s_j) - x_j(\lambda r_i + \mu s_i) = \lambda(x_i r_j - x_j r_i) + \mu(x_i s_j - x_j s_i) = \lambda a_{ij} + \mu b_{ij}.$$

(e) follows from (d).

(f) Let a and b be two points on l. Since π is injective, there exist two lines g and h of P_3 such that $a = \pi(g)$ and $b = \pi(h)$. Since l is contained in Q, by (b), g and h meet in a point x. Let E be plane generated by g and h. Then, by (e), we have $l = \pi(E_x)$.

(g) Let x be a point of P_3, and let g and h be two lines of G_x. If E is the plane generated by g and h, $\pi(E_x)$ is contained in $\pi(G_x)$. It follows that $\pi(G_x)$ is a subspace of P_5 contained in Q. Since $\pi(E_x)$ is a line and since π is injective, we have $\dim \pi(G_x) \geq 2$. By Theorem 2.3, we have $\dim \pi(G_x) \leq 2$. It follows that $\pi(G_x)$ is a plane of Q.

(h) follows analogously to (g).

(i) Let X be a plane of Q, and let a, b and c be three points of P_5 generating X. Since π is injective, there exist three lines g, h and l of P_3 such that $a = \pi(g)$, $b = \pi(h)$ and $c = \pi(l)$. By (b), the lines g, h and l meet pairwise. It follows that there exists a point x of P_3 such that $g \cap h \cap l = x$ or there is a plane E of P_3 such that $E = \langle g, h, l \rangle$. In the first case we have $X = \pi(G_x)$, in the second case we have $X = \pi(G_E)$.

(j) Let l be a line of Q. Then, by (f), there exists a plane E and a point x of E such that $l = \pi(E_x)$. It follows that $\pi(G_x)$ and $\pi(G_E)$ are the only planes of R, respectively of S through l.

(k) The assertions (α) to (δ) result from the fact that

$$G_x \cap G_y = xy \text{ if } x \neq y$$
$$G_E \cap G_F = E \cap F \text{ if } E \neq F$$
$$G_x \cap G_E = \varnothing \text{ if } x \notin E$$
$$G_x \cap G_E = E_x \text{ if } x \in E.$$

□

7.3 Theorem. *Let $P_3 = PG(3, K)$ and $P_5 = PG(5, K)$ be two projective spaces over the same commutative field K, and let Q be the Kleinian quadric of P_5. Furthermore, let S be the Kleinian polar space defined by P_3, and let S' be the polar space defined by the Kleinian quadric Q.*

(a) S and S' are isomorphic.

(b) With the notations of Theorem 7.2, the following correspondences are valid:

P_3	P_5	S
Lines of P_3	points of Q	points of S
E_x (lines of E through x)	line of Q	line of S
G_x (lines through x)	planes of $\mathscr{P}$ in Q	planes of $\mathscr{P}$ in S
G_E (lines in E)	planes of $\mathscr{E}$ in Q	planes of $\mathscr{E}$ in S

Proof. The proof follows from Theorem 7.2.   □

8   The Theorem of Segre

The Theorem of Segre [43] is a remarkable result about conics in finite projective planes. It says that every oval of a finite Desarguesian projective plane of odd order is a conic. The Theorem of Segre has initiated intensive research about finite projective spaces and finite geometries, in general. For the basic notions of finite geometries, we refer to Sect. 8 of Chap. I.

Definition. Let P be a projective plane, and let $\mathcal{O}$ be an oval of P. A line g of P is called a **secant of** $\mathcal{O}$ if g is incident with exactly two points of $\mathcal{O}$.

8.1 Theorem. *Let P be a finite projective plane of order q, and let $\mathcal{O}$ be an oval of P. The oval $\mathcal{O}$ has exactly $q + 1$ points.*

Proof. Let x be a point of $\mathcal{O}$.

Then, the point x is incident with exactly one tangent. All other lines through x are incident with exactly two points of $\mathcal{O}$ where the point x is one of these points. Since x is incident with exactly $q + 1$ lines, it follows that $|\mathcal{O}| = q + 1$.

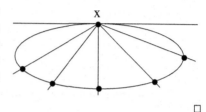

□

8.2 Theorem. *Let P be a finite projective plane of order q, and let $\mathcal{O}$ be an oval of P. Furthermore, let x be a point outside of $\mathcal{O}$.*

(a) If q is odd, x is either incident with no or with exactly two tangents of P.
(b) If q is even, there exists a point z such that all tangents of $\mathcal{O}$ intersect in z.

Proof. (a) Let x be a point outside of $\mathcal{O}$ which is incident with at least one tangent t. Let a be the number of lines disjoint to $\mathcal{O}$, let b be the number of tangents and let c be the number of secants through x. Then,

$$q + 1 = |\mathcal{O}| = b + 2c.$$

Since x is incident with the tangent t, we have $b \geq 1$. Since q is odd, $q + 1$ is even, it follows that $b + 2c$ is an even number. Since $b \geq 1$, it follows that $b \geq 2$. Hence, x is incident with at least one further tangent.

Since x is an arbitrary point outside of $\mathcal{O}$ on t, it follows that for any point y of t outside of $\mathcal{O}$, the point y is incident with t and with at least one further tangent of $\mathcal{O}$.

Any point of $\mathcal{O}$ is incident with exactly one tangent. By Theorem 8.1, there are exactly $q + 1$ points on $\mathcal{O}$. Hence, there are exactly q tangents of $\mathcal{O}$ different from t. It follows that every point on $t \setminus \mathcal{O}$ is incident with exactly one tangent different from t. In particular, the point x is incident with exactly two tangents.

(b) Step 1. Any point of P is incident with at least one tangent: Since $\mathcal{O}$ is an oval, any point of $\mathcal{O}$ is incident with exactly one tangent.

Let x be a point outside of $\mathcal{O}$, and let a be the number of lines disjoint to $\mathcal{O}$, let b be the number of tangents and let c be the number of the secants through x. Then,

$$q + 1 = |\mathcal{O}| = b + 2c.$$

Since q is even, $q + 1$ is odd, hence, $b \geq 1$. Thus, x is incident with at least one tangent.

Step 2. All tangents of $\mathcal{O}$ meet in a common point: Let t_1 and t_2 be two tangents, and let z be the intersection point of t_1 and t_2. Obviously, z is not contained in $\mathcal{O}$.

Assume that there is a tangent t intersecting the tangents t_1 and t_2 in two points different from z. Hence, there are exactly $q - 1$ points on t which are neither incident with t_1 nor with t_2. For each further tangent t', it holds: There are at most q points on t' outside of t_1 and t_2. Altogether, the $q + 1$ tangents of $\mathcal{O}$ cover at most

$$q + 1 \quad \text{(points on the tangent } t_1)$$

$$+q \quad \text{(points on the tangent } t_2 \text{ which are not on } t_1)$$

$$+q - 1 \quad \text{(points on the tangent } t \text{ which are neither on } t_1 \text{ nor on } t_2)$$

$$+(q - 2)q \quad \begin{cases} \text{(maximal number of points on the remaining q} - 2 \text{ tangents} \\ \text{which are neither on } t_1 \text{ nor on } t_2) \end{cases}$$

$$= q^2 + q. \quad \begin{cases} \text{(maximal number of points of } P \text{ which are incident with at} \\ \text{least one tangent)} \end{cases}$$

Since P has $q^2 + q + 1$ points, there is a point of P which is not incident with any tangent of $\mathcal{O}$, in contradiction to Step 1. $\qquad\qquad\square$

For the investigation of ovals in Desarguesian projective planes of finite order, we need the following Theorem of Wedderburn about finite fields.

8.3 Theorem (Wedderburn). *Let K be a finite (skew) field. Then, K is commutative.*

For a proof, the interested reader is referred to [33].

Definition. Let P be a projective plane, and let a, b and c be three non-collinear points. The **triangle** $\triangle abc$ is the set of the points on the lines ab, bc and ac.

In what follows, we will have to compute the coordinates of the lines joining two points and the coordinates of the intersection points of two lines in the projective plane $P = PG(2, K)$. Therefore, we recall the following obvious property:

8.4 Lemma. *Let $P = PG(2, K)$ be a projective plane over a commutative field K.*

(a) *If $g : \lambda_1 x_1 + \lambda_2 x_2 + \lambda_3 x_3 = 0$ and $h : \mu_1 x_1 + \mu_2 x_2 + \mu_3 x_3 = 0$ are two lines of P, the point $s = (s_1, s_2, s_3)^t$ is the intersection point of g and h if and only if $\lambda_1 s_1 + \lambda_2 s_2 + \lambda_3 s_3 = 0$ and $\mu_1 s_1 + \mu_2 s_2 + \mu_3 s_3 = 0$.*

(b) *The line $g : \lambda_1 x_1 + \lambda_2 x_2 + \lambda_3 x_3 = 0$ is the line joining the two points $a = (a_1, a_2, a_3)^t$ and $b = (b_1, b_2, b_3)^t$ if and only if $\lambda_1 a_1 + \lambda_2 a_2 + \lambda_3 a_3 = 0$ and $\lambda_1 b_1 + \lambda_2 b_2 + \lambda_3 b_3 = 0$.*

8.5 Lemma. *Let $P = PG(2, K)$ be a projective plane over a commutative field K. Let $a_1 := (1, 0, 0)^t$, $a_2 := (0, 1, 0)^t$ and $a_3 := (0, 0, 1)^t$.*

(a) The lines a_1a_2, a_1a_3 and a_2a_3 satisfy the equations $x_3 = 0$, $x_2 = 0$ and $x_1 = 0$, respectively.

(b) Let $c = (c_1, c_2, c_3)^t$ be a point of P which is not contained in the triangle $\triangle a_1a_2a_3$. Then, $c_1 \neq 0$, $c_2 \neq 0$ and $c_3 \neq 0$.

(c) Let $c = (c_1, c_2, c_3)^t$ be a point of P which is not contained in the triangle $\triangle a_1a_2a_3$. Then, the lines $g_1 := a_1c$, $g_2 := a_2c$ and $g_3 := a_3c$ are given by the equations

$$g_1 : x_2 = \lambda_1 \, x_3 \quad \text{with } \lambda_1 = c_2 c_3^{-1} \neq 0$$

$$g_2 : x_3 = \lambda_2 \, x_1 \quad \text{with } \lambda_2 = c_3 c_1^{-1} \neq 0$$

$$g_3 : x_1 = \lambda_3 \, x_2 \quad \text{with } \lambda_3 = c_1 c_2^{-1} \neq 0.$$

Furthermore, we have $\lambda_1 \lambda_2 \lambda_3 = 1$.

Proof. (a) follows from Lemma 8.4.

(b) follows from Part (a).

(c) The equations of the three lines is a consequence of Lemma 8.4 and Part (b). Furthermore, we have $\lambda_1 \lambda_2 \lambda_3 = c_2 c_3^{-1} \, c_3 c_1^{-1} \, c_1 c_2^{-1} = 1$. □

8.6 Theorem. *Let $K = GF(q)$ be a finite field with q elements, and let q be odd. We have*

$$\prod_{\lambda \in K, \lambda \neq 0} \lambda = -1.$$

Proof. It is well known that the multiplicative group K^* of K is a cyclic group of order $q - 1$.[20] Since q is odd, K^* is a cyclic group of even order, that is, there is exactly one element τ of K^* of order 2. Since $\lambda \neq \lambda^{-1}$ for all λ of $K^* \setminus \{\tau\}$, it follows that

$$\prod_{\lambda \in K^*, \lambda \neq \tau} \lambda = 1, \text{ that is, } \prod_{\lambda \in K^*} \lambda = \tau.$$

Since K is a field, we have $\tau = -1$. □

In the following, we will only consider Desarguesian projective planes of odd order.

8.7 Theorem. *Let $P = PG(2, q)$ be a finite Desarguesian projective plane of odd order, and let $a_1 := (1, 0, 0)^t$, $a_2 := (0, 1, 0)^t$ and $a_3 := (0, 0, 1)^t$. Furthermore, let $\mathcal{O}$ be an oval of P through the points a_1, a_2 and a_3.*

(a) Let t_1, t_2 and t_3 be the tangents of $\mathcal{O}$ at the points a_1, a_2 and a_3, respectively. Then, there exist elements $0 \neq \mu_1$, μ_2, μ_3 of K such that the lines t_1, t_2 and t_3 satisfy the following equations:

---

[20]See Jungnickel [33], Theorem 1.2.4.

$$t_1 : x_2 = \mu_1\, x_3$$

$$t_2 : x_3 = \mu_2\, x_1$$

$$t_3 : x_1 = \mu_3\, x_2.$$

(b) *We have* $\mu_1\, \mu_2\, \mu_3 = -1$.

(c) *Let* $z_1 := t_2 \cap t_3$, $z_2 := t_1 \cap t_3$ *and* $z_3 := t_1 \cap t_2$. *Then,* $z_1 = (\mu_3,\ 1,\ \mu_2\, \mu_3)^t$, $z_2 = (\mu_3\, \mu_1,\ \mu_1,\ 1)^t$ *and* $z_3 = (1,\ \mu_1\, \mu_2,\ \mu_2)^t$.

(d) *The lines* $g_1 := a_1 z_1$, $g_2 := a_2 z_2$ *and* $g_3 := a_3 z_3$ *satisfy the equations*

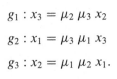
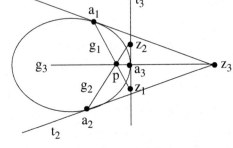

$$g_1 : x_3 = \mu_2\, \mu_3\, x_2$$

$$g_2 : x_1 = \mu_3\, \mu_1\, x_3$$

$$g_3 : x_2 = \mu_1\, \mu_2\, x_1.$$

(e) *The lines* g_1, g_2 *and* g_3 *meet in the point* $p = (1,\ \mu_1\, \mu_2,\ -\mu_2)^t$. *The point* p *is not contained in the triangle* $\triangle a_1 a_2 a_3$.

Proof. (a) Let $\lambda_1\, x_1 + \lambda_2\, x_2 + \lambda_3\, x_3 = 0$ be the equation defining t_1. Since a_1 is incident with t_1, it follows that $\lambda_1 = 0$. Since t_1 is a tangent of $\mathcal{O}$, we have $t_1 \neq a_1 a_2$. By Lemma 8.5 (a), it follows that $\lambda_3 \neq 0$.

Similarly, it follows from $t_1 \neq a_1 a_3$ that $\lambda_2 \neq 0$. Thus, t_1 satisfies the equation $x_2 = \mu_1\, x_3$ with $\mu_1 := -\lambda_3\, \lambda_2^{-1} \neq 0$.

The statements about the tangents t_2 and t_3 follow analogously.

(b) Step 1. Let $c \neq a_1,\ a_2,\ a_3$ be a point on $\mathcal{O}$. Then, the lines $g_1 := a_1 c$, $g_2 := a_2 c$ and $g_3 := a_3 c$ satisfy the equations

$$g_1 : x_2 = \lambda_1\, x_3 \qquad \text{with } 0 \neq \lambda_1 \neq \mu_1$$

$$g_2 : x_3 = \lambda_2\, x_1 \qquad \text{with } 0 \neq \lambda_2 \neq \mu_2$$

$$g_3 : x_1 = \lambda_3\, x_2 \qquad \text{with } 0 \neq \lambda_3 \neq \mu_3.$$

Furthermore, we have $\lambda_1\, \lambda_2\, \lambda_3 = 1$:

Since c is a point of $\mathcal{O}$, the point c is not contained in the triangle $\triangle a_1 a_2 a_3$. If $c = (c_1,\ c_2,\ c_3)^t$, by Theorem 8.5 (c), the lines g_1, g_2 and g_3 satisfy the equations

$$g_1 : x_2 = \lambda_1\, x_3 \qquad \text{with } \lambda_1 = c_2 c_3^{-1} \neq 0$$

$$g_2 : x_3 = \lambda_2\, x_1 \qquad \text{with } \lambda_2 = c_3 c_1^{-1} \neq 0$$

$$g_3 : x_1 = \lambda_3\, x_2 \qquad \text{with } \lambda_3 = c_1 c_2^{-1} \neq 0.$$

Furthermore, we have $\lambda_1 \lambda_2 \lambda_3 = 1$. Since the lines g_1, g_2 and g_3 are not tangents of $\mathcal{O}$, it follows that $\lambda_1 \neq \mu_1$, $\lambda_2 \neq \mu_2$, $\lambda_3 \neq \mu_3$.

Step 2. Conversely, let h_1, h_2 and h_3 be three lines of $\boldsymbol{P}$ with the equations

$$h_1 : x_2 = \lambda_1 x_3 \qquad \text{with } 0 \neq \lambda_1 \neq \mu_1$$
$$h_2 : x_3 = \lambda_2 x_1 \qquad \text{with } 0 \neq \lambda_2 \neq \mu_2$$
$$h_3 : x_1 = \lambda_3 x_2 \qquad \text{with } 0 \neq \lambda_3 \neq \mu_3.$$

Then, for $i = 1, 2, 3$, the line h_i is a secant of $\mathcal{O}$ through the point a_i intersecting $\mathcal{O}$ in a second point different from a_1, a_2, a_3:

Obviously, a_1 is a point on h_1. Furthermore, the lines through a_1 are exactly the lines

$$x_2 = \lambda x_3 \qquad \text{with } \lambda \in K \text{ and}$$
$$x_3 = 0.$$

The line defined by the equation $x_3 = 0$ is the line $a_1 a_2$. The line defined by the equation $x_2 = 0$ (that is, $\lambda = 0$) is the line $a_1 a_3$. The line defined by the equation $x_2 = \mu_1 x_3$ is the tangent of $\mathcal{O}$ through a_1. All other lines are secants of $\mathcal{O}$ through a_1 intersecting $\mathcal{O}$ in a second point different from a_2 and a_3. For the lines h_2 and h_3, the assertion can be seen analogously.

Step 3. We have $\mu_1 \mu_2 \mu_3 = -1$: For a point c of $\mathcal{O}$ with $c \neq a_1$, a_2, a_3, let $\lambda_1(c)$, $\lambda_2(c)$, $\lambda_3(c)$ be three elements of K such that the lines $g_1 := a_1 c$, $g_2 := a_2 c$ and $g_3 := a_3 c$ satisfy the following equations as indicated in Step 1:

$$g_1 : x_2 = \lambda_1(c) x_3$$
$$g_2 : x_3 = \lambda_2(c) x_1$$
$$g_3 : x_1 = \lambda_3(c) x_2.$$

By Step 2, we have $0 \neq \lambda_i(c) \neq \mu_i$ for $i = 1, 2, 3$. Furthermore, we have $\lambda_1(c) \lambda_2(c) \lambda_3(c) = 1$. It follows from Step 2 and Theorem 8.6 that

$$1 = \prod (\lambda_1(c)\lambda_2(c)\lambda_3(c) \mid c \in O, c \neq a_1, a_2, a_3)$$
$$= \prod (\lambda_1\lambda_2\lambda_3 \mid \lambda_1, \lambda_2, \lambda_3 \in K, 0 \neq \lambda_i \neq \mu_i (i = 1, 2, 3))$$
$$= \left(\prod (\lambda) \mid \lambda \in K, 0 \neq \lambda \neq \mu_1 \right) \left(\prod (\lambda) \mid \lambda \in K, 0 \neq \lambda \neq \mu_2 \right)$$
$$\left(\prod (\lambda) \mid \lambda \in K, 0 \neq \lambda \neq \mu_3 \right)$$

$$= \left(\prod_{\lambda \in K, \lambda \neq 0} \lambda \right) \mu_1^{-1} \left(\prod_{\lambda \in K, \lambda \neq 0} \lambda \right) \mu_2^{-1} \left(\prod_{\lambda \in K, \lambda \neq 0} \lambda \right) \mu_3^{-1}$$

$$= (-1)\,\mu_1^{-1}\,(-1)\,\mu_2^{-1}\,(-1)\,\mu_3^{-1}$$

$$= -\mu_1^{-1}\mu_2^{-1}\mu_3^{-1}.$$

The second equation follows from the fact that we compute the cardinality of the following set:

$$\{l \text{ secant } of \ \mathcal{O} \mid a_1 \in l, \ l \neq a_1 a_2, \ a_1 a_3\} \times \{l \text{ secant } of \ \mathcal{O} \mid a_2 \in l, \ l \neq a_1 a_2, \ a_2 a_3\}$$

$$\times \{l \text{ secant } of \ \mathcal{O} \mid a_3 \in l, \ l \neq a_1 a_3, \ a_2 a_3\}.$$

Hence, $\mu_1\,\mu_2\,\mu_3 = -1$.

(c) and (d) follow from Lemma 8.4.

(e) From $-\mu_2 = \mu_1\,\mu_2\,\mu_3\,\mu_2 = \mu_2\,\mu_3\,\mu_1\,\mu_2$, it follows that p is a point on g_1.
From $1 = -\mu_1\,\mu_2\,\mu_3 = \mu_3\,\mu_1\,(-\mu_2)$, it follows that p is a point on g_2.
From $\mu_1\,\mu_2 = \mu_1\,\mu_2$, it follows that p is a point on g_3.

It follows that $p = (1, \ \mu_1\,\mu_2, \ -\mu_2)^t$ is the intersection point of the lines g_1, g_2 and g_3. By Lemma 8.5 (a), it follows that the point p is not contained in the triangle $\Delta a_1 a_2 a_3$.   □

8.8 Theorem. *Let $P = PG(2, q)$ be a finite Desarguesian projective plane of odd order q, and let $\mathcal{O}$ be an oval of P. Let a_1, a_2, a_3 be three points on $\mathcal{O}$, and let t_1, t_2, t_3 be the tangents of $\mathcal{O}$ at a_1, a_2 and a_3, respectively. Finally, let $z_1 := t_2 \cap t_3$, $z_2 := t_1 \cap t_3$, $z_3 := t_1 \cap t_2$, $g_1 := a_1 z_1$, $g_2 := a_2 z_2$ and $g_3 := a_3 z_3$.*

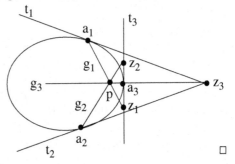

(a) *The lines g_1, g_2 and g_3 meet in a point p.*
(b) *The point p is not contained in the triangle $\Delta a_1 a_2 a_3$.*

Proof. Since P is Desarguesian, we can assume w.l.o.g. that $a_1 = (1, 0, 0)^t$, $a_2 = (0, 1, 0)^t$ and $a_3 = (0, 0, 1)^t$. Hence, the assertion follows from Theorem 8.7.   □

8.9 Theorem (Segre). *Let $P = PG(2, q)$ be a finite Desarguesian projective plane of odd order q. Then, every oval of P is a conic.*

Proof. Let $\mathcal{O}$ be an oval of P, and let a_1, a_2, a_3 be three points on $\mathcal{O}$. W.l.o.g., let $a_1 = (1, 0, 0)^t$, $a_2 = (0, 1, 0)^t$ and $a_3 = (0, 0, 1)^t$.

Let t_1, t_2, t_3 be the tangents of $\mathcal{O}$ at a_1, a_2 and a_3, respectively. Then, by Theorem 8.7, there exist elements μ_1, μ_2, $\mu_3 \in K \setminus \{0\}$ with

$$t_1 : x_2 = \mu_1 \, x_3$$

$$t_2 : x_3 = \mu_2 \, x_1$$

$$t_3 : x_1 = \mu_3 \, x_2.$$

Furthermore, by Theorem 8.7, for the points $z_1 := t_2 \cap t_3$, $z_2 := t_1 \cap t_3$ and $z_3 := t_1 \cap t_2$, we have $z_1 = (\mu_3, 1, \mu_2 \, \mu_3)^t$, $z_2 = (\mu_3 \, \mu_1, \mu_1, 1)^t$ and $z_3 = (1, \mu_1 \, \mu_2, \mu_2)^t$.

Finally, again by Theorem 8.7, the lines $a_1 z_1$, $a_2 z_2$ and $a_3 z_3$ meet in the point $p = (1, \mu_1 \, \mu_2, -\mu_2)^t$. Since the point p is not contained in the triangle $\Delta a_1 a_2 a_3$, we can assume w.l.o.g. that $p = (1, 1, 1)^t$.

It follows that $p = (1, 1, 1)^t = (1, \mu_1 \, \mu_2, -\mu_2)^t$, hence, $\mu_2 = -1$ and $\mu_1 = -1$. Since, by Theorem 8.7, the relation $\mu_1 \, \mu_2 \, \mu_3 = -1$ holds, it follows that $\mu_3 = -1$.

Let $c = (c_1, c_2, c_3)^t$ be an arbitrary point on $\mathcal{O}$ different from a_1, a_2, a_3. By Lemma 8.5, we have $c_1 \neq 0$, $c_2 \neq 0$ and $c_3 \neq 0$. Let b be the tangent of $\mathcal{O}$ at c with the equation

$$b : b_1 \, x_1 + b_2 \, x_2 + b_3 \, x_3 = 0.$$

Step 1. We have $b_1 \, c_1 + b_2 \, c_2 + b_3 \, c_3 = 0$: This assertion follows from the fact that the point c is incident with the line b.

Step 2. We have $b_1 \neq 0$, $b_2 \neq 0$ and $b_3 \neq 0$: Assume on the contrary that $b_1 = 0$. Then, the point $a_1 = (1, 0, 0)^t$ would be on the line b, in contradiction to the fact that b is a tangent of $\mathcal{O}$ at $c \neq a_1$. Similarly, it follows that $b_2 \neq 0$ and $b_3 \neq 0$.

Step 3. We have $-b_1 + b_2 + b_3 \neq 0$, $b_1 - b_2 + b_3 \neq 0$ and $b_1 + b_2 - b_3 \neq 0$: The point z_1 is incident with the two tangents t_2 and t_3. By Theorem 8.2, any point of $\boldsymbol{P}$ is incident with at most two tangents of $\mathcal{O}$. It follows that z_1 is not incident with b. Analogously, it follows that z_2 is not on b and that z_3 is not on b. Since $\mu_1 = \mu_2 = \mu_3 = -1$, we have $z_1 = (-1, 1, 1)^t$, $z_2 = (1, -1, 1)^t$ and $z_3 = (1, 1, -1)^t$. Since z_1 is not on b, it follows that

$$-b_1 + b_2 + b_3 \neq 0.$$

The two other equations follow from the fact the z_2 and z_3 are not on b.

Step 4. We have $b_2 \, (c_1 + c_2) = b_3 \, (c_1 + c_3)$:

(i) Let $r_1 := z_1 = t_2 \cap t_3$, $r_2 := b \cap t_3$ and $r_3 := b \cap t_2$. Then, by Theorem 8.4, we have $r_1 = (-1, 1, 1)^t$, $r_2 = (-b_3, b_3, b_1 - b_2)^t$ and $r_3 = (b_2, b_3 - b_1, -b_2)^t$.
(ii) Let $h_1 := r_1 c$, $h_2 := r_2 a_2$ and $h_3 := r_3 a_3$. Then, again by Lemma 8.4, the lines h_1, h_2 and h_3 satisfy the following equations:

$$h_1 : (c_3 - c_2) \, x_1 + (c_1 + c_3) \, x_2 - (c_1 + c_2) x_3 = 0$$

$$h_2 : (b_1 - b_2) \, x_1 + b_3 \, x_3 = 0$$

$$h_3 : (b_1 - b_3) \, x_1 + b_2 \, x_2 = 0.$$

(iii) By Theorem 8.8, it follows that the three lines h_1, h_2 and h_3 intersect in a point, that is, the system of equations defined by the lines h_1, h_2, h_3 has a non-trivial solution. It follows that

$$0 = \det \begin{pmatrix} c_3 - c_2 & c_1 + c_3 & -c_1 - c_2 \\ b_1 - b_3 & b_2 & 0 \\ b_1 - b_2 & 0 & b_3 \end{pmatrix}$$

$$= (c_3 - c_2)\, b_2\, b_3 + (c_1 + c_2)\, b_2\, (b_1 - b_2) + (c_1 + c_3)(b_3 - b_1)b_3$$

$$= (b_1 - b_2 - b_3)(b_2(c_1 + c_2) - b_3\, (c_1 + c_3)).$$

By Step 3, we have $b_1 - b_2 - b_3 \neq 0$, hence, we have $b_2\, (c_1 + c_2) = b_3\, (c_1 + c_3)$.

Step 5. By considering the points a_1, c, a_3 and a_1, a_2, c, we get as in Step 4 the equations

$$b_3\, (c_2 + c_3) = b_1\, (c_1 + c_2)$$

$$b_1\, (c_1 + c_3) = b_2\, (c_2 + c_3).$$

Step 6. We have $c_1\, c_2 + c_1\, c_3 + c_2\, c_3 = 0$: By Step 2, we have b_1, b_2, $b_3 \neq 0$. We have $c_1 + c_3 \neq 0$: Otherwise, it follows from $b_1\, (c_1 + c_3) = b_2\, (c_2 + c_3)$ and $b_2 \neq 0$ that $c_2 + c_3 = 0$. In the same way, it follows that $c_1 + c_2 = 0$ and $c_1 + c_3 = 0$. Hence, $c_1 = -c_3$ and $c_1 = -c_2$ implying that $c_2 = c_3$. It follows that $c_2 + c_2 = 0$. Since q is odd, we have $c_2 = 0$. Hence, $c_1 = c_2 = c_3 = 0$, a contradiction.

From Steps 4 and 5, it follows that

$$\frac{b_1}{b_2} = \frac{c_2 + c_3}{c_1 + c_3} \quad \text{and} \quad \frac{b_3}{b_2} = \frac{c_1 + c_2}{c_1 + c_3}.$$

By Step 1, we have $b_1\, c_1 + b_2\, c_2 + b_3\, c_3 = 0$. Thus, it follows that

$$0 = \frac{b_1}{b_2}c_1 + c_2 + \frac{b_3}{b_2}c_3$$

$$= \frac{c_2 + c_3}{c_1 + c_3}c_1 + c_2 + \frac{c_1 + c_2}{c_1 + c_3}c_3, \quad \text{hence,}$$

$$0 = 2(c_1\, c_2 + c_1\, c_3 + c_2\, c_3).$$

Since q is odd, the field K is of odd order,[21] thus, we have Char $K \neq 2$. Hence

$$c_1\, c_2 + c_1\, c_3 + c_2\, c_3 = 0.$$

---

[21] It is $|K| = q$.

Step 7. $\mathscr{O}$ is a conic: Let Q be the conic with the quadratic equation

$$x_1 \, x_2 + x_1 \, x_3 + x_2 \, x_3 = 0.$$

By Step 6, any point $c = (c_1, c_2, c_3)^t$ of $\mathscr{O}$ with $c \neq a_1, a_2, a_3$ satisfies the equation $c_1 \, c_2 + c_1 \, c_3 + c_2 \, c_3 = 0$, that is, c is a point of Q. Furthermore, the points a_1, a_2, a_3 are contained in Q. It follows that $\mathscr{O}$ is contained in Q.

Since every conic is an oval, it follows from Theorem 8.1 that $|Q| = q + 1$. Since $\mathscr{O}$ is contained in Q and since $|\mathscr{O}| = q + 1 = |Q|$, it follows that $\mathscr{O} = Q$.   □

8.10 Corollary (Barlotti, Panella). *Let Q be an ovoid in $P = PG(3, q)$, q odd. Then, Q is a quadric.*

Proof. From the Theorem of Segre (Theorem 8.9), it follows that $E \cap Q$ is a conic for all planes E such that $E \cap Q$ is an oval. By Theorem 6.3, Q is a quadric.   □

Corollary 8.10 was proven by Barlotti [5] and Panella [36] without using Theorem 6.3.

9   Further Reading

An introduction into the foundations of projective and polar spaces can only present a limited selection of results. As a consequence, a wide range of important and interesting results had to be omitted. Therefore, at the end of this book, I want to give some hints about further literature. These hints are by no means complete.

There are two encyclopaedic works about incidence geometry, namely the Handbook of Incidence Geometry [18] and the forthcoming book of Buekenhout and Cohen [20]. In both works, the interested reader will find a lot of information about projective and polar geometries and related topics.

The textbooks dedicated to projective and affine spaces are divided into books about projective and affine planes and about projective and affine spaces in general. For projective and affine planes see for example Hughes and Piper [31] and Pickert [41]. The interested reader will find in both books a lot of results about non-Desarguesian projective planes.

For projective and affine spaces in general see Artin [2], Baer [3], Beutelspacher and Rosenbaum [6], Cameron [22], Hilbert [27], Segre [44] and Veblen and Young [55]. All these books deal with the structure of projective spaces in detail, but with different priorities. In Beutelspacher and Rosenbaum [6], the reader will find applications of projective geometry for example to cryptography.

I only know about two textbooks concentrating on polar spaces, namely Buekenhout and Cohen [20] and Cameron [22]. Whereas Cameron [22] is an introduction into polar spaces, Buekenhout and Cohen [20] provides a detailed analysis of polar spaces including the Classification Theorem of Polar Spaces (Theorem 7.3 of Chap. IV). A detailed study of polar spaces is also included in Tits [50] where the

Classification Theorem of Polar Spaces is proved using the theory of buildings. In fact, Tits [50] also provides a classification of polar spaces of rank 3 whose planes are Moufang.

Since polarities stem from sesquilinear forms (see Theorem 5.11 of Chap. IV), the study of polarities is strongly related to the study of the classical groups. The interested reader is referred to Artin [2], Borel [8], Dieudonné [24], Garrett [26], Suzuki [45] and Taylor [46].

Projective and polar spaces are important classes of buildings. Further information about buildings can be found in Abramenko and Brown [1], Bourbaki [9], Brown [10], Garrett [26], Ronan [42], Tits [50], Tits [51] and Weiss [57, 58]. In Bourbaki [9], buildings are called systèmes de Tits. There are two main definitions for buildings, both introduced by Tits in [50, 51].

Generalized polygons have been introduced by Tits [49]. Two classes of generalized polygons are the projective planes and the generalized quadrangles. For the theory of generalized polygons see Payne and Thas [39], Thas [47], Tits and Weiss [52] and van Maldeghem [54]. Payne and Thas [39] and Thas [47] deal with finite generalized quadrangles, whereas Tits and Weiss [52] and van Maldeghem [54] deal with arbitrary generalized polygons.

Although diagram geometries are an important part of modern incidence geometry, I only know the textbook of Pasini [38], which is totally devoted to this subject. Buekenhout is one of the pioneers in diagram geometry. The interested reader is referred to [13, 15, 19].

Last but not least, there is a rich literature about finite geometries. The interested reader is referred to Batten [5], Dembowski [23], Hirschfeld [28], Hirschfeld [29], Hirschfeld and Thas [30]. Some contributions of the author to finite geometries can be found in [53].

Further interesting topics related to incidence geometry like topological geometry or geometry over rings can be found in the Handbook of Incidence Geometry [18].

References

1. Abramenko, P., Brown,K.: Buildings Theory and Applications. Springer, Berlin (2008)
2. Artin, E.: Algèbre Géométrique. Gauthier-villars, Paris (1978)
3. Baer, R.: Linear Algebra and Projective Geometry. Academic, New York (1952)
4. Barlotti, A.: Un' estensione del teorema di Segre-Kustaanheimo. Boll. Un. Mat. Ital. **11**, 96–98 (1955)
5. Batten, L.: Combinatorics of Finite Geometries, 2nd edn. Cambridge University Press, Cambridge (1997)
6. Beutelspacher, A., Rosenbaum, U.: Projective Geometry; From Foundations to Applications. Cambridge University Press, Cambridge (1998)
7. Birkhoff, G., Neumann, J.v.: The logic of quantum mechanics. Ann. Math. **37**(2), 823–843 (1936)
8. Borel, A.: Linear Algebraic Groups, 2nd enlarged edition. Springer, Berlin (1991)
9. Bourbaki, N.: Groupes et Algèbres de Lie, Chaps. 4–6. Hermann, Paris (1968)
10. Brown, K.S.: Buildings, 2nd edn. Springer, New York (1989)
11. Buekenhout, F.: Ensembles quadratiques des espaces projectifs. Math. Zeit. **110**, 306–318 (1969)
12. Buekenhout, F.: Une caractérisation des espaces affines basées sur la notion de droite. Math. Zeit. **111**, 367–371 (1969)
13. Buekenhout, F.: Foundations of one-dimensional projective geometry based on perspectivities. Abh. aus dem Math. Sem. d. Univ. Hamburg **43**, 21–29 (1975)
14. Buekenhout, F.: Diagrams for geometries and groups. J. Combin. Th. Ser. A **27**, 121–151 (1979)
15. Buekenhout, F.: The basic diagram of a geometry. In: Aigner, M., Jungnickel, D. (eds.) Geometries and Groups, Lecture Notes in Math., vol. 893, pp. 1–29. Springer, Berlin (1981)
16. Buekenhout, F.: On the foundations of polar geometry II. Geom. Dedicata **33**, 21–26 (1990)
17. Buekenhout, F.: A theorem of Parmentier characterizing projective spaces by polarities. In: De Clerck, F., et al. (eds.) Finite Geometry and Combinatorics, 69–71. Cambridge University Press, Cambridge (1993)
18. Buekenhout, F. (ed.): Handbook of Incidence Geometry; Buildings and Foundations. Elsevier, Amsterdam (1995)
19. Buekenhout, F., Buset, D.: On the foundations of incidence geometry. Geom. Ded. **25**, 269–296 (1988)
20. Buekenhout, F., Cohen, A.: Diagram Geometry. Springer, Berlin (to appear)
21. Buekenhout, F., Shult, E.: On the foundation of polar geometry. Geom. Ded. **3**, 155–170 (1973)
22. Cameron, P.J.: Projective and Polar Spaces. QMW Maths. Notes 13, 2nd edn. (2000)
23. Dembowski, P.: Finite Geometries, Ergebnisse der Math. und ihrer Grenzgebiete, vol. 44. Springer, Berlin (1968)

J. Ueberberg, *Foundations of Incidence Geometry*, Springer Monographs in Mathematics, 245
DOI 10.1007/978-3-642-20972-7, © Springer-Verlag Berlin Heidelberg 2011

24. Dieudonné, J.: La Géométrie des Groupes Classiques, Ergebnisse der Math. und ihrer Grenzgebiete, vol. 5. Springer, Heidelberg (1971)
25. Euclid: The Elements. In: Densmore, D. (ed.) Euclid's Elements. Green Lion Or (2002)
26. Garrett, P.: Buildings and Classical Groups. Chapman & Hall (1997)
27. Hilbert, D.: Grundlagen der Geometrie (Foundations of Geometry). Teubner, Stuttgart (1999)
28. Hirschfeld, J.W.P.: Projective Geometries over Finite Fields, 2nd edn. Oxford Math. Monographs, Clarendon, Oxford (1998)
29. Hirschfeld, J.W.P.: Finite Projective Geometries of Three Dimensions. Oxford University Press, Clarendon, Oxford (1986)
30. Hirschfeld, J.W.P., Thas, J.A.: General Galois Geometries, Oxford Math. Monographs. Clarendon, Oxford (1991)
31. Hughes, D.R., Piper, F.C.: Projective Planes, 2nd printing. Springer, Berlin (1982)
32. Johnson, P.: Polar spaces of arbitrary rank. Geom. Dedicata **35**, 229–250 (1990)
33. Jungnickel, D.: Finite Fields; Structure and Arithmetic. BI Wissenschaftsverlag (1993)
34. Lang, S.: Algebra, Springer Graduate Texts in Mathematics, vol. 211. Berlin (2002)
35. Neumaier, A.: Some sporadic geometries related to PG(3, 2). Arch. Math. **42**, 89–96 (1984)
36. Panella, G.: Caratterizzazione delle quadriche di uno spazio (tridimensionale) lineare sopra un corpo finito. Boll. Un. Mat. Ital. **10**, 507–513 (1955)
37. Parmentier, A.: Caractérisations des Polarités dans les Espace Projectifs et Linéaires. Master's Thesis, Université Libre de Bruxelles (1974)
38. Pasini, A.: Diagram Geometries. Oxford University Press, Oxford (1994)
39. Payne, S.E., Thas, J.A.: Finite Generalized Quadrangles, 2nd edn. Pitman (1984); EMS Series of Lect. in Math., Zürich (2009)
40. Percsy, N.: On the Geometry of Zara Graphs. J. Combin. Th. A **55**, 74–79 (1990)
41. Pickert, G.: Projektive Ebenen. Springer, Göttingen (1955)
42. Ronan, M.: Lectures on Buildings, 2nd edn. Academic, Boston (2009)
43. Segre, B.: Ovals in a fine projective plane. Canad. J. Math. **7**, 414–416 (1955)
44. Segre, B.: Lectures on Modern Geometry. Ed. Cremonese, Roma (1961)
45. Suzuki, M.: Group Theory I, Grundlehren der Math. Wissenschaften, vol. 247. Springer, Berlin (1982)
46. Taylor, D.E.: The Geometry of the Classical Groups, Sigma Series in Pure Math., vol. 9. Heldermann, Berlin (1992)
47. Thas, K.: Symmetry in Finite Generalized Quadrangles. Birkhäuser, Basel (2004)
48. Teirlinck, L.: On projective and affine hyperplanes. J. Combin. Th. A **28**, 290–306 (1980)
49. Tits, J.: Sur la trialité et certains groupes qui s'en déduisent. Inst. Hautes Etudes Scient. Publ. Math. **2**, 13–60 (1959)
50. Tits, J.: Buildings of spherical type and finite BN-Pairs, Lecture Notes in Math., vol. 386. Springer, Berlin (1974)
51. Tits, J.: A local approach to buildings. In: Davies, C., Grünbaum, B., Scherk, F.A. (eds.) The Geometric Vein (Coxeter-Festschrift), 519–547. Springer, Berlin (1981)
52. Tits, J., Weiss, R.: Moufang Polygons, Springer (2002)
53. Ueberberg, J.: Bögen, Blockaden und Baer-Unterräume in Endlichen Projektiven Räumen. Mitt. aus dem math. Sem. Giessen **205**, 1–91 (1991)
54. van Maldeghem, H.: Generalized Polygons, Monographs in Math., vol. 93. Birkhäuser, Basel (1998)
55. Veblen, O., Young, J.W.: Projective Geometry. Ginn & Blaisdell, Boston (1910)
56. Veldkamp, F.D.: Polar Geometry I–IV. Indag. Math. **21**, 412–551 (1959); **22**, 207–212 (1959)
57. Weiss, R.M.: The Structure of Spherical Buildings. Princeton University Press, Princeton (2003)
58. Weiss, R.M.: The Structure of Affine Buildings. Princeton University Press, Princeton (2009)

Index

$A(W)$, 98
Absolute, 156
Affine collineation, 105
Affine geometry, 31
Affine plane, 7, 24
Affine space, 24
Affine space over a vector space, 98
Affine subspace, 23
$AG(d, K)$, 98
Alternating, 164
Anti-automorphism, 158
Anti-hermitian, 164
Autocorrelation, 59
Automorphism, 58
Automorphism group, 58
Axis, 68

Basis, 12, 29
Bilinear form, 158

Central collineation, 68, 72
Centre, 68, 72
Chamber, 2
Codimension, 20
Collinear, 5, 124
Collineation, 61
Co-maximal, 4
Complementary, 20
Cone, 204
Conic, 170
Connected, 44
Corank, 2
Correlation, 59

Desargues configuration, 73
Desargues, Theorem of, 74

Desarguesian, 74
Diagram, 42
Dimension, 17, 30, 136
Double point, 186
Dual space, 153
Duality, 59, 154, 158

Elation, 68
Element of a geometry, 1

Finite, 52
Firm, 4
Fixed element, 67
Fixed line, 68
Fixed point, 68
Flag, 2
Frame, 93

Generalized digon, 42
Generalized polar space, 124
generalized projective plane, 6
generalized projective space, 10
Generalized quadrangle, 124
Geometry, 3

Hermitian, 163, 164
Homogeneous coordinates, 88
Homogenous quadratic polynomial, 192
Homology, 68, 72
Homomorphism, 58
Hyperbolic quadric, 203
Hyperplane, 5
Hyperplane at infinity, 29

J. Ueberberg, *Foundations of Incidence Geometry*, Springer Monographs in Mathematics,   247
DOI 10.1007/978-3-642-20972-7, © Springer-Verlag Berlin Heidelberg 2011

{i, j}-path, 46
Incidence graph, 44
Incidence relation, 1
Incident, 2
Independent, 12, 29
Isomorphic, 67
Isomorphism, 58

Kleinian polar space, 174
Kleinian quadric, 231

Line at infinity, 8
Linear, 47
Linear space, 4

Morphism, 57

Neumaier-geometry, 151
N-gon, 2
Nondegenerate, 158, 168
Non-singular, 168

Opposite field, 159
Opposite regulus, 199
Order, 53, 59
Origin, 102
Oval, 191
Ovoid, 204

P(V), 84
Parallel, 7, 23, 30, 98
Parallel axiom, 7
Parallel class, 7, 23
Parallelism, 23
Parallelism-preserving collineation, 63
Partial linear space, 124
Path of length r, 44
Perspective, 210
PG(d, K), 84
Plücker transformation, 231
Polar geometry, 144
Polar hyperplane, 179
Polar space, 124, 157, 171
Polarity, 59, 154
Pregeometry, 1

Projective closure, 9, 29, 65
Projective collineation, 92
Projective geometry, 21
Projective hyperplane, 126
Projective plane, 6
Projective space, 10
Projective space over a vector space, 84
Pseudo-quadratic form, 168
Pseudo-quadric, 170

Quadratic form, 168
Quadratic set, 186
Quadric, 170

Radical, 168, 186
Rank, 2, 125
Reflexive, 161
Regulus, 198
Residually connected, 45
Residue, 39

Secant, 235
Semilinear, 90
Sesquilinear form, 158
Set geometry, 3
Set pregeometry, 3
Singular subspace, 125
Skew, 198
Subgeometry, 4
Subspace, 5
Symmetric, 158, 164

Tangent, 186
Tangent set, 186
Tangent space, 186
Thick, 4
Thin, 4
Translation, 72
Transversal, 198
Type, 1, 2
Type function, 1

Veblen–Young, 10
Vector representation, 102
Vector space associated to an affine space, 102